NATIVE MACROMOLECULE-BASED 3D TISSUES REPAIR

NATIVE MACROMOLECULE-BASED 3D TISSUES REPAIR

Jin-Ye Wang

Shanghai Jiaotong University, China

W **World Scientific**

NEW JERSEY • LONDON • SINGAPORE • BEIJING • SHANGHAI • HONG KONG • TAIPEI • CHENNAI

Published by

World Scientific Publishing Co. Pte. Ltd.

5 Toh Tuck Link, Singapore 596224

USA office: 27 Warren Street, Suite 401-402, Hackensack, NJ 07601

UK office: 57 Shelton Street, Covent Garden, London WC2H 9HE

Library of Congress Cataloging-in-Publication Data
Wang, Jin-Ye, author.
    Native macromolecule-based 3D tissue repair / Jin-Ye Wang.
       p. ; cm.
    Includes bibliographical references and index.
    ISBN 978-9814551922 (hardcover : alk. paper)
    I. Title.
    [DNLM: 1. Polymers--therapeutic use. 2. Tissue Engineering. QT 37.5.P7]
    QP801.P64
    572'.33--dc23
                           2013030747

British Library Cataloguing-in-Publication Data
A catalogue record for this book is available from the British Library.

In-house Editor: Darilyn Yap

Typeset by Stallion Press
Email: enquiries@stallionpress.com

Contents

Preface   xiii

**Chapter 1   Native Polymer-based 3D Substitutes
                in Cardiovascular Tissue Engineering**   1
Hua-Jie Wang and Ying Cao

1. Introduction   1
2. Native Polymer-based Tissue-Engineered Heart Valves   2
   2.1. Collagen   5
   2.2. Fibrin   5
   2.3. Hyaluronic acid   6
   2.4. Other native polymers   8
3. Native Polymer-based Tissue-Engineered Blood Vessels   9
   3.1. Collagen   10
   3.2. Fibrin   11
   3.3. Elastin   12
   3.4. Silk fibroin   13
   3.5. Bacteria cellulose   14
   3.6. Other native polymers   15
4. Native Polymer-based Tissue-Engineered Myocardium   16
   4.1. *In-situ* myocardial tissue engineering   16
   4.2. *In-vitro* engineering of myocardium substitutes   19
5. Applications of Composite Materials in Cardiovascular
   Tissue Engineering   20
6. Summary and Future Directions   22
References   22

**Chapter 2   Native Polymer-based 3D Substitutes
for Nerve Regeneration**   35
Guo-Wu Wang and Jin-Ye Wang

1. Introduction   35
2. Design of Ideal Nerve Conduits   37
3. The Application of Bioengineered Natural Materials
   in Nerve Conduits   38
   3.1. Extracellular matrix components   38
        3.1.1. Bioengineered collagen and its derivatives   39
               3.1.1.1. Introduction   39
               3.1.1.2. Category and characteristic   40
        3.1.2. Bioengineered laminin and its derivatives   41
               3.1.2.1. Introduction   41
               3.1.2.2. Category and characteristic   42
        3.1.3. Bioengineered fibronectin & fibrin and their
               derivatives   43
               3.1.3.1. Introduction   43
               3.1.3.2. Category and characteristic   44
   3.2. Bioengineered gelatin and its derivatives   48
        3.2.1. Introduction   48
        3.2.2. Category and characteristic   49
   3.3. Bioengineered chitosan and its derivatives   49
        3.3.1. Introduction   49
        3.3.2. Category and characteristic   52
   3.4. Bioengineered silk and its derivatives   55
        3.4.1. Introduction   55
        3.4.2. Category and characteristic   56
   3.5. Bioengineered alginate and its derivatives   58
        3.5.1. Introduction   58
        3.5.2. Category and characteristic   60
4. Conclusions and Future Perspective   61
References   62

**Chapter 3 Native Polymer-Based 3D Substitutes
for Cartilage Repair** 75
*Huitang Xia, Yu Liu, Ran Tao, Chunlei Miao,
Shengjian Tang, Biaobing Yang, Guangdong Zhou*

1. Introduction 75
2. The Theory Basis of Native Polymer-Based 3D Substitutes
   for Cartilage Repair 77
   2.1. Native polymers mimic the molecular composition
        of cartilage ECM 78
        2.1.1. Molecular composition of cartilage ECM 78
        2.1.2. Native polymers mimic molecular composition
               of cartilage ECM 79
   2.2. Native polymers mimic the structure of cartilage ECM 81
        2.2.1. Structure of cartilage ECM 81
        2.2.2. Native polymers mimic the structure of
               cartilage ECM 82
   2.3. Native polymers mimic the function of cartilage ECM 84
        2.3.1. Regulation of cell fate by native adhesion ligand 84
        2.3.2. Regulation of chondrogenic differentiation
               and phenotypic maintenance 85
        2.3.3. Controlled delivery of biochemical factors 86
        2.3.4. Regulation of mechanical properties 87
        2.3.5. Integration of neocartilage with host tissues 89
3. The Main Native Polymers for Cartilage Regeneration 91
   3.1. Acellular matrix 91
        3.1.1. Properties of acellular matrix for cartilage
               regeneration 92
        3.1.2. A cellular matrix scaffolds for cartilage
               regeneration 93
   3.2. Collagen 95
        3.2.1. Properties of collagen for cartilage engineering 96
        3.2.2. Collagen scaffolds for cartilage regeneration 101
   3.3. Chitosan 102
        3.3.1. Properties of chitosan for cartilage regneration 103

3.3.2. Chitosan-based scaffolds for cartilage
regeneration   105
3.3.2.1. Modification of chitosan   105
3.3.2.2. Combination of chitosan with other
materials   107
3.3.2.3. Chitosan nanofibers   107
3.4. Hyaluronic acid   109
3.4.1. Properties of hyaluronic acid for cartilage
regeneration   110
3.4.2. Hyaluronic acid-based scaffolds for cartilage
regeneration   112
3.5. Other native polymers for cartilage regeneration   114
3.5.1. Silk   114
3.5.2. Fibrin   115
3.5.3. Alginate   117
3.5.4. Agarose   118
4. Cartilage Regeneration and Repair Based
on Native Polymers   119
4.1. Cartilage regeneration based on native polymers   119
4.1.1. Collagen   119
4.1.2. Gelatin   120
4.1.3. Fibrin   120
4.1.4. Alginate   121
4.1.5. Remarks and future directions   123
4.2. Cartilage repair based on native polymers   123
4.2.1. Articular cartilage repair based on native
polymers   123
4.2.2. Tracheal cartilage repair based on native
polymers   125
4.3. Challenges in cartilage repair   127
4.3.1. Tissue integration   127
4.3.2. The scale of cartilage defect repair   127
4.3.3. Defect design in animal models   128
4.4. Future directions in native polymer-based scaffolds
and cartilage regeneration   129
References   131

**Chapter 4   Native Polymer-based 3D Substitutes
for Bone Repair**   145
Yan Huang, Kerong Dai, Xiaoling Zhang

1. Introduction   145
2. Proteins   148
   2.1. Collagen   148
   2.2. Silk   152
   2.3. Zein   156
3. Polysaccharides   157
   3.1. Chitosan   159
   3.2. Hyaluronic acid   161
   3.3. Alginate   163
   3.4. Starch-based material   164
   3.5. Cellulose   166
   3.6. Dextran   170
4. Microbial Origin Polyesters   171
Acknowledgments   174
References   174

**Chapter 5   Native Polymer-based 3D Substitutes
in Plastic Surgery**   185
Jing Wang, Xiaoling Zhang, Qingfeng Li

1. Bioengineered Hyaluronic Acid and its Derivatives   187
   1.1. Introduction   187
   1.2. Category and characteristic   188
2. Bioengineered Collagen and Its Derivatives   190
   2.1. Introduction   190
   2.2. Category and characteristic   191
3. Bioengineered Poly-L-Lactic Acid (PLLA)   196
4. Clinical Indications   197
   4.1. Rhinoplasty   198
       4.1.1. Anatomy   198
       4.1.2. Clinical usage   199
   4.2. Nasolabial fold   200
       4.2.1. Anatomy   200
       4.2.2. Clinical usage   202

4.3. Glabellar rhytides   203
    4.3.1. Anatomy   203
    4.3.2. Clinical usage   204
4.4. Lip enhancement   204
4.5. Nasojugal grooves (tear troughs)   206
    4.5.1. Anatomy   206
    4.5.2. Clinical usage   207
5. Injection Techniques   208
    5.1. Tunneling technique   209
    5.2. Serial puncture   210
6. Complications   211
7. Summary   213
Reference   213

**Chapter 6   Nanofabrication Techniques in Native
            Polymer-based 3D Substitutes**   **221**
Yangchao Luo, Qin Wang

1. Introduction   221
2. Electrospinning   222
    2.1. Introduction of electrospinning technique   222
    2.2. Modifications in electrospinning   223
    2.3. Parameters affecting production of electrospin
         nanofibers   228
    2.4. Applications of native polymer-based
         electrospinning technique   229
    2.5. Challenges in electrospinning technique   232
3. Self-assembly   233
    3.1. Introduction of self-assembly technique   233
    3.2. Parameters affecting production of self-assembly
         nanostructures   234
    3.3. Applications of native polymer-based
         self-assembly technique   236
    3.4. Challenges of self-assembly technique   241
4. Phase Separation   242
    4.1. Introduction of phase separation technique   242

4.2. Parameters affecting production
      of phase separation nanostructures   242
4.3. Applications of native polymer-based
      phase separation technique   245
4.4. Challenges of phase separation technique   248
5. Nano-Patterning Techniques using Native Polymers   248
6. Concluding Remarks   251
References   251

**Chapter 7   Native Polymer-based 3D Substitutes
               as Alternatives with Slow-Release Functions**   257
Dongwei Guo, Benson J. Edagwa, Xin-Ming Liu

1. Introduction   257
2. Proteins   259
   2.1. Collagen   260
   2.2. Albumin   263
   2.3. Gelatin   266
   2.4. Zein   269
   2.5. Recombinant proteins and peptides   272
   2.6. Silk fibroin   274
   2.7. Fibrin   276
3. Polysaccharides   278
   3.1. Chitosan   279
   3.2. Starch   282
   3.3. Alginate   285
   3.4. Hyaluronan   288
   3.5. Chondroitin sulphate   291
4. Conclusion   294
References   295

Conclusions and Future Outlook   307

Index   311

Preface

Industries based on fiber, rubber, and plastic, including both natural and synthetic polymers, have undergone significant development within the last century. However, few types of petrochemical-based synthetic polymers decompose naturally, and the effect of the associated pollution on the environment is very difficult to resolve. Additionally, oil and coal resources are dwindling, with an energy crisis threatening the future of human society. Therefore, the research and development of recyclable and renewable green materials, which includes the constant development of new natural polymer materials and the expansion of applications of existing natural polymers, is imperative. The natural world encompasses a variety of animals and plants and is an inexhaustible repository for natural polymers, which can facilitate sustainable development of renewable resources. Plants can absorb solar energy through photosynthesis, consume carbon dioxide, which causes global warming, and produce useful substances. Using the natural world to produce renewable resources is an environmentally friendly and energy-saving strategy. The natural world is undoubtedly an ideal factory for fabricating medicines or materials.

Diseases, natural disasters, and wars cause significant damage to the health and organs of people throughout the world, such that relying on organ transplants alone cannot satisfy the high demand required in this area of medicine. We predict the use of animal- or plant-based factories to achieve cures to modern diseases and to develop organ repair technologies. By using genetic engineering techniques, substances that could not be produced by animals and plants before can now be developed; additionally, the properties of existing substances can be improved to meet the specific requirements of the various materials needed by human kind. In recent years, significant progress has been made in the research and application of natural polymers in the field of regenerative medicine. The number of publications in this area has rapidly increased. However, based on the current knowledge of the chief editor, no special relevant monograph has been published on this subject. To facilitate the readers' understanding of the current status of research progress in this field and to introduce research advancements to our colleagues, we have collected the latest associated research achievements and published this monograph. Additionally, we hope that more scholars and graduate students will pay attention to and participate in the studies within this field.

In the process of editing this book, we have received strong support from all authors involved. Several of the authors are chief physician specialists from different departments who have been working at the forefront of the relevant clinical practices. Their active participation, enthusiasm and serious contributions have made the publication of this monograph possible. Here, I would like to express my sincere thanks to these individuals!

Due to the editor's limited capabilities, it is inevitable that flaws and mistakes may occur in this work. Accordingly, we sincerely look forward to your kind corrections.

Editor
August 2013, Shanghai

Chapter 1

Native Polymer-based 3D Substitutes in Cardiovascular Tissue Engineering

Hua-Jie Wang* and Ying Cao

Key Laboratory of Green Chemical Media and Reactions
Ministry of Education
College of Chemistry and Chemical Engineering
Henan Normal University, P. R. China

1. Introduction

Cardiovascular and related diseases are one of the most frequently occurring disorders and a leading cause of death, although considerable advances have already been made in an attempt to discover therapies for acute and chronic cardiovascular disease using preventive cardiovascular medicines (Breuer, 2011). In the United States alone, the operative cost of cardiovascular disease in 2008 was estimated to be a staggering US$297.7 billion, which includes strategies such as valve replacement, coronary artery bypass graft surgery, and stenting, and it accounted for about 16% of the total health expenditure (Patra *et al.*, 2012a; Roger *et al.*, 2012). Autologous, allogeneic, and xenogeneic vessels have demonstrated only limited success in cardiovascular disease therapy because of the insufficient supply of autologous and allogeneic vessels and the

*Corresponding author. E-mail: wanghuajie972001@163.com

immunogenicity of allogeneic and xenogeneic vessels. Given the inadequacies of existing therapy strategies for cardiovascular disease, tissue engineering technology, one of the frontier sciences, offers new hope for better treatment (Fong *et al.*, 2006; Hopkins, 2006; Lichtenberg *et al.*, 2006; Naito *et al.*, 2011; Simon *et al.*, 2006). As one of the key factors of tissue-engineered substitutes, biomaterials undoubtedly play a crucial role during cardiovascular tissue repair, working as an artificial extracellular matrix (ECM) to provide physical and even biochemical support for both differentiated and progenitor cells (Chen *et al.*, 2013). Among the available biomaterials, the advantages of raw materials sources, easy fabrication, and adjustable biological activities give native polymer-based materials the potential to be applied across a wide scope of implants in cardiovascular disease treatments. More importantly, they can be degraded after tissue regeneration and metabolized into innoxious products, such as saccharides and amino acids. This section will provide a comprehensive review on native polymer-based biomaterials used as cardiovascular tissue engineering scaffolds in clinic or in research phases.

2. Native Polymer-based Tissue-Engineered Heart Valves

The physiologic purpose of the human heart valve is to maintain unidirectional and non-obstructed blood flow (Chen *et al.*, 2001; Fong *et al.*, 2006). Valvular dysfunction will occur when a valve has restricted valve opening (stenosis), valve leakage (regurgitation), or both. Congenital and rheumatic valvular diseases of the heart dominate the main cardiac surgery. For example, approximately 20,000 people die annually as a direct result of valvular dysfunction in the United States (Schoen, 1997). Valve replacement surgery is efficacious and approximately 100,000 of such operations are performed annually in the United States (Lloyd-Jones *et al.*, 2010; Pibarot and Dumesnil, 2009). In addition, it is estimated that over 850,000 patients will require heart valve replacement by 2050 (Yacoub and

Takkenberg, 2005). Generally speaking, the state-of-the-art valves used clinically can be divided into two groups, namely mechanical and biological valves (Pibarot and Dumesnil, 2009). Both of these valves could effectively improve blood flow kinetics, but they do not provide a definitive cure to the patient; they may even exchange the native valve disease for "prosthetic valve disease" (Chan *et al.*, 2006; Hermans *et al.*, 2013). For example, mechanical valves have a substantial risk of thromboembolic and thrombotic obstruction, and patients with such mechanical valves have to receive lifelong anti-thrombotic therapy, especially in the early postoperative phase (<three months) (Bonow *et al.*, 2006; Jamieson *et al.*, 2004; Zilla *et al.*, 2007). Due to a low risk of thromboembolism without anticoagulation, biological valves are superior to mechanical valves. However, they are troubled by calcific or non-calcific tissue deterioration (Farivar and Cohn, 2003; Schoen and Levy, 2005; Stein *et al.*, 2001). Besides, the failure to grow, repair, and remodel is another innegligible fundamental problem during the use of both prostheses (Schoen, 2011). Therefore, an alternative way to improve the quality of heart valve replacement is evidently pressing.

Heart valve tissue engineering is aimed at constructing heart valve grafts that retain the surgical and design advantages, while doing away with the biological disadvantages of allografts and which have or promote restoration of recellularization with appropriate host cells that should represent the ideal valve replacement (Fig. 1) (Neuenschwander and Hoerstrup, 2004). Up to now, the tissue-engineered valve has experienced three distinct eras and the achieved progress has been impressive. However, the existing valves are to some extent affected by calcification, immune, and inflammatory reactions, and ultimately structural failure (Chen *et al.*, 2001; Hoerstrup *et al.*, 2000; Hopkins, 2006). Furthermore, an FDA-approved viable biological valve with physiological function and biological activity is not available and still in the animal experimental stage (Hilbert *et al.*, 2004; Hopkins, 2006; Lichtenberg *et al.*, 2006; Zou and Dong, 2010). Valve scaffold is the basis of tissue-engineered valves and acts as a starting point to construct neotissue (Langer and Vacanti, 1993). Among the available

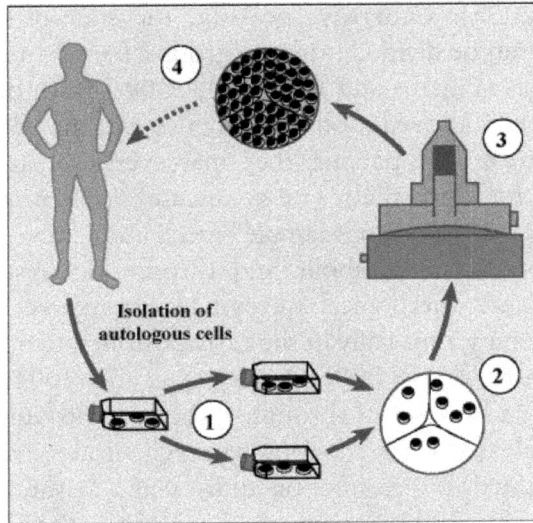

Fig. 1. Principle of tissue engineering of heart valves using autologous cells. (1) Isolation and expansion of autologous cells using standard monolayer culturing techniques; (2) cell seeding on a three-dimensional biodegradable scaffold; (3) "biomimetic" (mechanical) conditioning of the cells in a bioreactor; and (4) implantation of autologous living tissue-engineered heart valve (adapted from Neuenschwander and Hoerstrup, 2004).

scaffold materials, natural and biodegradable polymers derived from the organisms may be more biocompatible and less immunogenic than synthetic polymers. Moreover, they also can be engineered to meet early-stage biomechanical needs (Teebken *et al.*, 2000). Following that, the neotissue can develop its own biomechanical properties through the deposition of ECM on the temporary scaffold, which will be gradually degraded and absorbed. The currently used native polymers for artificially engineered valve scaffolds are mainly collagen, fibrin, hyaluronic acid, alginate, chitosan, and so on (Albanna *et al.*, 2012; Chow *et al.*, 2013; Daamen *et al.*, 2007; Mulder, 2002; Ramamurthi and Vesely, 2005; Taylor *et al.*, 2006b; Ye *et al.*, 2000). These materials are inherently biocompatible and biodegradable through enzymolysis, which enables cell-controlled tissue remodeling (Butcher *et al.*, 2011). In addition, the biodegradability can be adjusted by suitable cross-links according to the requirements (Masters *et al.*, 2005).

2.1. *Collagen*

Of the available native polymers, type I collagen, the major ECM protein of the native valve, contributes to the mechanical and tensile strength of the valve. It can only induce weak immunogenic reaction due to its homology across species (Chevallay and Herbage, 2000). Especially, collagens have the RGD peptide sequence (Arg–Gly–Asp), which supplies the cell attachment sites and drives the "RGD-dependent adhesion system" (Benoit *et al.*, 2012). Therefore, type I collagen is considered to be one of the most appropriate materials (Taylor *et al.*, 2002; Taylor *et al.*, 2006a). Many research groups have reported that different cardiovascular cells (e.g., human valve interstitial cells and vascular cells, myofibroblasts, and smooth muscle cells) tended to attach, migrate, and proliferate on/in type I collagen-based scaffolds. Meanwhile, the scaffolds could enhance the capacity of cells to express their native phenotype and synthesize ECM proteins, including highly organized collagen and proteoglycans (Taylor *et al.*, 2002; Rothenburger *et al.*, 2001; Rothenburger *et al.*, 2002a; Rothenburger *et al.*, 2002b). However, a choice has to be made between non-cross-linked and fully cross-linked collagen-based scaffolds, because the former will degrade too quickly to exert its functions and the latter will reduce the cell infiltration and remodeling capacity, and even prevent tissue regeneration (Filova *et al.*, 2009).

2.2. *Fibrin*

Fibrin is one of the most promising native polymers because it can be easily obtained from the patient's blood. Thus, it corresponds to an autologous scaffold and will not produce any inflammatory reactions (Jockenhoevel *et al.*, 2001; Ye *et al.*, 2000). Additionally, fibrin could also be chemically derivatized or cross-linked into stable hydrogels or scaffolds (Ibrahim *et al.*, 2011; Segura *et al.*, 2005). Flanagan *et al.* (2007) reported tissue-engineered heart valves

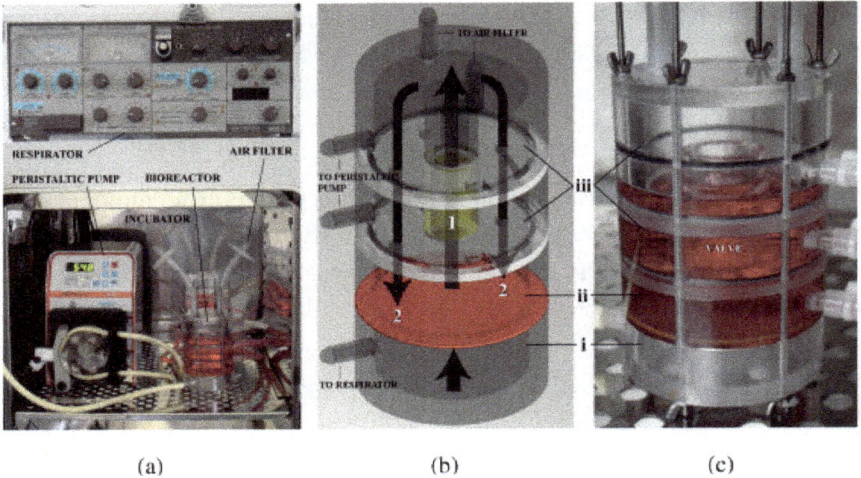

Fig. 2. (a) The bioreactor system consists of a respirator, peristaltic pump, standard cell culture incubator, and customized bioreactor. (b,c) The bioreactor consists of three chambers: (i) air chamber, (ii) ventricular chamber, and (iii) recirculation chamber. During systole, the silicone membrane separating the air and ventricular chambers is periodically displaced by a respirator [small arrow in (b) shows direction of displacement], with culture medium propelled through the lumen of the valve (1). During diastole, pressure is released by the respirator, and the culture medium is drawn back by gravity, closing the valve and opening silicone valves to the ventricular chamber for refilling (2) (adapted from Flanagan *et al.*, 2007).

constructed with autologous fibrin as the scaffold. Ovine carotide artery-derived cells were co-cultured with the fibrin-based scaffolds in a custom-designed bioreactor system. After a low-pressure and dynamic culture, the attachment, alignment, and expression of α-smooth muscle actin of seed cells within fibrin-based scaffolds were obtained (Figs. 2 and 3). It also enhanced the deposition of ECM proteins, including type I and II collagen, fibronectin, laminin, and chondroitin sulfate.

2.3. *Hyaluronic acid*

Hyaluronic acid, a non-sulfated polysaccharide, is an attractive candidate scaffold material for tissue-engineered valves (Masters *et al.*,

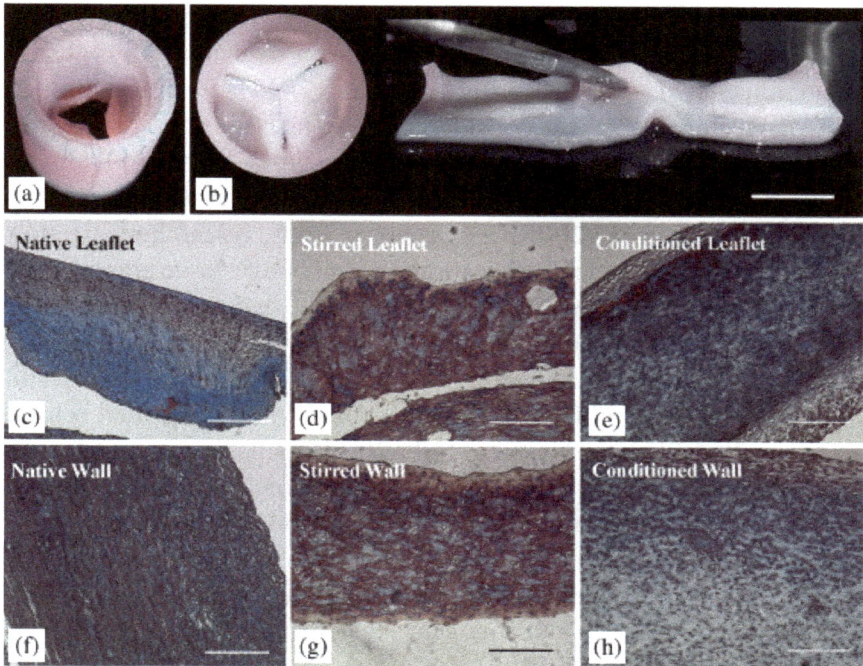

Fig. 3. (a) Gross appearance of the tissue-engineered heart valve immediately after removal from the bioreactor. The temporary silicone support cylinder is attached to two PVDF mesh rings (positioned at either end of the conduit and infiltrated by the gel) using polypropylene sutures. (b) The fibrin-based valves demonstrated intact, flexible tissue consisting of three leaflets continuous with a conduit wall (left: outflow side of closed valve; right: opened conduit cut through conduit wall following removal of the silicone cylinder). Scale: 10 mm. (c–h) Histological micrographs of trichrome-stained samples revealed an extensive collagenous component (blue) within the native ovine aortic valve leaflet and aortic wall (c,f), and also the conditioned leaflet and wall (e,h), with less staining in stirred control tissue (d,g). The majority of cells in stirred tissue samples were detached from fibrin scaffold and surrounded by lacunae (d,g), while attachment and alignment of a more dense population of cells was evident in conditioned samples (e,h). Scale: 250 μm (adapted from Flanagan *et al.*, 2007).

2004; Masters *et al.*, 2005). It is the major glycosaminoglycan of the native valve leaflet and plays a key role during cardiac morphogenesis (Lagendijk *et al.*, 2013; Rodriguez *et al.*, 2011; Torii and Bashey, 1966). Hyaluronic acid is considered to be non-immunogenic, non-thrombogenic, viscoelastic, and can be cross-linked to form hydrogel scaffolds, with controllable biodegradable and polymerization properties (Masters *et al.*, 2005). For these reasons, researchers are now exploiting hyaluronic acid applications in heart valve tissue engineering. Several groups have prepared hyaluronic acid hydrogels scaffolds by the photopolymerizable technology and chemical cross-linking reaction. Clearly, the preponderance of evidence suggested that hyaluronic acid-based scaffolds were suitable for valvular interstitial cells to attach and proliferate (Masters *et al.*, 2005; Ramamurthi and Vesely, 2005). Moreover, the degradation products of hyaluronic acid also could significantly promote the cell proliferation of valvular interstitial cells and endothelial cells, and angiogenesis and secretion of the ECM, which will be beneficial to enhancing the mechanical and biological properties of hyaluronic acid scaffolds (Masters *et al.*, 2005; Ramamurthi and Vesely, 2005).

2.4. *Other native polymers*

Other native polymers, such as alginate and chitosan, are also applied in engineered heart valves. Although the studies are limited, they still exhibit a great potential (Mulder, 2002). Alginate is a naturally derived linear polysaccharide and has excellent biocompatibility and biodegradability properties (Huang *et al.*, 2010; Jork *et al.*, 2002). For example, Kanakis *et al.* (2001) found that sodium alginate could effectively reduce the calcification process of porcine and human heart valves, and the rates were decreased to 53% and 45%, respectively. Kuo and Ma (2001) tried to control the gelation rate, mechanical propertries, and 3D structure of alginate hydrogel scaffolds by using ionic cross-linking methods. They demonstrated that the ratio of $CaCO_3$ and $CaSO_4$, calcium content, polymer concentration, and gelation temperature were the key factors to affect the uniform structure, stronger mechanical properties, better structural integrity, and inductivity on new tissue formation and function of alginate-based scaffolds. However, the deficiency of specific interactions between mammalian cells and alginate scaffolds is one of the drawbacks. Due to the negative charge property of alginate scaffolds, the electrostatic repulsion between scaffolds and cells has to be considered before their application as valve scaffolds.

Chitosan, linear copolymers of D-GlcN and D-GluNAc residues, has exhibited its unlimited application potential in the field of tissue engineering due to its desirable biological properties (Prashanth and Tharanathan, 2007). However, its use in tissue-engineered valves is limited due to low mechanical properties. Thus, this drawback has garnered much attention. Albanna *et al.* (2012) accomplished an improvement on the mechanical properties of chitosan-based heart valve scaffolds using chitosan fibers, and the tensile strength of the obtained fiber-reinforced heart valve scaffold could be comparable even to human pulmonary valve leaflets. They demonstrated that the key factors affecting the mechanical properties of scaffolds included fiber/scaffold mass ratio, fiber mechanical properties, and fiber length.

3. Native Polymer-based Tissue-Engineered Blood Vessels

Blood vessel diseases such as atherosclerosis and arteritis, chronic venous insufficiency, and thrombosis remain as key global vascular problems (Nemeno-Guanzon *et al.*, 2012; Rosamond *et al.*, 2007), the treatment of which results in a major economic burden (Roger *et al.*, 2011). An ideal tissue-engineered vessel is biologically and functionally stable with the appropriate healing properties similar to the native vascular tissue (Thomas *et al.*, 2003). The thrombogenicity and poor mechanical properties are the typical challenges in front of tissue-engineered vessels (Nemeno-Guanzon *et al.*, 2012). The former can be solved by seeding endothelial cells onto the grafts (Ku and Park, 2010). However, the latter is a key block for native polymer-based engineered vessels (L'Heureux *et al.*, 2006; Thomas *et al.*, 2012). To date, native polymer-based engineered vessels are still in the infant stage and the involved native polymers mainly contain collagen, fibrin, hyaluronic acid, alginate, chitosan, etc. (Boccafoschi *et al.*, 2005; Krenning *et al.*, 2008).

3.1. *Collagen*

Collagen is one of the most plentiful proteins found in the native blood vessel, working as a strength supporter in vessels (Sell *et al.*, 2009). The study on the application of collagen in vascular tissue engineering dates back to 1986. Weinberg and Bell (1986) firstly constructed a collagen-based scaffold with multilayered structure that resembled an artery. Bovine endothelial cells, smooth muscle cells, and fibroblasts were embedded within it. The expression of von Willebrand factor and prostacyclin could be detected in the endothelial cells lining the lumen. Moreover, the smooth cells in the wall were healthy and well differentiated. However, the model was not stable in itself, and its strength depended on its multiple layers of collagen and an integrated Dacron mesh. Following that, L'Heureux *et al.* (1993) also fabricated a human collagen-based scaffold using similar techniques. Unfortunately, they encountered the same mechanical limitations when human vascular cells were cultured on the scaffold and the burst pressures were only 120 mmHg. The type I collagen-based tissue engineered blood vessel construction reported by Hirai *et al.* (1994) and Hirai and Matsuda (1996) also ended up in failure when it was grafted alone in a canine model and ruptured within few days of implantation. Mechanical properties of collagen-based tissue-engineered blood vessels are generally demonstrated to be poor and thus various approaches have begun to address these issues. For example, Tranquillo's group (1996) used a strong magnetic field to spatially align collagen, and this circumferential orientation mimicked natural arteries. The results revealed that the magnetic field increased the stiffness of media equivalents with reduced creep in the circumferential direction. Girton *et al.* (1999) also successfully applied naturally occurring glycation cross-linking to stiffen and strengthen tissue equivalents and increased their resistance to collagenolytic degradation. Dynamic mechanical stimulation to augment cell-mediated remodeling of the collagen-based scaffolds is also considered to be successful and it can be attributed to the circumferential orientation of collagen fibers (Seliktar *et al.*, 2000). Berglund *et al.* (2003) successfully increased the mechanical strength of type I collagen-based tissue-engineered vessels by cross-linking with glutaraldehyde, ultraviolet, or dehydrothermal treatments, and the burst pressure rose to 650 mmHg. Additionally, Berglund *et al.* (2004) also incorporated elastin into collagen-based tissue-engineered vessels, which better mimicked arterial physiology. Moreover, this modification could increase tensile strength and linear stiffness moduli of collagen-based tissue-engineered blood vessels, and improve their viscoelastic properties.

3.2. *Fibrin*

Fibrin contains binding sites for endothelial cells, and smooth muscle cells and fibrin gel-based scaffolds could provide high seeding efficiency and uniform cell distribution by gelation entrapment (Aper *et al.*, 2004). At the same time, the secretion of reinforcing ECM proteins by seeded cells could be stimulated by fibrin. The biodegradation rate also could be adjusted by the loading of protease inhibitors, e.g., tranexamic acid or aprotinin (Cholewinski *et al.*, 2009; Ye *et al.*, 2000). Thus, fibrin has been developed as a vascular tissue-engineered scaffold material for applications in several settings, such as coronary artery and peripheral artery bypass procedures, and arteriovenous access grafts for hemodialysis patients (Tschoeke *et al.*, 2009). However, fibrin alone also does not possess sufficient mechanical properties to resist the high pressure in the body. Subsequently, the application of fibrin-based tissue-engineered blood vessels is the same as that of collagen, and it is always supported by a bio-absorbable, macroporous mesh, working as a temporary supporting system upon implantation. Tschoeke *et al.* (2008) fabricated a fibrin-based vascular graft seeded with carotid myofibroblasts and the graft was supported by a polyvinylidene fluoride mesh. After 14 days of dynamic culture at a flow rate of 250 mL/min, cell growth and tissue development were excellent within the fibrin gel. Furthermore, the mean suture retention strength and mean burst strength of grafts

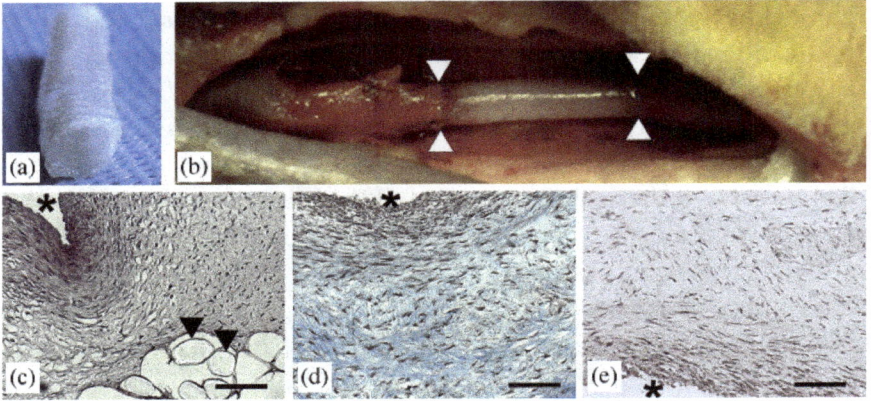

Fig. 4. (a) Pre-implanted tissue-engineered vascular composite graft; (b) vascular composite grafts were implanted as an end-to-end anastomosis (arrowheads) to the common carotid artery; (c) H&E staining of pre-implanted grafts revealed a homogenous cell distribution within the fibrin-based cell carrier, which was supported by P(L/D)LA 96/4 fibers (arrowheads); (d) Gomori's trichrome staining illustrated abundant collagen (blue) within the pre-implanted grafts; (e) absence of calcific deposits prior to implantation was evidenced by van Kossa stain (black). Images c–e (graft mid-section). Graft lumen is indicated by *. Scale bars: 200 μm (adapted from Koch *et al.*, 2010).

increased to 642 g and 236±26 mmHg, respectively. Following that, Koch *et al.* (2010) implanted fibrin-based grafts supported by a polylactide mesh in the arterial circulation of an ovine carotide model for up to six months (Fig. 4). The grafts showed no evidences for thrombus formation, aneurysm, calcification, and infection, and functioned for six months. Meanwhile, histological analysis indicated the remodeling of the fibrin scaffold with autologous connective tissue elements.

3.3. *Elastin*

Elastin is the dominant ECM protein deposited in the arterial wall during early postnatal life (Kielty, 2006; Kielty *et al.*, 2002). It confers blood vessels with the critical properties of elastic recoil and prevents dynamic tissue creep (Patel *et al.*, 2006). Moreover, elastin possesses specific cell attachment motifs, such as VGVAPG (Mochizuki *et al.*, 2002). They can mediate cell attachment, migration, survival, proliferation, phenotype, and differentiation (Nivison-Smith and Weiss, 2011). All these drive the development of elastin as the vascular engineering scaffold (Kielty *et al.*, 2007). Chuang *et al.* (2009) fabricated tubular elastin-based tissue-engineered vascular grafts, which were stabilized by penta-galloyl glucose, an established polyphenolic elastin-stabilizing agent. The results revealed that the scaffolds could effectively resist elastase digestion and maintain their original shapes *in vitro*. The scaffolds obtained a high resistance to burst pressures and it could get to 800 mmHg. The *in-vivo* tests also demonstrated that the penta-galloyl glucose modification reduced the rate of elastin biodegradation and controlled cell infiltration, but did not hamper new collagen and proteoglycan deposition and secretion of matrix-degrading proteases after eight weeks of implantation. Moreover, no calcium salts were detected at four weeks, and at eight weeks, little was deposited on the penta-galloyl glucose-stabilized elastin scaffolds. However, due to the extensive cross-linking and consequential insolubility of elastin, it is difficult for elastin to be manipulated (Daamen *et al.*, 2007). Therefore, the feasibility of hydrolyzed elastin as vascular tissue-engineered scaffold material was discussed (Nivison-Smith and Weiss, 2011). Hydrolyzed elastin could be obtained by hydrolysis with oxalic acid, potassium hydroxide, or proteinases. This type of elastin is water-soluble and can achieve polymerization through temperature-induced aggregation (coacervation). Furthermore, it still retains the regulation activity on smooth muscle cell and fibroblast phenotype. However, the poor mechanical properties are still a key disadvantage for elastin when it is independently applied as tissue-engineered blood vessels, and subsequently, a synthetic polymer scaffold has to be used as a mechanical support (Smith and McClure, 2008).

3.4. *Silk fibroin*

The application of silk fibroin as a biomaterial is currently a hot spot in tissue engineering due to their excellent biocompatibility, adaptable biodegradability, good oxygen/water vapor permeability, and remarkable mechanical properties (Kundu *et al.*, 2012; Zhang *et al.*, 2009). Besides, silk fibroin is proved to be anti-thrombotic and has good resistance to high shear stress and blood flow pressure (Enomoto *et al.*, 2010; Marelli *et al.*, 2010; Zhang *et al.*, 2009). All these advantages provide some insight about the potential of application of silk fibroin in vascular tissue engineering (Causin *et al.*, 2011; Leonardi *et al.*, 2011; Lovett *et al.*, 2007). Soffer *et al.* (2008) applied the electrospinning technique to fabricate a silk fibroin nano-fibrous-based tubular structure, which had 3 mm diameter and 0.15 mm thickness. Besides the good biocompatibility with human endothelial cells and smooth muscle cells, the engineered blood vessel also had impressive mechanical properties, including a burst strength of 811 ± 77.2 mmHg, an average tensile strength of 2.42 ± 0.48 mPa, and a linear modulus of 2.45 ± 0.47 mPa. In order to enhance the anti-thrombotic lining, sulfation of silk fibroin was suggested and proved to be effective. For instance, the electrospinning technique was also applied by Liu *et al.* (2011) and they fabricated a sulfated silk fibroin nano-fibrous-based engineered blood vessel. This scaffold was suitable for endothelial cells and smooth muscle cells to attach on, proliferate, and express some phenotype-related marker genes and proteins. Moreover, it was shown to be anticoagulant. The *in-vivo* evaluation of fibroin-based grafts in the abdominal aorta of rats for one year indicated better patency than that of poly(tetrafluoroethylene) (PTFE) grafts (85.1% *vs.* 30%) (Enomoto *et al.*, 2010).

3.5. *Bacteria cellulose*

Bacteria cellulose, a polysaccharide, is produced by *Acetobacter xylinum* and well-known in the field of tissue engineering due to its high strength and stiffness, high water content, high crystallinity, biodegradability and renewability, and ultra-fine, highly pure fibrous structure similar to that of collagen (Backdahl *et al.*, 2006; Czaja *et al.*, 2007; Klemm *et al.*, 2001). In particular, bacteria cellulose can be molded into almost any size and shape during its synthesis without changing its physical properties, and hence, studies about the application of bacteria cellulose in vascular tissue engineering became more active since the

1990s (Backdahl *et al.*, 2011; Bodin *et al.*, 2007; Charpentier *et al.*, 2006; Klemm *et al.*, 2001; Pooyan *et al.*, 2012; Yamanaka *et al.*, 1990). Helenius *et al.* (2006) performed a systematic evaluation on the *in-vivo* biocompatibility of bacterial cellulose and found that bacterial cellulose elicited negligible foreign body or imflammatory response after 12 weeks of subcutaneous implantation in rats. Moreover, the lower thrombogenicity of bacterial cellulose than that of expanded PTFE (ePTFE) and poly(ethyleneterephtalate) (PET) had been proven by Fink *et al.* (2010). In addition, the works of Klemm *et al.* (2001; 2003; 2005) are also typical. They fabricated a cellulose-based vascular graft named BASYC, with an i.d. of 1 mm, a length of about 5 mm, and wall thickness of 0.7 mm. The maximal tensile strength of this graft rose to about 800 mN and was able to withstand the blood pressure of a rat (0.02 mPa). After four weeks of implantation in place of the carotid artery of a rat, the graft was covered with connective tissue and its inner surface was completely covered with properly oriented endothelial cells.

3.6. *Other native polymers*

Other native polymers, including hyaluronan, fibronectin, and chitosan, have also been developed to act as the vascular tissue engineering scaffolds (Turner *et al.*, 2004; Zhu *et al.*, 2005). For example, Lepidi *et al.* (2006) fabricated a hyaluronan-based tissue engineering blood vessel with 2 mm in diameter and successfully induced completely artery regeneration after four months of implantation in a rat experimental model. Additionally, Zavan *et al.* (2008) implanted the hyaluronan-based tubular scaffold of 4 mm in diameter and 5 cm in length into a porcine model. The results indicated that there were three cases of partial or complete occlusion due to inimal hyperplasia and thrombosis. After five months, the remaining grafts were completely degraded and replaced by a neoartery segment, including mature smooth cells, collagen, and elastin fibers, organized in layers. Turner *et al.* (2004) conjugated fibronectin onto highly porous 3D poly(carbonate) urethane scaffolds through grafted poly(acrylic acid) spacers on the urethane backbone. This scaffold possessed better attachment and infiltration depth than those of human coronary artery smooth muscle cells. As for chitosan-based vascular tissue engineering scaffolds, Haque *et al.* (2001) found it was able to elicit an intense thrombotic and foreign body reaction despite heparin bonding in a pig model. However, they considered that the alteration of the base material or addition of bioactive side chains could reduce the thrombogenic and antigenic qualities of chitosan-based scaffold. Zhu *et al.* (2005) proved this opinion by using azidebenzoic to modify chitosan and fabricated a chitosan–heparin complex surface-modified with a ePTFE vascular graft. The graft markedly reduced platelets adhesion *in vitro* and retained patency after two weeks of implantation into the saphenous vein in the abdomen of a dog.

4. Native Polymer-based Tissue-Engineered Myocardium

Myocardial infarction and end-stage heart failure following myocardial infarction are major cardiac diseases, with the characters of cardiac muscle damage (Wang and Guan, 2010). Especially for adults, their cardiac muscle are terminally differentiated cells and difficult to regenerate. In 1998, the National Institutes of Health initiated the Bioengineered Autologous Tissue program, which aims to develop a living piece of heart muscle (Zandonella, 2003). Myocardial tissue engineering is different from heart valves or blood vessels tissue engineering as myocardial tissue is unable to be replaced and relies on the regeneration of the myocardium (Leor *et al.*, 2005). Native polymer-based myocardial tissue engineering has seen many attempts to find solutions for myocardium regeneration and mainly involves *in-situ* myocardial tissue engineering and *in-vitro* engineering of myocardium substitutes (Bouten *et al.*, 2011; Wang and Guan, 2010; Zimmermann *et al.*, 2004).

4.1. *In-situ myocardial tissue engineering*

In-situ myocardial tissue engineering involves the application of injectable biomaterials, which are combined with seeding cells and delivered directly into the infracted myocardium (Reffelmann and Kloner, 2003). The experimental data have shown that injectable biomaterials could improve cell retention with the aid of its higher viscosity and is suitable for angiogenesis (Ryu *et al.*, 2005). Such native polymers mainly contain collagen, fibrin, alginate, chitosan, and so on (Christman *et al.*, 2004; Landa *et al.*, 2008; Lu *et al.*, 2008; van Amerongen *et al.*, 2006). For example, Dai *et al.* (2005) injected zyderm collagen into Fischer rats with one-week-old myocardial infarcts. Six weeks later, they found that collagen injection significantly increased scar thickness, prevented paradoxical systolic bulging, increased the end-diastolic and decreased end-systolic volumes. Ryu *et al.* (2005) induced an infarction model in rat myocardium by cryo-injury and investigated the potential of injectable fibrin and bone marrow mononuclear cells at enhancing neovascularization in infarcted myocardium (Fig. 5). The results revealed that the implanted cells survived in the infarcted myocardium after eight weeks of implantation and this treatment could cause more extensive tissue regeneration and neovascularization (Fig. 6). Leor *et al.* (2009) tried to reverse left ventricular remodeling in swine after myocardial infarction through *in-situ* formation of alginate hydrogel. The alginate solution could be delivered effectively into the infracted myocardium by intracoronary injection and formed a calcium cross-linked alginate gel. Sixty days later, this treatment significantly prevented and even reversed left ventricular enlargement and increased scar thickness (Leor *et al.*, 2009). Moreover, the gel was replaced by myofibroblasts and collagen. Recently, Ruvinov *et al.* (2011) successfully delivered IGF-1 and HGF into the infarct in a Sprague–Dawley rat model of acute

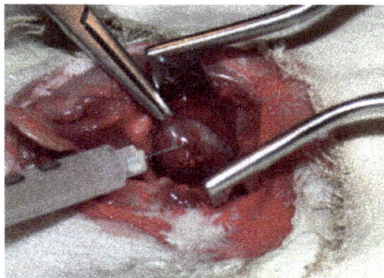

Fig. 5. Photograph of cell injection into infarcted myocardium. Thrombin solution containing BMMNCs and fibrinogen solution were injected into infarcted myocardium through a device designed for the simultaneous injection of the fibrinogen and thrombin solutions (adapted from Ryu *et al.*, 2005).

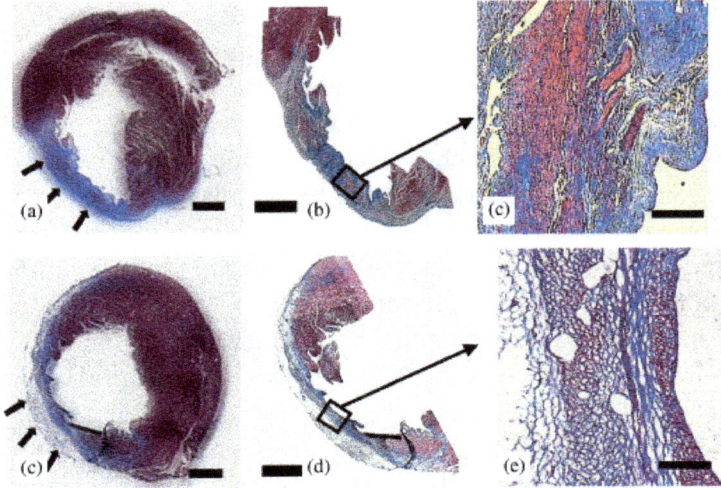

Fig. 6. Masson's trichrome staining of the infarcted myocardium eight weeks after the treatments. The arrows indicated infarction sites. (a,b,c) BMMNC implantation with fibrin matrix (scale bar = (a,b) 2 mm, and (c) 100 μm). (d,e,f) BMMNC implantation without matrix (scale bar = (d,e) 2 mm, and (f) 100 μm). A larger of viable tissues (appearing as a red color) and a smaller amount of fibrous tissues (appearing as a blue color) were present in the group of BMMNC implantation with fibrin matrix than the group of BMMNC implantation without matrix (adapted from Ryu *et al.*, 2005).

myocardial infarction with the application of an injectable alginate. This treatment obviously preserved scar thickness, attenuated infarct expansion, and reduced scar fibrosis after four weeks, concomitantly with increased angiogenesis and mature blood vessel formation at the infarct. Other recent researches have also proved that cell adhesion peptides-based modification and microencapsulation techniques could effectively improve the therapeutical efficacy of alginate in myocardial infarction (Tsur-Gang *et al.*, 2009; Yu *et al.*, 2010). Co-injection of embryonic or adipose-derived mesenchymal stem cells with chitosan hydrogel also proved to be effective at enhancing stem cell engraftment, survival and homing, and improving heart function, wall thickness, and microvessel densities within the infarcted area in the rat infarction models (Liu *et al.*, 2012; Lu *et al.*, 2008).

4.2. *In-vitro engineering of myocardium substitutes*

In-vitro engineering of myocardium substitutes involves the application of transportable biomaterials, which are firstly co-cultured *in vitro* with seeding cells and subsequently implanted on the infarcted myocardium. Due to the difference of seeding cells, the *in-vitro* engineering of engineered myocardium substitutes can be divided into two routes, namely *in-vitro* engineering of beating cardiomyocytes-containing tissue constructs, and engineering of stem cell-containing tissue constructs (Wang and Guan, 2010). Type I collagen to some extent has produced some success (van Amerongen *et al.*, 2006). For example, Zimmermann *et al.* (2002) mixed cardiac myocytes from neonatal rats with type I collagen from rat tails in circular casting molds. After seven days of culture *in vitro*, a unidirectional cyclic stretch was performed on the reconstitution mixture for another seven days. By comparison with native heart tissue from six-day-old and adult rats, cardiac cells in the engineered myocardium substitutes intensively interconnected, longitudinally oriented, and exhibited morphological features and contractile characteristics of adult rats. Following this finding, they continued to implant the large (thickness/diameter, 1–4 mm/15 mm), force-generating, and type I collagen/cardiac myocyte-based reconstitution mixtures on myocardial infarcts in immune-suppressed rats (Zimmermann *et al.*, 2006). Twenty-eight days later, engineered heart tissue prevented further dilation, induced systolic wall thickening of infarcted myocardial segments, improved fractional area shortening of infarcted hearts, and showed undelayed electrical coupling to the native myocardium without evidence of arrhythmia induction. The reports from Zhao *et al.* (2005), Kofidis *et al.* (2002), and Radisic *et al.* (2004) also proved the potential of collagen as a compatible scaffold for myocardium substitutes. Of course, other native polymers, e.g., alginate and silk fibroin, were also investigated and excellent experimental data were obtained (Dar *et al.*, 2002; Patra *et al.*, 2012b; Sakai *et al.*, 2001). Dar *et al.* (2002) prepared a tissue-engineered cardiac graft composed of cardiomyocytes seeded within a porous alginate scaffold. By a centrifugal seeding technique, a uniform cell distribution could be realized. Moreover, the forming cell aggregates within the scaffold contracted spontaneously. Recently, the modification on alginate with RGD peptide was performed and succeeded in improving survival and cell organization (Shachar *et al.*, 2011). Patra *et al.* (2012b) found that the silk fibroin was suitable for cardiomyocytes to attach without affecting their response to extracellular stimuli. Cardiomyocytes grew well on silk fibroin, exhibited aligned sarcomers, and coupled electrically with each other, resulting in synchronous beating. In addition, the stable and spontaneously beat form the generated cardiac tissues which can last for at least 20 days.

Nowadays, there is growing evidence that stem cells are the major key for heart muscle regeneration (Karam *et al.*, 2012). When the myocardium is damaged, the resident cardiac stem cells will be activated. Moreover, a stem cell population from other tissues also can be recruited (Beltrami *et al.*, 2003). The stem cells then differentiate under the myocardial environment to form cardiomyocytes and endothelial cells to improve cardiac function. Therefore, several groups have started research in the engineering of stem cell-containing tissue constructs for the repair of infracted myocardium. Fiumana *et al.* (2013) recently fabricated a mesenchymal stem cells-/hyaluronan-based graft. After two weeks of implantation in the infarcted area of the heart, the grafted cells moved toward the border zone, while the scaffold kept stable. Moreover, it promoted vascularization and reduced fibrosis in the infarcted area. Shi *et al.* (2011) fabricated a smart stem cell-capturing collagen. By conjugating with a stem cell-specific antibody (anti-Sca-1 antibody), the collagen scaffold could enrich more stem cells both *in vitro* and *in vivo*. Furthermore, the functional collagen scaffold successfully promoted cardiac tissue regeneration in the mouse.

5. Applications of Composite Materials in Cardiovascular Tissue Engineering

Besides the above-mentioned artificial engineered cardiovascular scaffolds composed of a sole native material, the composition of two or more materials is another alternative to act as tissue-engineered cardiovascular scaffolds. This composite matrix may actually combine the

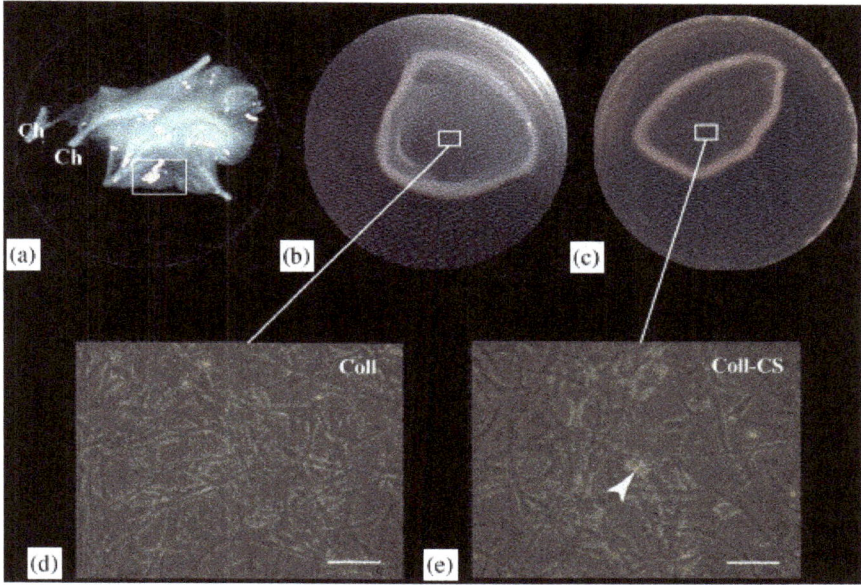

Fig. 7. Gross structure of porcine mitral valve leaflet, collagen, and collagen–chitosan constructs. (a) Ventricular surface of the mitral valve anterior leaflet showing attached chordae tendineae (Ch). The tissue composing the free edge of the leaflet (white box) bears a macroscopic resemblance to the porcine mitral valve gel constructs. (b,c) Collagen (b) and collagen–chitosan (c) gel constructs seeded with mitral VICs and VECs. Each gel has contracted to form a tissue-like structure. (d,e) Phase contrast micrographs of collagen (d) and collagen–chitosan (e) gel constructs. Focal sectioning throughout the full thickness of the constructs revealed an extensive VIC network with homogeneous cell distribution. Mitotic cells were also evident deep within the constructs (e, arrowhead). Scale bar: 100 μm (adapted from Flanagan *et al.*, 2006).

advantages of different materials. For example, Flanagan *et al.* (2006) synthesized a tissue-engineered heart valve with type I collagen–chondroitin sulfate hydrogels (Fig. 7). Clearly, evidences showed that the composition scaffold was more beneficial for porcine mitral valve endothelial cells to grow than when only collagen scaffolds were used. Besides, more ECM, including elastin and laminin, could be secreted by valve interstitial cells. More significantly, they detected the expression of the vasoactive molecule, eNOS, but contrary to collagen scaffolds. In 2013, Chi *et al.* used chitosan–hyaluronan/silk fibroin patches to repair myocardium in the myocardial infarction model of rats. The implantation treatment significantly reduced the dilation of the inner diameter of left ventricle, increased wall thickness, improved the fractional shortening, and promoted the secretion of paracrine factors.

6. Summary and Future Directions

Native polymers have received considerable interest for cardiovascular tissue engineering applications. Nevertheless, a conclusion is still yet to be reached regarding which material is the most optimal for cardiovascular tissue engineering scaffolds due to the complexity of tissue structure. In contrast to the promising *in-vitro* and *in-vivo* results with native polymer-based cardiovascular tissue-engineered substitutes, many critical issues remain to be addressed and be solved for the realization of a tissue-engineered angiocarpy. For example: how can the compliance match between the scaffold and native blood vessels be optimized? How can the native alignment of different cells in the scaffold be improved to form a functional substitute? How can the biodegradation rate of the scaffold and its functionalization be balanced? How can the mechanical properties of substitutes be enhanced to sustain long-term systemic pressure? Can a ready-to-use substitute be synthesized for emergency patients? All these questions pose as a basis to drive subsequent research and development of native polymer-based 3D cardiovascular tissue engineering.

References

Albanna MZ, Bou-Akl T, Walters III HL, Matthew HWT. Improving the mechanical properties of chitosan-based heart valve scaffolds using chitosan fibers. *J Mech Behav Biomed Mater* 2012;5:171–180.

Aper T, Teebken OE, Steinhoff G, Haverich A. Use of a fibrin preparation in engineering of a vascular graft model. *Eur J Vasc Endovasc Surg* 2004;28: 296–302.

Backdahl H, Helenius G, Bodin A, Nannmark U, Johansson BR, Risberg B, Gatenholm P. Mechanical properties of bacterial cellulose and interactions with smooth muscle cells. *Biomaterials* 2006;27:2141–2149.

Backdahl H, Risberg B, Gatenholm P. Observations on bacterial cellulose tube formation for application as vascular graft. *Mater Sci Eng* 2011;31:14–21.

Beltrami AP, Barlucchi L, Torella D, Baker M, Limana F, Chimenti S, Kasahara H, Rota M, Musso E, Urbanek K, Leri A, Kajstura J, Nadal-Ginard B, Anversa P. Adult cardiac stem cells are multipotent and support myocardial regeneration. *Cell* 2003;114:763–776.

Benoit YD, Groulx JF, Gagne D, Beaulieu JF. RGD-dependent epithelial cell-matrix interactions in the human intestinal crypt. *J Signal Transduction* 2012;248759.

Berglund JD, Mohseni MM, Nerem RM, Sambanis A. A biological hybrid model for collagen-based tissue engineered vascular constructs. *Biomaterials* 2003;24:1241–1254.

Berglund JD, Nerem RM, Sambanis A. Incorporation of intact elastin scaffolds in tissue-engineered collagen-based vascular grafts. *Tissue Eng* 2004;10(9–10):1526–1535.

Boccafoschi F, Habermehl J, Vesentini S, Mantovani D. Biological performances of collagen-based scaffolds for vascular tissue engineering. *Biomaterials* 2005;26:7410–7417.

Bodin A, Backdahl H, Fink H, Gustafsson L, Risberg B, Gatenholm P. Influence of cultivation conditions on mechanical and morphological properties of bacterial cellulose tubes. *Biotechnol Bioeng* 2007;97:425–434.

Bonow RO, Carabello BA, Kanu C, de Leon AC Jr, Faxon DP, Freed MD, Gaasch WH, Lytle BW, Nishimura RA, O'Gara PT, O'Rourke RA, Otto CM, Shah PM, Shanewise JS, Smith SC Jr, Jacobs AK, Adams CD, Anderson JL, Antman EM, Faxon DP, Fuster V, Halperin JL, Hiratzka LF, Hunt SA, Lytle BW, Nishimura R, Page RL, Riegel B. ACC/AHA 2006 guidelines for the management of patients with valvular heart disease: A report of the American College of Cardiology/American Heart Association Task Force on Practice Guidelines. *Circulation* 2006;114:e84–e231.

Bouten CVC, Dankers PYW, Driessen-Mol A, Pedron S, Brizard AMA, Baaijens FPT. Substrates for cardiovascular tissue engineering. *Adv Drug Delivery Rev* 2011;63:221–241.

Breuer CK. The development and translation of the tissue-engineered vascular graft. *J Pediatr Surg* 2011;46:8–17.

Butcher JT, Mahler GJ, Hockaday LA. Aortic valve disease and treatment: the need for naturally engineered solutions. *Adv Drug Delivery Rev* 2011;63:242–268.

Causin F, Pascarella R, Pavesi G, Marasco R, Zambon G, Battaglia R, Munari M. Acute endovascular treatment (<48 hours) of uncoilable ruputured aneurysms at non-branching sites using silk flow-diverting devices. *Interv Neuroradiol* 2011;17(3):357–364.

Chan V, Jamieson WR, Germann E, Chan F, Miyagishima RT, Burr LH, Janusz MT, Ling H, Fradet GJ. Performance of bioprostheses and mechanical prostheses assessed by composites of valve-related complications to 15 years after aortic valve replacement. *J Thorac Cardiovasc Surg* 2006;131(6):1267–1273.

Charpentier PA, Maguire A, Wan WK. Surface modification of polyester to produce a bacterial cellulose-based vascular prosthetic device. *Appl Surf Sci* 2006;252: 6360–6367.

Chen Q, Liang S, Thouas GA. Elastomeric biomaterials for tissue engineering. *Prog Polym Sci* 2013;38:584–671.

Chen XW, Chi YF, Niu ZZ, Hou WM, Sun ZD, Sun Y. Application and safety evaluation of different types of heart valve biomaterials. *J Clin Rehabil Tissue Eng Res* 2001;15(12):2257–2260.

Chevallay B, Herbage D. Collagen-based biomaterials as 3D scaffold for cell cultures: Applications for tissue engineering and gene therapy. *Med Biol Eng Comput* 2000;38(2):211–218.

Chi NH, Yang MC, Chung TW, Chou NK, Wang SS. Cardiac repair using chitosan-hyaluronan/silk fibroin patches in a rat heart model with myocardial infarction. *Carbohydr Polym* 2013;92:591–597.

Cholewinski E, Dietrich M, Flanagan TC, Schmitz-Rode T, Jockenhoevel S. Tranexamic acid — an alternative to a protinin in fibrin-based cardiovascular tissue engineering. *Tissue Eng Part A* 2009;15(11):3645–3653.

Chow JP, Simionescu DT, Warner H, Wang B, Patnaik SS, Liao J, Simionescu A. Mitigation of diabetes-related complications in implanted collagen and elastin scaffolds using matrix-binding polyphenol. *Biomaterials* 2013;34:685–695.

Chuang TH, Stabler C, Simionescu A, Simionescu DT. Polyphenol-stabilized tubular elastin scaffolds for tissue engineered vascular grafts. *Tissue Eng Part A* 2009;15(10):2837–2851.

Christman KL, Vardanian AJ, Fang Q, Sievers RE, Fok HH, Lee RJ. Injectable fibrin scaffold improves cell transplant survival, reduces infarct expansion, and induces neovasculature formation in ischemic myocardium. *J Am Coll Cardiol* 2004;44:654–660.

Czaja WK, Young DJ, Kawecki M, Brown RB Jr. The future prospects of microbial cellulose in biomedical application. *Biomacromolecules* 2007;8(1):1–12.

Daamen WF, Veerkamp JH, van Hest JC, van Kuppevelt TH. Elastin as a biomaterial for tissue engineering. *Biomaterials* 2007;28(30):4378–4398.

Dai W, Wold LE, Dow JS, Kloner RA. Thickening of the infarcted wall by collagen injection improves left ventricular function in rats: A novel approach to preserve cardiac function after myocardial infarction. *J Am Coll Cardiol* 2005;46(4):714–719.

Dar A, Shachar M, Leor J, Cohen S. Optimization of cardiac cell seeding and distribution in 3D porous alginate scaffolds. *Biotechnol Bioeng* 2002;80(3): 305–312.

Enomoto S, Sumi M, Kajimoto K, Nakazawa Y, Takahashi R, Takabayashi C, Asakura T, Sata M. Long-term patency of small-diameter vascular graft made from fibroin, a silk-based biodegradable material. *J Vasc Surg* 2010;51:155–164.

Farivar RS, Cohn LH. Hypercholesterolemia is a risk factor for bioprosthetic valve calcification and explantation. *J Thorac Cardiovasc Surg* 2003;126:969–975.

Filova E, Straka F, Mirejovsky T, Masin J, Bacakova L. Tissue-engineered heart valves. *Physiol Res* 2009;58(Suppl. 2):S141–S158.

Fink H, Faxalv L, Molnar GF, Drotz K, Risberg B, Lindahl TL, Sellborn A. Real-time measurements of coagulation on bacterial cellulose and conventional vascular graft materials. *Acta Biomater* 2010;6:1125–1130.

Fiumana E, Pasquinelli G, Foroni L, Carboni M, Bonafe F, Orrico C, Nardo B, Tsivian M, Neri F, Arpesella G, Guarnieri C, Caldarera CM, Muscari C. Localization of mesenchymal stem cells grafted with a hyaluronan-based scaffold in the infarcted heart. *J Surg Res* 2013;179:E21–E29.

Flanagan TC, Cornelissen C, Koch S, Tschoeke B, Sachweh JS, Schmitz-Rode T, Jockenhoevel S. The *in vitro* development of autologous fibrin-based tissue-engineered heart valves through optimised dynamic condition. *Biomaterials* 2007;28(23):3388–3397.

Flanagan TC, Wilkins B, Black A, Jockenhoevel S, Smith TJ, Pandit AS. A collagen-glycosaminoglycan co-culture model for heart valve tissue engineering applications. *Biomaterials* 2006;27:2233–2246.

Fong P, Shin'oka T, Lopez-Soler PI, Breuer C. The use of polymer-based scaffolds in tissue-engineered heart valves. *Prog Pediatr Cardiol* 2006;21:193–199.

Girton TS, Oegema TR, Tranquillo RT. Exploiting glycation to stiffen and strengthen tissue equivalents for tissue engineering. *J Biomed Mater Res* 1999;46(1):87–92.

Haque MI, Beekley AC, Gutowska A, Reardon RA, Groo P, Murray SP, Andersen C, Azarow K. Bioabsorption quatilities of chitosan-absorbable templates. *Curr Surg* 2001;58(1):77–80.

Helenius G, Backdahl H, Bodin A, Nannmark U, Gatenholm P, Risberg B. *In vivo* biocompatibility of bacterial cellulose. *J Biomed Mater Res A* 2006;76(2):431–438.

Hermans H, Vanassche T, Herijgers P, Meuris B, Herregods MC, Van de Werf F, Verhamme P. Antithrombotic therapy in patients with heart valve prostheses. *Cardiol Rev* 2013;21(1):27–36.

Hilbert SL, Yanagida R, Souza J, Wolfinbarger L, Jones AL, Krueger P, Stearns G, Bert A, Hopkins RA. Prototype anionic detergent technique used to decellularized allograft valve conduits evaluated in the right ventricular outflow tract in sheep. *J Heart Valve Dis* 2004;13:831–840.

Hirai J, Kanda K, Oka T, Matsuda T. Highly oriented, tubular hybrid vascular tissue for a low-pressure circulatory system. *ASAIO J* 1994;40:M383–M388.

Hirai J, Matsuda T. Venous reconstruction using hybrid vascular tissue composed of vascular cells and collagen-tissue regeneration process. *Cell Transplant* 1996;5:93–105.

Hoerstrup SP, Sodian R, Daebritz S, Wang J, Bacha EA, Martin DP, Moran AM, Guleserian KJ, Sperling JS, Kaushal S, Vacanti JP, Schoen FJ, Mayer JE Jr.

Functional living trileaflet heart valves grown *in vitro*. *Circulation* 2000;102(Suppl.):III44–49.

Hopkins R. From cadaver harvested homograft valves to tissue-engineered valve conduits. *Prog Pediatr Cardiol* 2006;21:137–152.

Huang SB, Wu MH, Lee GB. Microfluidic device utilizing pneumatic micro-vibrators to generate alginate microbeads for microencapsulation of cells. *Sens Actuators B* 2010;147:755–764.

Ibrahim S, Kothapalli CR, Kang QK, Ramamurthi A. Characterization of glycidyl methacrylate-crosslinked hyaluronan hydrogel scaffolds in corporating elasto-genic hyaluronan oligomers. *Acta Biomater* 2011;7:653–665.

Jamieson WR, Cartier PC, Allard M, Boutin C, Burwash IG, Butany J, de Varennes B, Del Rizzo D, Dumesnil JG, Honos G, Houde C, Munt BI, Poirier N, Rebeyka IM, Ross DB, Siu SC, Williams WG, Rebeyka IM, David TE, Dyck JD, Feindel CM, Fradet GJ, Human DG, Lemieux MD, Menkis AH, Scully HE, Turpie AG, Adams DH, Berrebi A, Chambers J, Chang KL, Cohn LH, Duran CM, Elkins RC, Freedman R, Huysman HA, Jue J, Perier P, Rakowski H, Schaff HV, Schoen FA, Shah P, Thompson CR, Warnes C, Westaby S, Yacoub MH. Surgical management of valvular heart disease 2004. *Can J Cardiol* 2004;20(Suppl. E):7E–120E.

Jockenhoevel S, Zund G, Hoerstrup SP, Chalabi K, Sachweh JS, Demircan L, Messmer BJ, Turina M. Fibrin gel-advantages of a new scaffold in cardiovascular tissue engineering. *Eur J Cardiothorac Surg* 2001;19(4):424–430.

Jork A, Thurmer F, Cramer H, Zimmermann G, Gessner P, Hamel K, Hof G. Biocompatible alginate from freshly collected laminaria pallida for implantation. *Appl Microbiol Biotechnol* 2002;53:224–229.

Kanakis J, Malkaj P, Petroheilos J, Dalas E. The crystallization of calcium carbonate on porcine and human cardiac valves and the antimineralization effect of sodium alginate. *J Cryst Growth* 2001;223:557–564.

Karam JP, Muscari C, Montero-Menei CN. Combining adult stem cells and poly-meric devices for tissue engineering in infarcted myocardium. *Biomaterials* 2012;33(23):5683–5695.

Kielty CM. Elastic fibres in health and disease. *Expert Rev Mol Med* 2006;8:1–23.

Kielty CM, Sherratt MJ, Shuttleworth CA. Elastic fibres. *J Cell Sci* 2002;115:2817–2828.

Kielty CM, Stephan S, Sherratt MJ, Williamson M, Shuttleworth CA. Applying elas-tic fibre biology in vascular tissue engineering. *Phil Trans R Soc B* 2007;362:1293–1312.

Klemm D, Heublein B, Fink HP, Bohn A. Cellulose: Fascinating biopolymer and sustainable raw material. *Angew Chem Int Ed* 2005;44:3358–3393.

Klemm D, Schumann D, Udhardt U, Marsch S. Bacterial synthesized cellulose-artificial blood vessels for microsurgery. *Prog Polym Sci* 2001;26:1561–1603.

Klemm D, Udhardt U, Marsch S, Schumann D. Method and device for producing shaped microbial cellulose for use as biomaterial, especially for microsurgery. U.S. Patent 0013163A1, 2003.

Koch S, Flanagan TC, Sachweh JS, Tanios F, Schnoering H, Deichmann T, Ella V, Kellomaki M, Gronloh N, Gries T, Tolba R, Schmitz-Rode T, Jockenhoevel S. Fibrin-polylactide-based tissue-engineered vascular graft in the arterial circulation. *Biomaterials* 2010;31(17):4731–4739.

Kofidis T, Akhyari P, Boublik J, Theodorou P, Martin U, Ruhparwar A, Fishcher S, Eschenhagen T, Kubis HP, Leyh R, Haverich A. *In vitro* engineering of heart muscle: Artificial myocardial tissue. *J Thorac Cardiovasc Surg* 2002;124(1):63–69.

Krenning G, Moonen J-RAJ, van Luyn MJA, Harmsen MC. Vascular smooth muscle cells for use in vascular tissue engineering obtained by endothelial-to-mesenchymal transdifferentiation (EnMT) on collagen matrices. *Biomaterials* 2008;29:3703–3711.

Ku SH, Park CB. Human endothelial cell growth on mussel-inspired nanofiber scaffold for vascular tissue engineering. *Biomaterials* 2010;31:9431–9437.

Kundu B, Rajkhowa R, Kundu SC, Wang X. Silk fibroin biomaterials for tissue regenerations. *Adv Drug Delivery Rev* doi: 10.1016/j.addr.2012.09.043, 2012.

Kuo CK, Ma PX. Ionically crosslinked alginate hydrogels as scaffolds for tissue engineering: Part 1. Structure, gelation rate and mechanical properties. *Biomaterials* 2001;511–521.

Lagendijk AK, Szabo A, Merks RMH, Bakkers J. Hyaluronan: A critical regulator of endothelial-to-mesenchymal transition during cardiac valve formation. *Trends Cardiovasc Med* doi: 10.1016/j.tcm.2012.10.002, 2012.

Landa N, Miller L, Feinberg MS, Holbova R, Schachar M, Freeman I, Cohen S, Leor J. Effect of injectable alginate implant on cardiac remodeling and function after recent and old infarcts in rat. *Circulation* 2008;117:1388–1396.

Langer R, Vacanti JP. Tissue engineering. *Science* 1993;260:920–926.

Leonardi M, Cirillo L, Toni F, Dall'Olio M, Princiotta C, Stafa A, Simonetti L, Agati R. Treatment of intracranial aneurysms using flow-diverting silk stent (BALT): A single centre experience. *Interv Neuroradiol* 2011;17(3): 306–315.

Leor J, Amsalem Y, Cohen S. Cells, scaffolds, and molecules for myocardial tissue engineering. *Pharmacol Ther* 2005;105:151–163.

Leor J, Tuvia S, Guetta V, Manczur F, Castel D, Willenz U, Petnehazy O, Landa N, Feinberg M, Konen E, Goitein O, Tsur-Gang O, Shaul M, Klapper L, Cohen S. Intracoronary injection of *in situ* forming alginate hydrogel reverses left ventricular remodeling after myocardial infarction in swine. *J Am Coll Cardiol* 2009;54(11):1014–1023.

Lepidi S, Abatangelo G, Vindigni V, Deriu GP, Zavan B, Tonello C, Cortivo R. *In vivo* regeneration of small-diameter (2 mm) arteries using a polymer scaffold. *FASEB J* 2006;20:103–105.

Lepidi S, Grego F, Vindigni V, Zavan B, Tonello C, Deriu GP, Abatangelo G, Cortivo R. Hyaluronan biodegradable scaffold for small-caliber artery grafting: Preliminary results in an animal model. *Eur J Vasc Endovasc Surg* 2006;32:411–417.

L'Heureux N, Dusserre N, Konig G, Victor B, Keire P, Wight TN, Chronos NA, Kyles AE, Gregory CR, Hoyt G, Robbins RC, McAllister TN. Human tissue-engineered blood vessels for adult arterial revascularization. *Nat Med* 2006;12:361–365.

L'Heureux N, Germain L, Labbe R, Auger FA. *In vitro* construction of a human blood vessel from cultured vascular cells: A morphologic study. *J Vasc Surg* 1993;17:499–509.

Lichtenberg A, Breymann T, Cebotari S, Haverich A. Cell seeded tissue engineered cardiac valves based on allograft and xenograft scaffold. *Prog Pediatr Cardiol* 2006;21:211–217.

Liu H, Li X, Zhou G, Fan H, Fan Y. Electrospun sulfated silk fibroin nanofibrous scaffolds for vascular tissue engineering. *Biomaterials* 2011;32: 3784–3793.

Liu Z, Wang H, Wang Y, Lin Q, Yao A, Cao F, Li D, Zhou J, Duan C, Du Z, Wang Y, Wang C. The influence of chitosan hydrogel on stem cell engraftment, survival and homing in the ischemic myocardial microenvironment. *Biomaterials* 2012;33:3093–3106.

Lloyd-Jones D, Adams RJ, Brown TM, Carnethon M, Dai S, De Simone G, Ferguson TB, Ford E, Furie K, Gillespie C, Go A, Greenlund K, Haase N, Hailpern S, Ho PM, Howard V, Kissela B, Kittner S, Lackland D, Lisabeth L, Marelli A, McDermott MM, Meigs J, Mozaffarian D, Mussolino M, Nichol G, Roger VL, Rosamond W, Sacco R, Sorlie P, Roger VL, Thom T, Wasserthiel-Smoller S, Wong ND, Wylie-Rosett J. Heart disease and stroke statistics — 2010 update: A report from the American Heart Association. *Circulation* 2010;121:e46–e215.

Lovett M, Cannizzaro C, Daheron L, Messmer B, Vunjak-Novakovic G, Kaplan DL. Silk fibroin microtubes for blood vessel engineering. *Biomaterials* 2007;28:5271–5279.

Lu WN, Lu SH, Wang HB, Li DX, Duan CM, Liu ZQ, Hao T, He WJ, Xu B, Fu Q, Song YC, Xie XH, Wang CY. Functional improvement of infarcted heart by co-injection of embryonic stem cells with temperature-responsive chitosan hydrogel. *Tissue Eng Part A* 2008;15(6):1437–1447.

Marelli B, Alessandrino A, Fare S, Freddi G, Mantovani D, Tanzi MC. Compliant electrospun silk fibroin tubes for small vessel bypass grafting. *Acta Biomater* 2010;6:4019–4026.

Masters KS, Shah DN, Leinwand LA, Anseth KS. Crosslinked hyaluronan scaffolds as a biologically active carrier for valvular interstitial cells. *Biomaterials* 2005;2517–2525.

Masters KS, Shah DN, Walker G, Leinwand LA, Anseth KS. Designing scaffolds for valvular interstitial cells: Cell adhesion and function on naturally derived materials. *J Biomed Mater Res A* 2004;71(1):172–180.

Mochizuki S, Brassart S, Hinek A. Signaling pathways transduced through the elastin receptor facilitate proliferation of arterial smooth muscle cells. *J Biol Chem* 2002;277:44854–44863.

Mulder L. Cell adhesion on alginate scaffolds for tissue engineering of aortic valve — a review. Faculty Biomedical Engineering, Eindhoven University of Technology. 2002;22–34.

Naito Y, Shinoka T, Duncan D, Hibino N, Solomon D, Cleary M, Rathore A, Fein C, Church S, Breuer C. Vascular tissue engineering: Towards the next generation vascular grafts. *Adv Drug Delivery Rev* 2011;63:312–323.

Nemeno-Guanzon JG, Lee S, Berg JR, Jo YH, Yeo JE, Nam BM, Koh YG, Lee JI. Trends in tissue engineering for blood vessels. *J Biomed Biotechnol* doi: 10.1155/2012/956345, 2012.

Neuenschwander S, Hoerstrup SP. Heart valve tissue engineering. *Transplant Immunol* 2004;12:359–365.

Nivison-Smith L, Weiss A. Elastin based constructs. In: Eberli D, ed. *Regenerative Medicine and Tissue Engineering — Cells and Biomaterials.* InTech, 2011, pp. 323–340.

Patel A, Fine B, Sandig M, Mequanint K. Elastin biosynthesis: The missing link in tissue-engineered blood vessels. *Cardiovasc Res* 2006;71:40–49.

Patra C, Ricciardi F, Engel FB. The functional properties of nephronectin: An adhesion molecule for cardiac tissue engineering. *Biomaterials* 2012a;33(17): 4327–4335.

Patra C, Talukdar S, Novoyatleva T, Velagala SR, Muhlfeld C, Kundu B, Kundu SC, Engel FB. Silk protein fibroin from *Antheraea mylitta* for cardiac tissue engineering. *Biomaterials* 2012b;33:2673–2680.

Pibarot P, Dumesnil JG. Prosthetic heart valves: Selection of the optimal prosthesis and long-term management. *Circulation* 2009;119:1034–1048.

Pooyan P, Tannenbaum R, Garmestani H. Mechanical behaviour of a cellulose-reinforce scaffold in vascular tissue engineering. *J Mech Behav Biomed Mater* 2012;7:50–59.

Prashanth KVH, Tharanathan RN. Chitin/chitosan: Modifications and their unlimited application potential — an overview. *Trends Food Sci Technol* 2007;18:117–131.

Radisic M, Park H, Shing H, Consi T, Schoen FJ, Langer R, Freed LE, Vunjak-Novakovic G. Functional assembly of engineered myocardium by electrical stimulation of cardiac myocytes cultured on scaffolds. *Proc Natl Acad Sci USA* 2004;101(52):18129–18134.

Ramamurthi A, Vesely I. Evaluation of the matrix-synthesis potential of crosslinked hyaluronan gels for tissue engineering of aortic heart valves. *Biomaterials* 2005;26(9):999–1010.

Reffelmann T, Kloner RA. Cellular cardiomyoplasty-cardiomyocytes, skeletal myoblasts, or stem cells for regenerating myocardium and treatment of heart failure? *Cardiovasc Res* 2003;58:358–368.

Rodriguez KJ, Piechura LM, Masters KS. Regulation of valvular interstitial cell phenotype and function by hyaluronic acid in 2-D and 3-D culture environments. *Matrix Biol* 2011;30(1):70–82.

Roger VL, Go AS, Lloyd-Jones DM, Adams RJ, Berry JD, Brown TM, Carnethon MR, Dai S, de Simone G, Ford ES, Fox CS, Fullerton HJ, Gillespie C, Greenlund KJ, Hailpern SM, Heit JA, Ho PM, Howard VJ, Kissela BM, Kittner SJ, Lackland DT, Lichtman JH, Lisabeth LD,Makuc DM, Marcus GM, Marelli A, Matchar DB, McDermott MM, Meigs JB, Moy CS, Mozaffarian D, Mussolino ME, Nichol G, Paynter NP, Rosamond WD, Sorlie PD, Stafford RS, Turan TN, Turner MB, Wong ND, Wylie-Rosett J, American Heart Association Statistics Committee and Stroke Statistics Subcommittee. Heart disease and stroke statistics — 2011 update: A report from the American Heart Association. *Circulation* 2011;123:e18–e209.

Roger VL, Go AS, Lloyd-Jones DM, Benjamin EJ, Berry JD, Borden WB, Bravata DM, Dai S, Ford ES, Fox CS, Fullerton HJ, Gillespie C, Hailpern SM, Heit JA, Howard VJ, Kissela BM, Kittner SJ, Lackland DT, Lichtman JH, Lisabeth LD, Makuc DM, Marcus GM, Marelli A, Matchar DB, Moy CS, Mozaffarian D, Mussolino ME, Nichol G, Paynter NP, Soliman EZ, Sorlie PD, Sotoodehnia N, Turan TN, Virani SS, Wong ND, Woo D, Turner MB, American Heart Association Statistics Committee and Stroke Statistics Subcommittee. Heart disease and stroke statistics — 2012 update: A report from the American Heart Association. *Circulation* 2012;125:e2–e220.

Rosamond W, Flegal K, Friday G, Furie K, Go A, Greenlund K, Haase N, Ho M, Howard V, Kissela B, Kittner S, Lloyd-Jones D, McDermott M, Meigs J, Moy C, Nichol G, O'Donnell CJ, Roger V, Rumsfeld J, Sorlie P, Steinberger J, Thom T, Wasserthiel-Smoller S, Hong Y, American Heart Association Statistics Committee and Stroke Statistics Subcommittee. Heart disease and stroke statistics — 2007 update: A report from the American Heart Association Statistics Committee and Stroke Statistics Subcommittee. *Circulation* 2007;115(5):e69–e171.

Rothenburger M, Vischer P, Volker W, Glasmacher B, Berendes E, Scheld HH, Deiwick M. *In vitro* modeling of tissue using isolated vascular cells on a synthetic collagen matrix as a substitute for heart valves. *Thorac Cardiovasc Surg* 2001;49(4):204–209.

Rothenburger M, Volker W, Vischer P, Glasmacher B, Scheld HH, Deiwick M. Ultrastructure of proteoglycans in tissue-engineered cardiovascular structures. *Tissue Eng* 2002a;8(6):1049–1056.

Rothenburger M, Volker W, Vischer JP, *et al.* Tissue engineering of heart valves: Formation of a three-dimensional tissue using porcine heart valve cells. *ASAIO J* 2002b;48(6):586–591.

Ruvinov E, Leor J, Cohen S. The promotion of myocardial repair by the sequential delivery of IGF-1 and HGF from an injectable alginate biomaterial in a model of acute myocardial infarction. *Biomaterials* 2011;32:565–578.

Ryu JH, Kim IK, Cho SW, Cho MC, Hwang KK, Piao H, Piao S, Lim SH, Hong YS, Choi CY, Yoo KJ, Kim BS. Implantation of bone marrow monouclear cells using injectable fibrin matrix enhances neovascularization in infarcted myocardium. *Biomaterials* 2005;26:319–326.

Sakai T, Li RK, Weisel RD, Mickle DA, Kim ET, Jia ZQ, Yau TM. The fate of a tissue-engineered cardiac graft in the right ventricular outflow tract of the rat. *J Thorac Cardiovasc Surg* 2001;121(5):932–942.

Shachar M, Tsur-Gang O, Dvir T, Leor J, Cohen S. The effect of immobilized RGD peptide in alginate scaffolds on cardiac tissue engineering. *Acta Biomater* 2011;7:152–162.

Schoen FJ. Aortic valve structure-function correlations: Role of elastic fibers no longer a stretch of the imagination. *J Heart Valve Dis* 1997;6(1):1–6.

———. Heart valve tissue engineering: *Quo vadis? Curr Opin Biotechnol* 2011;22:698–705.

Schoen FJ, Levy RJ. Calcification of tissue heart valve substitutes: Progress toward understanding and prevention. *Ann Thorac Surg* 2005;79:1072–1080.

Segura T, Anderson BC, Chung PH, Webber RE, Shull KR, Shea LD. Crosslinked hyaluronic acid hydrogels: A strategy to functionalize and pattern. *Biomaterials* 2005;26:359–371.

Seliktar D, Black RA, Vito RP, Nerem RM. Dynamic mechanical conditioning of collagen-gel blood vessel constructs induces remodeling *in vitro*. *Ann Biomed Eng* 2000;28:351–362.

Sell SA, McClure MJ, Garg K, Wolfe PS, Bowlin GL. Electrospinning of collagen/biopolymers for regenerative medicine and cardiovascular tissue engineering. *Adv Drug Delivery Rev* 2009;61:1007–1019.

Shi C, Li Q, Zhao Y, Chen W, Chen B, Xiao Z, Lin H, Nie L, Wang D, Dai J. Stem-cell-capturing collagen scaffold promotes cardiac tissue regeneration. *Biomaterials* 2011;32:2508–2515.

Simon P, Kasimir MT, Rieder E, Weigel G. Tissue engineering of heart valves-immunologic and inflammatory challenges of the allograft scaffold. *Prog Pediatr Cardiol* 2006;21(2):161–165.

Smith MJ, McClure MJ. Suture-reinforced electrospun polydioxanone-elastin small-diameter tubes for use in vascular tissue engineering: A feasibility study. *Acta Biomater* 2008;4(1):58–66.

Soffer L, Wang X, Zhang X, Kluge J, Dorfmann L, Kaplan DL, Leisk G. Silk-based electrospun tubular scaffolds for tissue-engineered vascular grafts. *J Biomater Sci Polym Ed* 2008;19:653–664.

Stein PD, Alpert JS, Bussey HI, Dalen JE, Turpie AG. Antithrombotic therapy in patients with mechanical and biological prosthetic heart valves. *Chest* 2001;119(Suppl. 1):220S–227S.

Taylor PM, Allen SP, Dreger SA, Yacoub MH. Human cardiac valve interstitial cells in collagen sponge: A biological three-dimensional matrix for tissue engineering. *J Heart Valve Dis* 2002;11(3):298–306.

Taylor PM, Cass AEG, Yacoub MH. Extracellular matrix scaffolds for tissue engineering heart valves. *Prog Pediatr Cardiol* 2006a;21:219–225.

Taylor PM, Sachlos E, Dreger SA, Chester AH, Czernuszka JT, Yacoub MH. Interaction of human valve interstitial cells with collagen matrices manufactured using rapid prototyping. *Biomaterials* 2006b;27:2733–2737.

Teebken OE, Bader A, Steinhoff G, Haverich A. Tissue engineering of vascular grafts: Human cell seeding of decellularised porcine matrix. *Eur J Vasc Endovasc Surg* 2000;19:381–386.

Thomas AC, Campbell GR, Campbell JH. Advances in vascular tissue engineering. *Cardiovasc Pathol* 2003;12:271–276.

Thomas LV, Lekshmi V, Nair PD. Tissue-engineered vascular grafts — preclinical aspects. *Int J Cardiol* doi: 10.1016/j.ijcard.2012.09.069, 2012.

Torii S, Bashey R. High content of hyaluronic acid in normal human heart valves. *Nature* 1966;209:506–507.

Tranquillo RT, Girton TS, Bromberek BA, Triebes TG, Mooradian DL. Magnetically oriented tissue-equivalent tubes: Application to a circumferentially orientated media-equivalent. *Biomaterials* 1996;17(3):349–357.

Tschoeke B, Flanagan TC, Cornelissen A, Koch S, Roehl A, Sriharwoko M, Sachweh JS, Gries T, Schmitz-Rode T, Jockenhoevel S. Development of a composite degradable/nondegradable tissue-engineered vascular graft. *Artif Organs* 2008;32(10):800–809.

Tschoeke B, Flanagan TC, Koch S, Harwoko MS, Deichmann T, Ella V, Sachweh JS, Kellomaki M, Gries T, Schmitz-Rode T, Jockenhoevel S. Tissue-engineered small-caliber vascular graft based on a novel biodegradable composite fibrin-polylactide scaffold. *Tissue Eng Part A* 2009;15(8):1909–1918.

Tsur-Gang O, Ruvinov E, Landa N, Holbova R, Feinberg MS, Leor J, Cohen S. The effects of peptide-based modification of alginate on left ventricular remodeling and function after myocardial infarction. *Biomaterials* 2009;30:189–195.

Turner NJ, Kielty CM, Walker MG, Canfield AE. A novel hyaluronan-based biomaterial (Hyaff-11®) as a scaffold for endothelial cells in tissue-engineered vascular grafts. *Biomaterials* 2004;25:5955–5964.

Vahanian A, Baumgartner H, Bax J, Butchart E, Dion R, Filippatos G, Flachskampf F, Hall, Iung B, Kasprzak J, Nataf P, Tornos P, Torracca L, Wenink A. Guidelines on the management of valvular heart disease: The Task Force on the Management of Valvular Heart Disease of the European Society of Cardiology. *Eur Heart J* 2007;28:230–268.

Van Amerongen MJ, Harmsen MC, Petersen AH, Kors G, van Luyn MJA. The enzymatic degradation of scaffolds and their replacement by vascularized extracellular matrix in the murine myocardium. *Biomaterials* 2006;27:2247–2257.

Wang F, Guan J. Cellular cardiomyoplasty and cardiac tissue engineering for myocardial therapy. *Adv Drug Delivery Rev* 2010;62:784–797.

Weinberg CB, Bell E. A blood vessel model constructed from collagen and cultured vascular cells. *Science* 1986;231(4736):397–400.

Yacoub MH, Takkenberg JJ. Will heart valve tissue engineering change the world? *Nat Clin Pract Cardiovasc Med* 2005;2:60–61.

Yamanaka S, Ono E, Watanabe K, Kusakabe M, Suzuki Y. Hollow microbial cellulose, process for preparation thereof, and artificial blood vessel formed of said cellulose. European Patent No. 0396344, 1990.

Ye Q, Zund G, Benedikt P, Jockenhoevel S, Hoerstrup SP, Sakyama S, Hubbell JA, Turina M. Fibrin gel at a three-dimentional matrix in cardiovascular tissue engineering. *Eur J Cardiothoral Surg* 2000;17:587–591.

Yu J, Du KT, Fang Q, Gu Y, Mihardja SS, Sievers RE, Wu JC, Lee RJ. The use of human mesenchymal stem cells encapsulated in RGD-modified alginate microspheres in the repair of myocardial infarction in the rat. *Biomaterials* 2010;31: 7012–7020.

Zandonella C. Tissue engineering: The beat goes on. *Nature* 2003;421:884–886.

Zavan B, Vindigni V, Lepidi S, Iacopetti I, Avruscio G, Abatangelo G, Cortivo R. Neoarteries growth *in vivo* using a tissue-engineered hyaluronan-based scaffold. *FASEB J* 2008;22(8):2853–2861.

Zhang X, Reagan MR, Kaplan DL. Electrospun silk biomaterials scaffolds for regenerative medicine. *Adv Drug Delivery Rev* 2009;6:988–1006.

Zhao YS, Wang CY, Li DX, Zhang XZ, Qiao Y, Guo XM, Wang XL, Dun CM, Dong LZ, Song Y. Construction of a unidirectionally beating 3-dimensional cardiac muscle construct. *J Heart Lung Transplant* 2005;24(8):1091–1097.

Zhu AP, Ming Z, Jian S. Blood compatibility of chitosan/heparin complex surface modified ePTFE vascular graft. *Appl Surf Sci* 2005;241:485–492.

Zilla P, Brink J, Human P, Bezuidenhout D. Prosthetic heart valves: Catering for the few. *Biomaterials* 2008;29:385–406.

Zimmermann WH, Melnychenko I, Eschenhagen T. Engineered heart tissue for regeneration of diseased hearts. *Biomaterials* 2004;25(9):1639–1647.

Zimmermann WH, Melnychenko I, Wasmeier G, Didie M, Naito H, Nixdorff U, Hess A, Budinsky L, Brune K, Michaelis B, Dhein S, Schwoerer A, Ehmke H, Eschenhagen T. Engineered heart tissue grafts improve systolic and diastolic function in infarcted rat hearts. *Nat Med* 2006;12(4):452–458.

Zimmermann WH, Schneiderbanger K, Schubert P, Didie M, Munzel F, Heubach JF, Kostin S, Neuhuber WL, Eschenhagen T. Tissue engineering of a differentiated cardiac muscle construct. *Circ Res* 2002;90(2):223–230.

Zou MH, Dong NG. Research and progress of scaffold materials for tissue engineered heart valves. *Zhongguo Zuzhi Gongcheng Yanjiu yu Linchuang Kangfu Zazhi* 2010;14(29):5471–5474.

Chapter 2

Native Polymer-based 3D Substitutes for Nerve Regeneration

*Guo-Wu Wang and Jin-Ye Wang**

School of Biomedical Engineering
Shanghai Jiaotong University
800 Dongchuan Road, Shanghai 200240, P. R. China

1. Introduction

With the increase in the number of patients suffering from peripheral nerve injury as a result of traffic jams, natural disasters, and war damages, the repair and the regeneration of peripheral nerves have drawn much attention. Unlike the central nervous system, peripheral nerves are able to regenerate after being damaged. After the peripheral nerve suffers an injury, it will undergo a series of molecular and cellular events called Wallerian degeneration, where the distal stump degenerates while the proximal stump survives and promotes regeneration (Hsu *et al.*, 2011). The peripheral nerve system has an intrinsic regenerative capacity after peripheral nerves are transected (Fig. 1) (Allodi *et al.*, 2012). First, large amounts of axons along the distal stump are reduced to amorphous debris and granular and following the myelin sheath, are then transformed toward the short segment (Chaudhry *et al.*, 1992). Then, the axon debris and myelin are removed by macrophages and monocytes recruited to the degenerating nerve stumps.

* Corresponding author: E-mail: jinyewang@sjtu.edu.cn

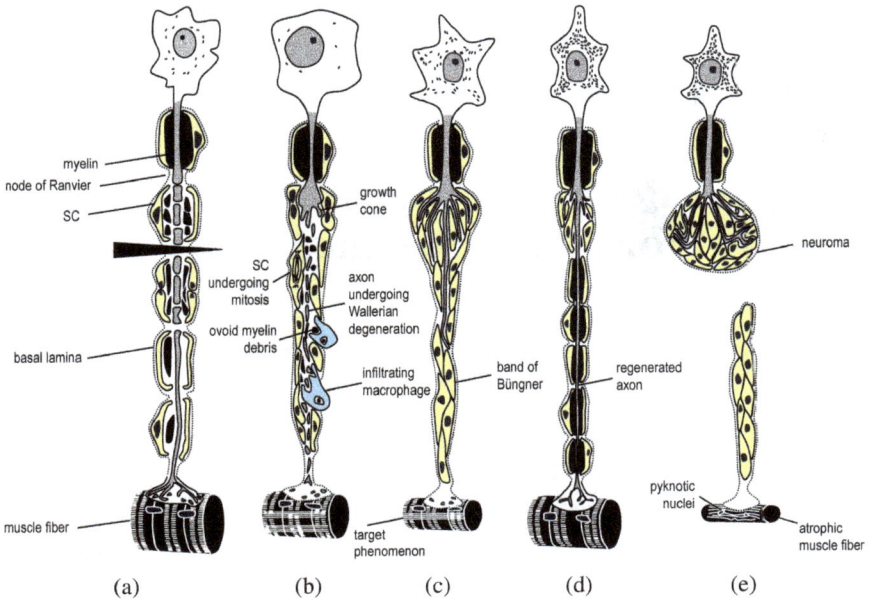

Fig. 1. Schematic representation of the degenerative and regenerative events associated with peripheral nerve injury. a: During the early phase (first few days) after axonal injury (arrowhead), local degenerative events are accompanied by both retrograde and antero-grade degeneration of axon and myelin. b: During the intermediate phase (a few days to weeks), the anterograde pattern of Wallerian degeneration proceeds to completion with infiltrating macrophages contributing to the removal of tissue debris and SC undergoing mitosis. The axotomised neuronal cell body undergoes reactive, chromatolytic changes and the severed proximal end of the axon develops regenerative axonal sprouts. c: Of the numerous axonal sprouts that successfully traverse the injury site (during the first few weeks to months), some re-enter appropriate endoneurial tubes and continue to extend through the distal nerve stump, supported by SC in the bands of Büngner. The target organ/tissue (in this case skeletal muscle) undergoes disuse atrophy. d: Successful axon regeneration through the bands of Büngner and the re-establishment of neurotransmis-sion at the neuromuscular junction results in the retraction or dying-back of unsuccessful axon sprouts, the reversal of muscle fiber atrophy, of neuronal cell body chromatolysis and the establishment of maturing SC-axon interactions (including reduced internodal spacing). e: Failure of regenerating axonal sprouts to cross the injury site (possibly due to the formation of a physical scarring barrier or the loss of a large segment of nerve) results in neuroma formation. The permanently denervated muscle fibers demonstrate severe atrophy, loss of their characteristic striations and pyknotic nuclei. The figure is adapted from (Deuments *et al.*, 2010).

At the same time, the Schwann cells grow into longitudinal cell columns, called Bands of Büngner (Stoll *et al.*, 1989). Schwann cells will stimulate the proximal stump to regenerate new axons by releasing extracellular matrix components and neurotrophic factors (Webber and Zochodne, 2010; Morris *et al.*, 1972). Functional reinnervation usually requires the regenerating axons to elongate towards their synaptic target under the mediation of growth cones. However, once the nerve suffers a complete lesion, the growth cone is unable to correctly reach the target organ and make a functional recovery (Lundborg, 2000; Gordon *et al.*, 2003). For a long time, physicians and scientists have attempted to search methods to repair the injury peripheral nerve. Since the 17th century, the suture technique was used to repair injury nerve (Artico *et al.*, 1996). With the progress in medical technology and scientific research, two types of surgical techniques, manipulative nerve operations and bridge operations, were developed to restore the peripheral nerve gaps. As one of the manipulative nerve operations, end to end coaptation is usually an appropriate treatment for the nerve lesion with a short gap (Sanders, 1942; di Summa *et al.*, 2010). However, the implantation of a graft is often necessary to bridge the proximal and distal nerve stumps for promoting nerve regeneration when a nerve defect or gap is longer (Lin *et al.*, 2008). Autologous nerve grafts have been used as an optimal choice, which is the golden standard in clinical surgery for nerve reconstructions. It still has several limitations, such as the limited availability of donor tissue, sacrifice of functional nerve and potential formation of neuroma (Cao *et al.*, 2011). Considering these defects, the tubulization technique has been developed. This technique is based on the application of hollow cylindrical conduit made of natural or synthetic materials, which provides a good microenvironment for the elongation of the growth cone (Lundborg *et al.*, 1982; Chamberlain *et al.*, 1988).

2. Design of Ideal Nerve Conduits

There are many requirements for ideal nerve conduits. An appropriate nerve conduit must be biodegradable and must exhibit good biocompatibility with extremely low inflammatory, immunogenic, and cytotoxic responses (Cunha *et al.*, 2011). Nerve conduits should also retain adequate mechanical strength and flexibility to support surgical suture and protect the regenerating axons from invasion by the surrounding connective tissue (Sundback *et al.*, 2003). In addition, the mechanical properties of the nerve conduit must guarantee that it does not collapse during the patient's movements but, at the same time, sufficiently elastic to avoid tension in the lesion site (Cunha *et al.*, 2011). The microstructures of the nerve conduits are also very important. Nerve conduits with micro-porous walls are not only permeable to the entry of nutrients and oxygen into the conduit lumen but also provide a necessary barrier against the infiltration of fibrous tissues into the conduit.

The materials selected for fabricating nerve conduits play an important role in the regeneration of injured nerve. At present, a variety of biomaterials, mainly including natural and synthetic materials, have been employed to prepare neural conduits of different structures and properties. Compared with synthetic materials, natural materials have good biocompatibility and are non-toxic. They not only stimulate the adhesion and migration of cells, but also promote the proliferation and growth of cells. Moreover, natural biomaterials are biodegradable and the degradation products are always non-toxic.

3. The Application of Bioengineered Natural Materials in Nerve Conduits

3.1. *Extracellular matrix components*

The extracellular matrix (ECM) is an important part of the extracellular environment. In addition to some proteins and non-proteoglycan polysaccharides, ECM is composed of a variety of glycoproteins and proteoglycans. The main functions of ECM are providing structural support and regulating the activity of cells. Glycoproteins, which mainly include laminin and fibronectin, play important roles in the elongation and growth of axons (Grimpe and Silver, 2002; Rutishauser, 1993). Proteoglycans and glycosaminoglycans can stimulate or inhibit neural activity and neuritis extension (Asher *et al.*, 2001; Bovolenta and Fernaud-Espinosa, 2000). ECM scaffolds have become a reliable and good biomaterial candidate as a result of these properties and good biocompatibility (Cozzolino *et al.*, 1999).

3.1.1. *Bioengineered collagen and its derivatives*

3.1.1.1. Introduction

As a major component in the extracellular matrix, collagen has been widely used as biomedical scaffolds in tissue engineering. Collagen is the most abundant protein in the extracellular matrix, making up more than one-third the weight of body protein tissue (Patino *et al.*, 2002). Collagen is derived from animal tissues, such as porcine skin, bovine deep flexor tendon and bone (Pati *et al.*, 2012). Collagen molecule has a triple-helix structure, formed by three chains in which two chains (α_1) are associated with a third chain (α_2) (Fratzl, 2003). There are about 28 types of collagen that have been identified up till now (Gelse *et al.*, 2003). Every family member contains domains of the proline–rich tripeptide Gly–X–Y that are involved in the formation of triple-helices. Different collagen types are characterized by their distinct considerable complexity and diversity in structure, splice variants, the presence of additional non-helical domains and functions (Gelse *et al.*, 2003). Collagen plays important roles in the process of organ development, wound healing and tissue repairing because it makes great contributions to the entrapment, storage and delivery of growth factors and cytokines (Yamaguchi *et al.*, 1990). Collagen has many advantages such as abundance, high porosity, easy processing, hydrophilicity, biodegradability, good biocompatibility and low antigenicity (Ferreira *et al.*, 2012). Therefore, collagen has been widely used in the field of biomedicine including: sutures, hemostatic agents, tissue replacement and regeneration (bone, cartilage, skin, blood vessels, etc.), cosmetic surgery, dental composites, membrane oxygenators, corneal bandage and drug delivery (Meena *et al.*, 1999; Scheibel, 2005; Pannone, 2007). In collagen families, the most abundant collagens are types I, II, and III collagen, especially type I collagen which has been widely used in tissue engineering (Kadler *et al.*, 2007). With about 49% of the total proteins in the nerves being type I and type III collagen, collagen has been widely used as the biomaterial for fabricating nerve conduits (Bunge *et al.*, 1989).

3.1.1.2. Category and characteristic

Since the 1990s, different forms of collagen nerve conduits have been applied to repair the injury peripheral nerve and the results are justified (Archibald *et al.*, 1991; Li *et al.*, 1992; Yoshii *et al.*, 2002). Alluin *et al.* (2009) used nerve conduits fabricated by type I and III collagen to bridge a 1 cm nerve gap and found that the empty collagen tube supported the motor axonal regeneration and promoted the locomotor recovery. In the rat sciatic nerve crush injury model, Sun *et al.* (2009) found that the collagen conduit binding nerve growth factors (NGF) could enhance functional recovery after nerve damage. Yao *et al.* (2010) fabricated the collagen into multi-channel nerve guidance to repair a 1 cm gap of sciatic nerve. The result showed that multi-channel guidance could control the dispersion of axonal regeneration but did not decrease the quantitative results of regeneration (Yao *et al.*, 2010) (Fig. 2). Besides rats, other animal models such as

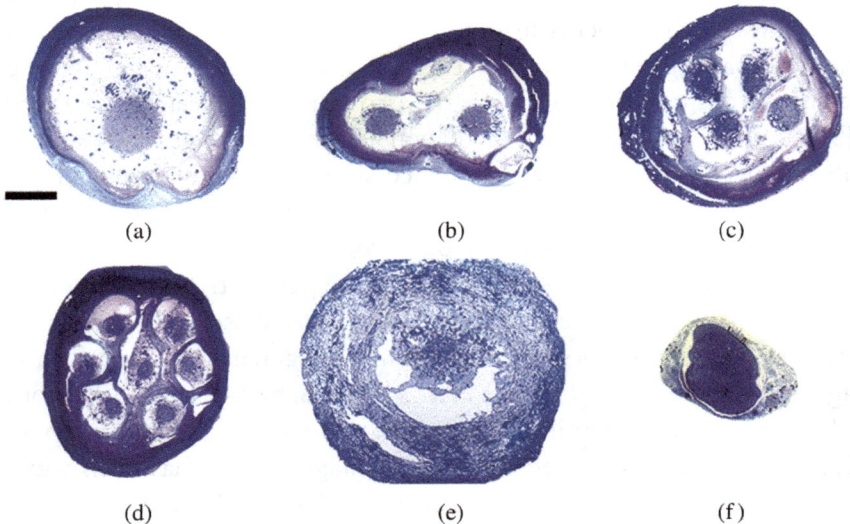

Fig. 2. Microscopic images (5× magnification) of sections stained with toluidine blue taken through the middle of an (a) 1-channel, (b) 2-channel, (c) 4-channel, and (d) 7-channel collagen conduit, (e) a NeuraGen® single channel conduit and (f) autograft. Scale bar, 500 μm. The figure is adapted from (Yao *et al.*, 2010).

cats and monkeys are also used to investigate the role that collagen tube plays in repairing the peripheral nerve defects (Kitahara *et al.*, 1999; Kitahara *et al.*, 2000; Archibald *et al.*, 1995). In Felix's and Keilhoff's works, the collagen conduit is rapidly integrated into the host tissue and the revascularization is soon observed after implantation (Stang *et al.*, 2005; Keilhoff *et al.*, 2003).

Although collagen is a good candidate for fabricating nerve conduits, it has poor mechanical properties due to high hydrophilicity. The rapid degradation rate may limit its application in repairing large size nerve gaps (Ferreira *et al.*, 2012; Alluin *et al.*, 2009; Harley *et al.*, 2004). To enhance the mechanical property and control the degradation time of the collagen conduit, many cross-linking methods which include chemical cross-linking and physical cross-linking have been used. These methods are reported to improve mechanical properties of collagen nerve conduits significantly (Alluin *et al.*, 2009; Itoh *et al.*, 2002; Ahmed *et al.*, 2005; Hu *et al.*, 2009). With the improvement, comparable functional recovery to autografts has been achieved (Li *et al.*, 1992; Kitahara *et al.*, 2000; Archibald *et al.*, 1995; Waitayawinyu *et al.*, 2007). Today, commercially available collagen nerve conduits such as NeuraGen®, Neuroflex®, NeuroMatrix® have been put into clinical use and have achieved good therapeutic effects (Archibald *et al.*, 1995; Farole and Jamal, 2008; Lohmeyer *et al.*, 2009). In conclusion, collagen is a very promising biomaterial in the application of fabricating nerve conduit to repair nerve injury.

3.1.2. *Bioengineered laminin and its derivatives*

3.1.2.1. Introduction

As a family member of glycoproteins of the extracellular matrix (ECM), laminin is the major constituent of basement membranes (Colognato and Yurchenco, 2000). At least 15 isoforms of laminin have been identified so far (Mochizuki *et al.*, 2003), each of which being large heterotrimers that consists of α, β, and γ chains including many subunits (Allodi *et al.*, 2012). Laminin not only is able to self-assemble but also binds to other matrix macro-molecules. Laminin was first isolated from the mouse tumor and the parietal yolk sac carcinoma cells (Timpl *et al.*, 1979; Chung *et al.*, 1979). Laminin is the first expressed ECM protein during embryogenesis which plays an important role in the early embryonic development (Gu *et al.*, 2011; Li *et al.*, 2002). Laminin has unique and shared cell interactions which are mediated by integrins, dystroglycan and other receptors (Colognato and Yurchenco, 2000). Through these interactions, laminin can make contribution to stimulate cell growth and differentiation, mediate cell communication and promote the survival of tissues (Beck *et al.*, 1990). Presently, it has exhibited bioactivity with many cell types, especially the neuronal cells, and has been widely used in neural tissue regeneration (Kleinman *et al.*, 2006; Tate *et al.*, 2009; Jurga *et al.*, 2011).

3.1.2.2. Category and characteristic

In the peripheral nervous system, laminin mainly originates from Schwann cells and disperses widely (Longo *et al.*, 1984; Lander *et al.*, 1985). Laminin plays an important role in promoting adhesion, migration, proliferation of the Schwann cells and guiding neurite outgrowth (Deister *et al.*, 2007). Once the peripheral nerves are injured, laminin is significantly upregulated at the injury site which will sustain the regeneration of the axons (Martini, 1994; Fu and Gordon, 1997). After nerve injury, the regenerated nerve fibers may disperse in the hollow nerve conduits and the misdirection of regenerating axons to inappropriate muscles can lead to poor functional recovery (Brushart *et al.*, 1995; Hamilton *et al.*, 2011). Therefore, it is necessary to control the dispersion and improve the directed axonal regeneration. Many attempts have been made, such as fabricating multichannel conduits, filling the conduits with microfilaments and luminal fillers and applying electrical stimulation (Yao *et al.*, 2010; Ngo *et al.*, 2003; Chen *et al.*, 2006; Yoshii *et al.*, 2009; Al-Majed *et al.*, 2000). Laminin is a common luminal filler which is widely used to repair the injury peripheral nerve. Matsumoto *et al.* (2000) used PGA–collagen nerve conduits filled with laminin-coated collagen fibers to bridge an 80 mm gap of dogs and observed favorable functional recovery (Matsumoto *et al.*, 2000). Madison *et al.* (1985) used nerve guide lumens containing laminin gel to achieve an increased number of myelinated axons when compared with empty lumens. In another study, similar conclusions were also drawn which shows that the presence of a laminin-containing gel increased the rate of axons crossing a transection site (Madison *et al.*, 1987). The concentration of laminin gels used may influence the regeneration of axons. High concentration of laminin gels will impede the axonal outgrowth (Labrador *et al.*, 1998). Besides, the permeability of the nerve conduits may impact the organization of the matrix and consequently affect the nerve regeneration (Valentini *et al.*, 1987). Therefore, these potential factors that may influence the regeneration of nerve should be considered when using laminin as luminal fillers to bridge a nerve defect.

3.1.3. *Bioengineered fibronectin & fibrin and their derivatives*

3.1.3.1. Introduction

Fibronectin (Fn) is a large extracellular matrix glycoprotein with a molecular weight of approximately 440 kDa. The quantity of fibronectin is only to collagen in the ECM. Fibronectins are dimers, consisting of two monomers, and they have many isoforms owing to the alternative mRNA splicing. Fibronectin mainly have two forms, one of which is soluble and the other one is insoluble. Fibronectin is soluble in plasma but not soluble in the ECM (Pankov and Yamada, 2002). Soluble isoform contains two nearly identical subunits which are joined by disulfide bonds (Roy *et al.*, 2011). The primary structure of each subunit contains three types of repeating homologous units (I, II, and IIII) (Koide *et al.*, 1998). The tripeptide sequence Arg-Gly-Asp (RGD) is rich in fibronectin, which plays an important role in cell adhesion via the $\alpha_5 \beta_1$ integrin (Yurchenco *et al.*, 1994). Fibronectin can interact with other ECM components such as collagen, fibrin, and cell surface receptors of the integrin family (De Olivera, 2004). Fibronectin has exhibited adhesive and chemotactic properties for many cell types, such as endothelial cells and fibroblasts, which are important to the wound-healing process (Mcpherson and Badylak, 1998). Therefore, fibronectin plays important roles in wound healing which include cell adhesion, migration and angiogenesis (Harding *et al.*, 2002). Also, fibronectin expressed at an early stage of developing embryos is critical for normal development (Badylak, 2004). The friendly biocompatibility of fibronectin has made it an attractive material to use as a substrate for cell culture and coatings of various synthetic scaffold materials to enhance their biocompatibility (Badylak, 2002). Fibronectin can promote cell adhesion and migration (Yamada, 1983). Fibronectin have also been used as conduits to regenerate peripheral nerves (Ahmed *et al.*, 2004).

Fibrin, a self-assembling polypeptide, forms in the natural blood coagulation cascade. It is derived from the fibrinogen which is a 340 KDa hexamer composed of three polypeptide chains termed Aα, Bβ, and γ (Doolittle, 1984). Fibrinogen contains an E region and two distal D regions. Three chains are joined together by disulfide bridges and form a helical coiled structure from the N-terminal E domain to the distal D regions (Standeven *et al.*, 2005). The first step of the fibrin formation process is the production of soluble fibrin monomers (α, β, γ). The fibrin monomers form through the cleavage of fibrinopeptides A and B from the fibrinogen chain by thrombin. Then, the fibrin is covalently cross-linked in the presence of factor XIII and form insoluble cross-linked fibrin (Silver *et al.*, 1995). The formation process of fibrin is delineated in Fig. 3. Fibrin plays an important role in blood coagulation and wound healing. Fibrin sealant has been used clinically for many years due to its good biocompatibility (Ahmed *et al.*, 2008). Besides, it has also been used as a delivery system of cells, growth factors, gene, and drugs in tissue engineering (Gorodetsky *et al.*, 2004; Bensaid *et al.*, 2003; Ehrbar *et al.*, 2008; Kidd *et al.*, 2012; Spicer and Mikos, 2010). Fibrin has also been tested to repair the primary nerve injury in humans (Bertelli and Mira, 1993).

3.1.3.2. Category and characteristic

Fibronectin is widely expressed in peripheral nervous system and can promote the development and regeneration of nerve after injury (McGrath, 2012). Fibronectin is mainly secreted by Schwann cells in nerves which will be up-regulated after peripheral nerve injury (Baron-Van evercooren *et al.*, 1986; Lefcort *et al.*, 1992). Whitworth *et al.* (1995) demonstrated that fibronectin conduit could support the growth of axons and the amount of regenerated axons and Schwann

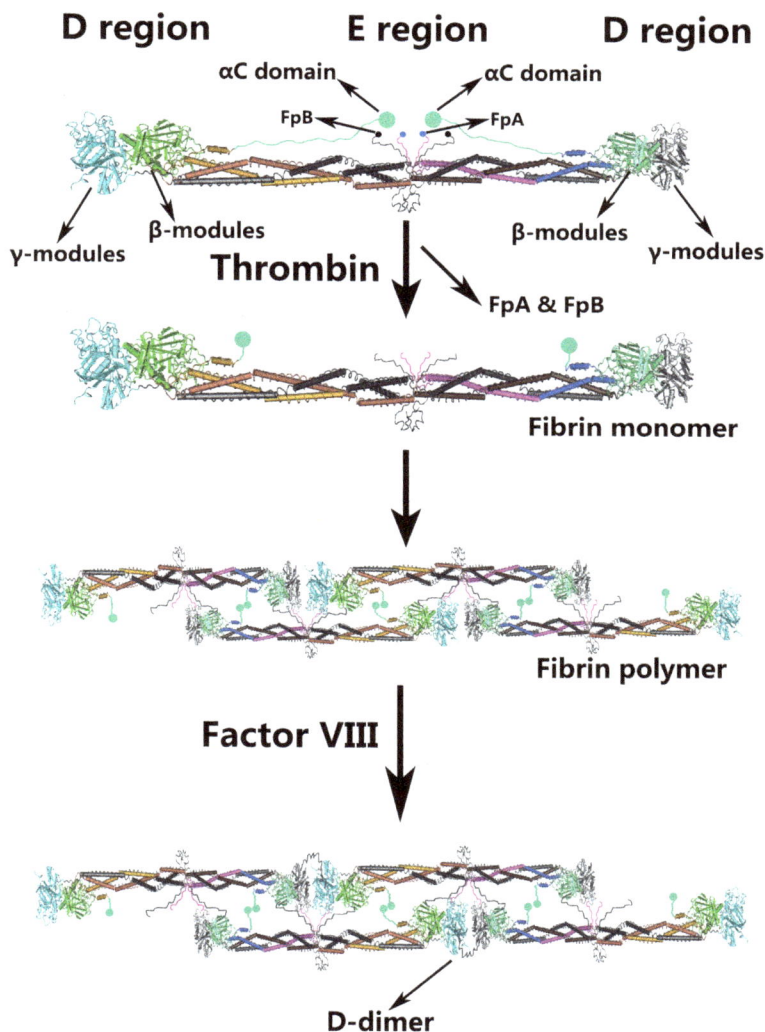

Fig. 3. The process of the fibrin formation. Fibrinogen consists of three regions including two D-regions and one E-region, which is composed of 2Aα-, 2Bβ-, and 2γ-chains. The D-regions contain the distal portions of the coiled-coil and the C-terminal β- and γ-modules. The E-region contains the central N-terminal part of the molecule and the proximal portions of both sets of coiled-coils. The blue and black circles represent fibrinopeptides A (FpA) and B (FpB), respectively. The first step of the fibrin formation process is the production of soluble fibrin monomers. The fibrin monomers form through the cleavage of fibrinopeptides A and B from the fibrinogen chain by thrombin. Then, the monomers are polymerized to form fibrin polymer. At last, the fibrin is covalently cross-linked in the presence of factor XIII and form D-dimer.

cells could be comparable to nerve grafts. In Phillips's study, the nerve tubes fabricated with plasma fibronectin were used to bridge a 5 mm interstump gap and the growth of axons through the tube was observed six weeks post-surgery in rats (Phillips *et al.*, 2002). Filling fibronectin into hollow conduits can improve nerve regeneration across a 18 mm gap of rats but the number of regenerated axons, sensory neurons and motor neurons were fewer than filling the mixture of fibronectin with laminin into conduits (Bailey *et al.*, 1993). In addition, fibronectin mats have been developed and successfully used to promote the neurite growth and the regeneration of injured peripheral nerve (Ahmed *et al.*, 1999; King *et al.*, 2003; Phillips *et al.*, 2004; Brown *et al.*, 1994). It has been suggested that fibronectin promotes cell proliferation and differentiation by undergoing conformational changes and forming fibril networks (Mao and Schwarzbauer, 2005).

Fibrin has shown good capacity to repair injured nerves (Kalbermatten *et al.*, 2009), and has been successfully used as sealant on transected nerve defect in adult rabbits in early studies (Palazzi *et al.*, 1995; Zeng *et al.*, 1995). Pettersson *et al.* (2010) investigated the long-term effects of a fibrin conduit by assessing the recovery of muscle weight, axonal sprouting and neuronal regeneration after peripheral nerve injury. Compared with the autograft, the fibrin conduit promoted regeneration of 86% of myelinated axons and induced more than 80% of muscles weight recovery (Pettersson *et al.*, 2010). However, the fibrin conduit can only effectively bridge a short nerve gap (10 mm for adult rats) (Pettersson *et al.*, 2011). To repair long nerve defects, further modifications, such as supplemented with growth factors of regenerative cells, to the conduit is required. Di Summa *et al.* (2011) used different cells, which include primary Schwann cells (SC), bone marrow-derived mesenchymal stem cells (dMSC) and adipose-derived stem cells seeding on fibrin conduit to repair a 1 cm nerve gap of adult rats. Compared with the empty fibrin conduit, the conduit seeding with these cells reduced muscle atrophy and improved the nerve regeneration. The authors considered that adipose derived stem cells could provide a clinically new method to enhance injury peripheral nerve regeneration (Di Summa *et al.*, 2011) (Fig. 4). Fibrin glue can also be used as a vehicle for the delivery of neurotrophic factors and

Cross section   **Light microscopy**

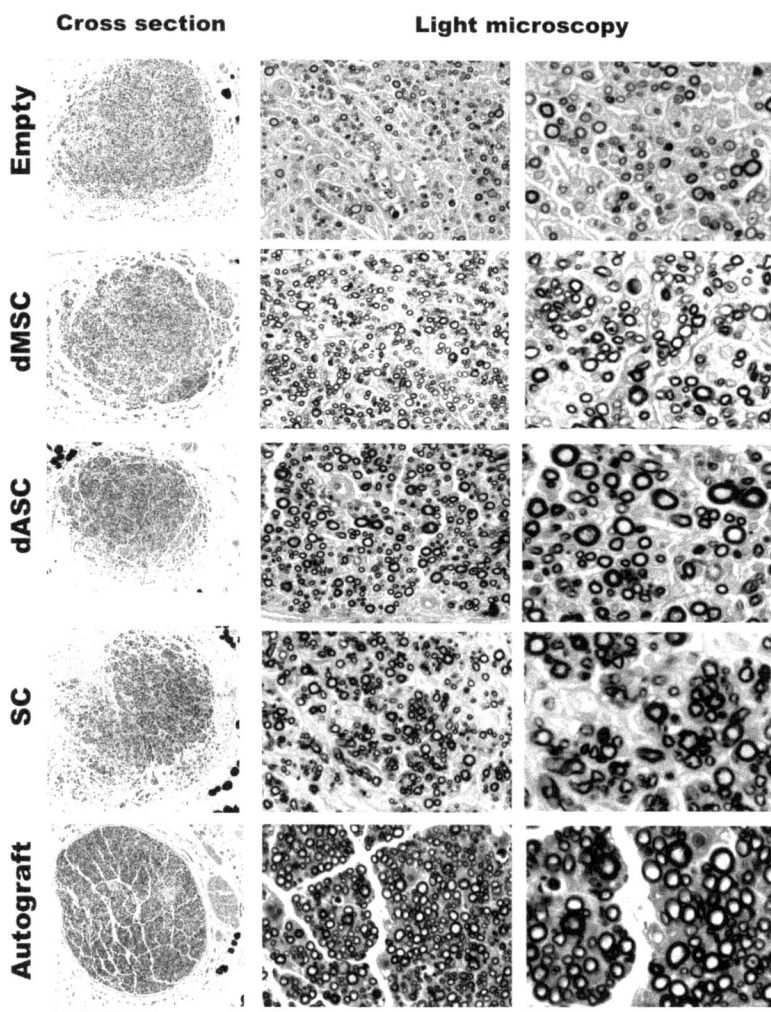

Fig. 4. Evaluation of sciatic nerve regeneration 16 weeks after total axotomy with simple empty fibrin conduit treatment, fibrin conduit seeded with SC-like differentiated MSC (dMSC), SC-like differentiated ASC (dASC), primary Schwann cells (SC) and autografts (rows from top to bottom). Columns from left to right: the general appearance of whole nerve cross sections (scale bar: 500 μm), light microscopy representative pictures of myelinated axons in the distal stump, 1 mm distal to the suture site (scale bars: 10 and 20 μm). The figure is adapted from (Di Summa *et al.*, 2011).

growth factors (Ehrbar *et al.*, 2008; Pandit *et al.*, 2000). Yin *et al.* (2001) reported that fibrin glue mixed with neurotrophin-4 significantly improved the myelin thickness, axonal diameter and the number of regenerated axons compared with the single fibrin glue.

3.2. *Bioengineered gelatin and its derivatives*

3.2.1. *Introduction*

Gelatin, a denatured collagen, is mainly derived from animal's bones, cartilages and skins. Therefore, the source, age of the animal and the collagen type have influences on the properties of the gelatin (Gómez-Guillén *et al.*, 2011). While the pH value, temperature, and extraction time may also influence the degree of collagen conversion into gelatin (Gómez-Guillén *et al.*, 2011). There are two types of gelatin known as type A and type B, whose precursors are pre-treated with acid and alkali respectively (Eysturskarð *et al.*, 2009). Gelatin aqueous solution can be transformed into sol when it is heated to 40°C and it will change into gel when cooled to room temperature. The gelatin chains will experience a conformational disorder-order transition and tend to recover the collagen triple-helix structure when it is transformed from sol to gel (Bigi *et al.*, 2001). Geltatin has many advantages: firstly, it is very cheap and easy to obtain; secondly, its physicochemical properties can be modulated according to the requirements; more importantly, it does not express antigenicity and can be completely degraded *in vivo* (Bigi *et al.*, 2002). Taking these advantages into account, gelatin has been applied in the field of food, cosmetic, photographic and pharmaceutical industries (Sarbon *et al.*, 2013). In the food field, it is usually used to make confections, low-fat spreads, baked goods and meat products (Karim and Bhat, 2009). While in the pharmaceutical and medical fields, gelatin has been used as drug delivery microspheres, hard and soft capsules, sealants for vascular prostheses, wound dressing and adsorbent pads for surgery (Bigi *et al.*, 2001). Recently, gelatin-based biomaterials have been used to make artificial skin, bone substitutes and extracellular matrix for peripheral nerve regeneration in tissue engineering (Lien *et al.*, 2009; Chang *et al.*, 2003; Yao *et al.*, 2004; Liu *et al.*, 2004).

3.2.2. *Category and characteristic*

Gelatin may be a good candidate for fabrication of nerve conduits attributing to its good biodegradability, plasticity, adhesion and bio-compatibility. However, the mechanical property of gelatin is poor as a result of its water solubility and the conduit may collapse and rapidly be cleared after implantation. Thus, gelatin is often cross-linked with chemical agents such as genipin (Chang *et al.*, 2009), proanthocyanidin (Liu, 2008), and 1-ethyl-3-(3-dimethylaminopropyl) carbodiimide (EDC)/N-hydroxysuccinimide (NHS) (Chang *et al.*, 2007) to improve its characteristics for supporting peripheral nerve regeneration. Genipin cross-linked gelatin conduit (GGC) has been widely studied and the permeable GGC can accelerate favorable nerve functional recovery compared with the non-porous GGC (Chang *et al.*, 2009; Chen *et al.*, 2005). In fact, the mechanical strength of genipin cross-linked gelatin is not satisfactory under physiological conditions which may limit its application. Therefore, Yang *et al.* (2011) enhanced the mechanical property of GGC conduit by doping the tricalcium phosphate in the fabricating process. The compression modulus of conduits was increased from 14.02 ± 1.22 MPa to 252.70 ± 61.24 MPa after tricalcium phosphate was added and this kind of conduit can well promote the peripheral nerve regeneration. Shen *et al.* (2012) seeded undifferentiated adipose tissue-derived stem cells into genipin cross-linked gelatin-tricalcium phosphate conduit to repair a 10 mm nerve gap of rats and achieved similar functional recovery compared with the autograft (Table 1).

3.3. *Bioengineered chitosan and its derivatives*

3.3.1. *Introduction*

Since Rouget paid attention to the deacetylated form of chitosan in 1859, a lot of works have been reported about this natural polymer and its application in different fields (Dodane and Vilivalam, 1998). Chitosan, deacetylated derivative of chitin, is consisted of d-glucosamine and N-acetyl-d-glucosamine units linked by β (1–4) glycosidic

Table 1. Morphometric parameters from the regenerated nerve tissue at 8-weeks after implantation of the normal nerve, GGT, GGT/ADSCs nerve conduits, and autografts.

Morphometric parameters	Normal nerve	GGT	GGT/ADSCs	Autografts
Density of nerve fiber number (no./mm^2)	5537 ± 890[a]	3660 ± 152	4181 ± 266[a]	4,510 ± 478[a]
Nerve fiber diameter (μm)	10.11 ± 1.17[a]	8.89 ± 0.59	9.10 ± 1.28	9.30 ± 0.87
Axon diameter (μm)	5.19 ± 0.84[a]	2.91 ± 0.32	4.31 ± 0.76[a]	4.06 ± 0.35[a]
Myelin sheath thickness (μm)	4.07 ± 0.61[a]	2.43 ± 0.29	2.80 ± 0.48	2.95 ± 0.24[a]
Medial nerve area (μm^2)	2,170,000 ± 3500[a]	51,330 ± 2238	914,818 ± 106,742[a]	841,598 ± 116,385[a]

[a] Significantly ($p < 0.05$) greater than the GGT group. The table is adapted from (Shen et al., 2012).

bond, which is the most abundant natural polymer second to cellulose (Siemionow *et al.*, 2010). Chitin is a white, hard, inelastic, nitrogenous polysaccharide, which only has a different group (an acetamido group) from cellulose (hydroxyl) at C2 position in structure (Ravi, 2000). While in chitosan, the acetamido group of chitin at position C2 is replaced by amino (Fig. 5) (Croisier and Jérôme, 2013). Chitin is usually obtained from the skeleton crab or shrimp shells and the cell wall of fungal mycelia (Jayakumar *et al.*, 2011). The process of obtaining chitosan from chitin can be described as: (1) removing the

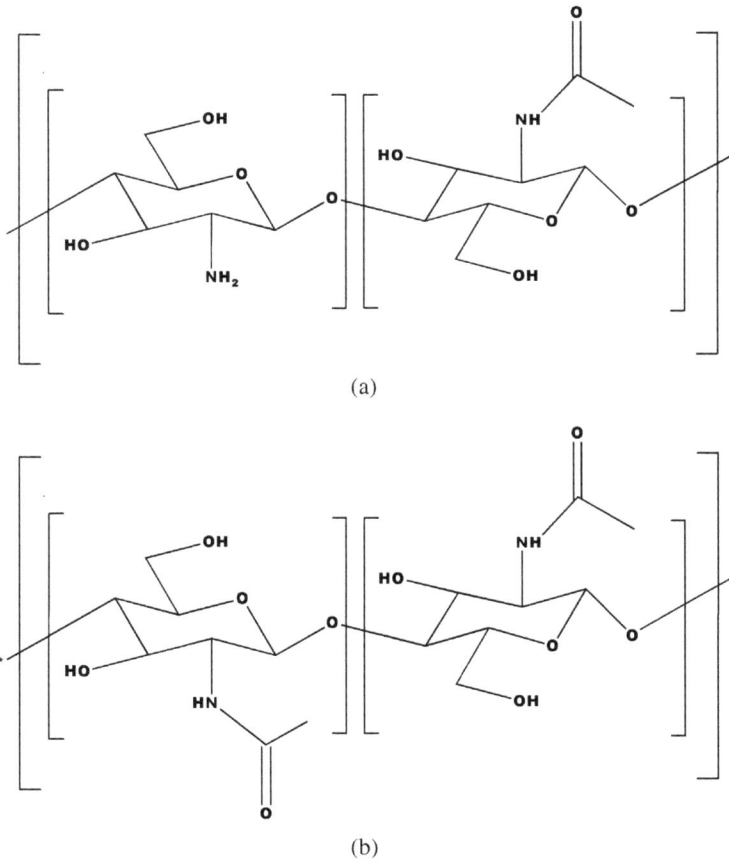

(a)

(b)

Fig. 5. Structures of chitosan and chitin: a, chitosan; b, chitin. The figure is adapted from (Croisier and Jérôme, 2013).

proteins and the dissolved calcium carbonate of crustacean shells; (2) deacetylating the resulting chitin in 40% sodium hydroxide at 120°C for 1–3 hours to produce 70% deacetylated chitosan. The molecular weight of chitosan ranges from 300 to over 1000 kDa with a degree of deacetylation from 30% to 95% (Kim et al., 2008). Chitosan is normally insoluble at neutral and alkaline pH but it becomes soluble in acids (Ilium, 1998). The characteristics of chitosan are influenced by the molecular weight and degree of deacetylation, which are determined by the conditions during preparation (Zhu et al., 2012). The desired properties such as the mechanical or biological properties of chitosan can be achieved through chemical modification so that it can be used for specific applications. Chitosan has been used in various fields including fisheries, textiles, food, ecology, agriculture, cosmetics, biotechnology, pharmaceutical and medicine (Agrawal et al., 2010; Aranaz et al., 2009; Harish and Tharanathan, 2007).

3.3.2. Category and characteristic

As a natural biomaterial, chitosan has many favorable properties such as low cost, biocompatibility, biodegradability, non-toxicity and characteristic physicochemical and biological activities. These properties offer powerful potential for its applications in biomedical field (Khor and Lim, 2003; VandeVord et al., 2001; Cheng et al., 2003; Yan et al., 2001). Chitosan has played important roles in tissue engineering and has been widely used as artificial organs such as skin, blood vessel, cartilage, bone, liver, and nerve (Kim et al., 2008). In vitro studies suggest that chitosan is biocompatible to Schwann cells and it can facilitate the attachment, differentiation and growth of nerve cells (Cheng et al., 2003; Yuan et al., 2004). Raisi et al. (2010) used single-lumen chitosan conduit to bridge a 10 mm peripheral nerve defect in sciatic nerve transection model of rats in vivo. The nerve regeneration was assessed based on walking track analysis, electrophysiological study, muscle mass measurement, histomorphometric and immuohistochemical criteria at the time of four, eight, and 12 weeks post-surgery. Morphometric and immunohistochemical indices indicated that nerve regeneration was promoted. Functional and electrophysiological analysis showed that nerve function was significantly improved (Raisi *et al.*, 2010). Permeability of nerve conduits plays an important role in the process of axons regeneration and porous chitosan nerve conduit has been developed to repair the peripheral nerve defects. Yang *et al.* (2004) constructed a porous chitin/chitosan nerve conduit using a mold-casting/lyophilization method and the pore size of conduits was ranged from 40 to 250 um. After being implanted into animals, the porous conduits did not swell and were compatible with surrounding tissue. However, the tensile strength of single-lumen porous chitosan conduit (0.75–0.95 MPa) was much lower than non-porous chitosan conduit (8.5–10.5 MPa). Therefore, the porous hollow chitosan conduit may collapse after being implanted *in vivo* as a result of its poor mechanical strength. Besides, single-lumen conduits may lead to inappropriate target reinnervation which can impede the functional recovery of injured nerve (Pfister *et al.*, 2007; de Ruiter *et al.*, 2008). Thus, the hollow chitosan conduit is filled with additives such as biodegradable matrix fibers to enhance the mechanical property, which has achieved satisfactory results (Cheng *et al.*, 2007; Wang *et al.*, 2007). Chitosan conduits filled with 2,000 longitudinally aligned PGA fibres were used to bridge a 30 mm dog sciatic nerve defect (Wang *et al.*, 2005). Six months post-operation, the analysis of fluorogold retrograde tracing, histological and immunohistochemical assessment as well as morphometric analysis revealed that the nerve trunks were reconstructed with restoration of nerve continuity and functional recovery (Fig. 6). Moreover, multi-channel chitosan conduits have also been designed to enhance the mechanical strength and reduce the dispersion of axonal branches and misdirection rate of regenerated axons (Wan *et al.*, 2012). To enhance functional recovery of regenerated peripheral nerve, biochemical cues such as growth factors and supporting cells are usually incorporated into the nerve conduits. Ciliary neurotrophic factor-coated polylactic–polyglycolic acid chitosan (PLGA/chitosan–CNTF) nerve conduits were used to bridge a 25 mm long canine tibial nerve defect. The conduction velocity of regenerated nerve and the number of myelinated axons (4684 ± 1007 *vs.* 6175 ± 816 axons per section) in PLGA/chitosan–CNTF group were significantly higher

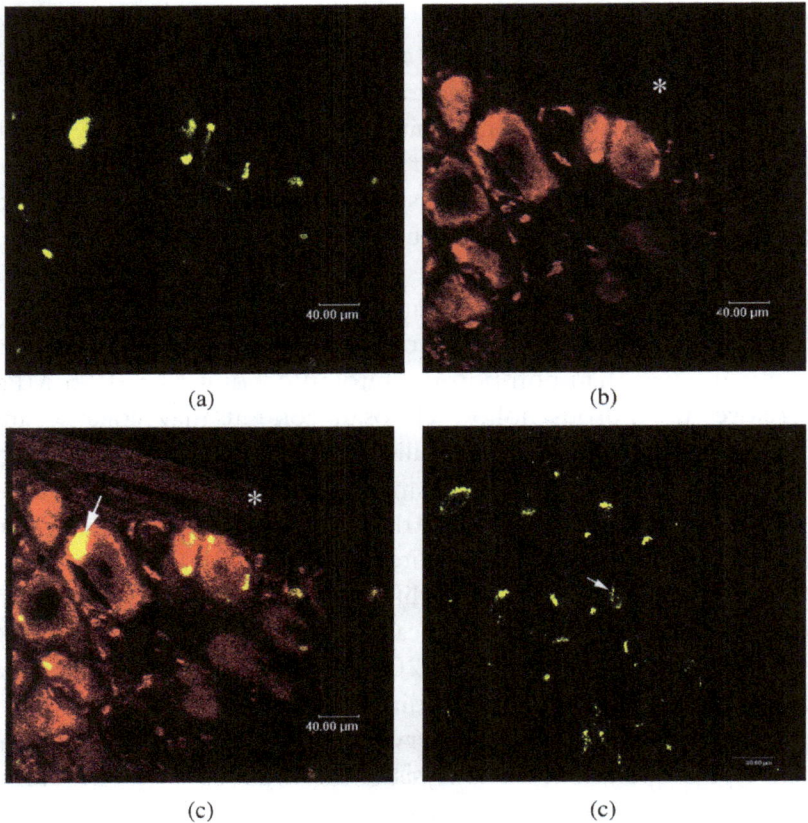

Fig. 6. FluoroGold retrograde tracing carried out 6 months after implantation of the chitosan/PGA graft. (a) DRG neurons labelled by FG (b) Immimostained for NF-200. (c) Overlay of (a) and (b). FG fluorescent particles aggregated at one side within cell bodies (arrow); an asterisk indicates the DRG capsule. (d) FG-labelled motoncurons in the anterior horn of grey matter in the spinal cord. FG fluorescent particles also aggregated at one side within cell bodies (arrow). Scale bar: 40 μm for a, b and c; 80 μm for d. The figure is adapted from (Wang *et al*, 2005).

than in PLGA/chitosan group which suggested that the use of ciliary neurotrophic factor is favorable for the regeneration of nerve (Shen *et al.*, 2010). Zheng *et al.* (2012) used multi-channel chitosan conduits seeded with bone marrow mesenchymal stem cells (BMSCs) to treat an 8 mm critical defect in peripheral nerves of rats. After six weeks post-operation, the sciatic nerve function index, average regenerated fiber density, and fiber diameter in nerves bridged with multi-channel chitosan conduits seeded with BMSCs were similar to those treated with autograft, but significantly higher than those bridged with single chitosan conduits (Zheng and Cui, 2012). These results demonstrated that the incorporation of BMSCs into chitosan conduits can promote the regeneration of injured nerve. In Chen's study, BMSCs were filled into the chitosan conduits to bridge the gap in the transected spinal cord of rats and the similar conclusion (that axon growth and remyelination in the regenerated nerve was promoted by the addition of BMSCs) was drawn, which suggests that BMSCs-loaded chitosan conduits have a promising future in the application of repairing nerve injury (Chen *et al.*, 2011).

3.4. *Bioengineered silk and its derivatives*

3.4.1. *Introduction*

Silk is a natural structure protein, mainly derived from silk worms or spiders. Silk farming has a history of more than 5,000 years in China and silk was traded with the west in ancient times (Lubec *et al.*, 1993). To date, the two best known types of silks are from the silkworm *Bombyx mori* and the spider *Nephila clavipes*. The silk from the silk-worm is largely used in textile industry and its production is 1,000 metric tons annually. The silk from *Bombyx mori* has a core-shell struc-ture containing two proteins, the heavy chain (~ 370 kDa) and light chain (~25 kDa), which are held together by a glue-like proteins called sericin (Nazarov *et al.*, 2004). The heavy chain of silk fibroin contains alternating hydrophobic and hydrophilic domains. The hydrophobic block consists of highly conserved sequence repeats of GAGAGS and less conserved repeats of GAGAGX (X is V or Y) which form β-sheet in the crystalline regions, while the hydrophilic part is non-repetitive and very short (Wenk *et al.*, 2011). Compared with the H-chain, the L-chain and sericin are more hydrophilic and relatively elastic, in which sericin is related to immune response but easily removed by boiling silk in alkaline solution (Altman *et al.*, 2003; Kaplan *et al.*, 1998). Unlike silkworm, spiders are difficult to be domesticated for textile applications because spiders have a cannibalistic nature and the production quantity of silk from spiders is very low. There are several species of spider and more than 34,000 species have been identified. The most widely studied spider silk is dragline silk from the spider *Nephila clavipes*, which is a mixture of two proteins whose molecular weights range from 70–700 kDa depending on sources (Numata and Kaplan, 2010). These proteins have similar structures with silk from silkworm. They are composed of large hydrophobic blocks and small hydrophilic blocks. The hydrophobic blocks have highly conserved repetitive sequences consisting of glycine and alanine, while the sequence of hydrophilic blocks is complex and they consists of bulkier side-chains and charged amino acids (Winkler *et al.*, 2000; Bini *et al.*, 2004). The hydrophobic blocks also form tight β-sheets in silk fibers (Jelinski, 1998). The process to obtain silk from both spiders and insects are complex and various (Vollrath and Knight, 1999). Presently, boiling the cocoons in Na_2CO_3 solution is widely used to isolate silk fibroin from silkworms. Also, genetically engineered silk fibroin analogues are also available and the property can be developed for special use (Meinel and Kaplan, 2012).

3.4.2. *Category and characteristic*

Silk is gradually capturing researchers' attentions and playing more important roles in tissue engineering owing to its good biocompatibility, processing versatility, oxygen and water vapor permeability, impressive mechanical performance and tailorable degradability (Zhang *et al.*, 2009; Oliveira *et al.*, 2012). Silk has been used for the engineering of bone, vascular, skin, cartilage, ligaments, bladder, tendons, cardiac, ocular, and neural tissues (Kundu *et al.*, 2012). Here, the application of silk in nerve regeneration will be discussed in detail. Chen *et al.* (2004) seeded human fibroblasts on the silk modified poly (ε-caprolactone) (PCL) scaffold *in vitro* and found that silk modification improved the biocompatibility of PCL scaffold (Chen *et al.*, 2004). Besides, Yang *et al.* (2007) also evaluated the biocompatibility of silk fibroin with peripheral nerve tissues and cells *in vitro*. The result suggests that silk fibroin has good biocompatibility with rat dorsal root ganglia and is beneficial to the survival of Schwann cells without cytotoxic effects on their morphologies and functions (Yang *et al.*, 2007). Yan *et al.* (2009) assessed biological safety of the silk fibroin-based nerve guidance conduits (SF-NGC) both *in vitro* and *in vivo*. The results of subcutaneous implantation in rabbits indicated that SF-NGC has a good compatibility with the surrounding tissues which provides beneficial support for nerve regeneration (Yan *et al.*, 2009). In Yang's study, silk fibroin-based nerve grafts were used to bridge a 10 mm long sciatic nerve defect of adult rats (Yang *et al.*, 2008). Six months post-implantation, the functional recovery was evaluated by a combination of electrophysiological assessment, Fluorogold retrograde tracing and histological investigation, and these showed that silk fibroin grafts could promote peripheral nerve regeneration. Ghaznavi *et al.* (2011) used silk fibroin conduits to examine cell inflammatory responses and functional recovery in a sciatic nerve defect model. The result suggests that the silk guides only induce slight inflammatory response and promote the nerve regeneration from the proximal to distal nerve stumps when compared with an autograft and collagen nerve guide (Ghaznavi *et al.*, 2011). The degradation behaviors of SF-NGCs have also been studied, in which they were incubated in the protease XIV solution or to be subcutaneously implanted in rabbits. SF-NGCs were subjected to light and electron microscopy, mass loss assessment, gel electrophoresis for testing the dynamic course of *in vitro* or *in vivo* degradation. The results indicated that SF-NGCs degraded at an appropriate rate and could meet the requirements of peripheral nerve regeneration, which provides an advantage for its application in nerve repair. A recent study used *Bombyx mori* regenerated silk conduits containing different numbers of luminal Spidrex fibers to repair an 8 mm nerve gap of rats (Huang *et al.*, 2012). At 4 weeks, conduits containing 200 luminal Spidrex fibers (PN200) supported similar axon growth and displayed similar Schwann cell support and macrophage response compared with the autologous nerve graft control (Fig. 7). At 12 weeks, rats implanted with PN 200 conduits showed similar numbers of myelinated axons, gastrocnemius muscle innervation and hindpaw stance assessed by Catwalk footprint analysis to the autologous nerve graft control (Fig. 8). This study demonstrated that Spidrex conduits could promote excellent axonal

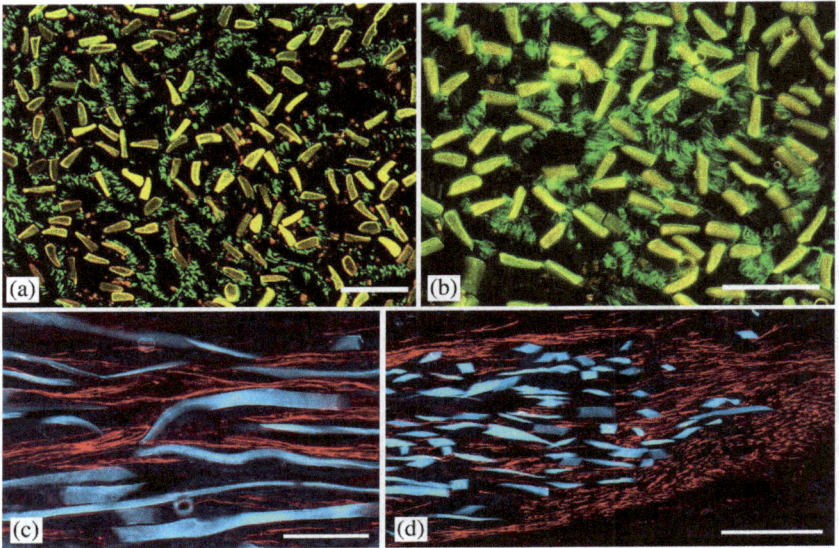

Fig. 7.   Examples of axon growth and Schwann cell migration through a 10 mm PN200 conduit, revealed using neurofilament (a,c,d) and p75 (b) immunofluorescence. Transverse sections (a,b) taken mid-conduit show the even distribution of regenerated axons (green, a) and Schwann cell processes (green, b) among luminal silk fibres (yellow). Longitudinal sections taken mid-conduit (c) and at the distal interface (d) show that axons (red) lie on and between the luminal silk fibres (blue), and regenerate out of the conduit into the distal nerve. Scale bars = 100 μm (a–c) and 200 μm (d). (For interpretation of the references to colour in this figure legend, the reader is referred to the web version of this article.) The figure is adapted from (Huang *et al.*, 2012).

regeneration and functional recovery. As a natural biomaterial, silk may have the potential for clinical application in nerve repair.

3.5. *Bioengineered alginate and its derivatives*

3.5.1. *Introduction*

Alginate is a biodegradable polysaccharide extracted from brown seaweed, which is made up of 1,4-linked β-D-mannuronate (M) and 1,4-linked α-L-guluronate (G) residues in variable proportions (Tan *et al.*, 2012). The sequence of alginate is comprised of GG, MG,

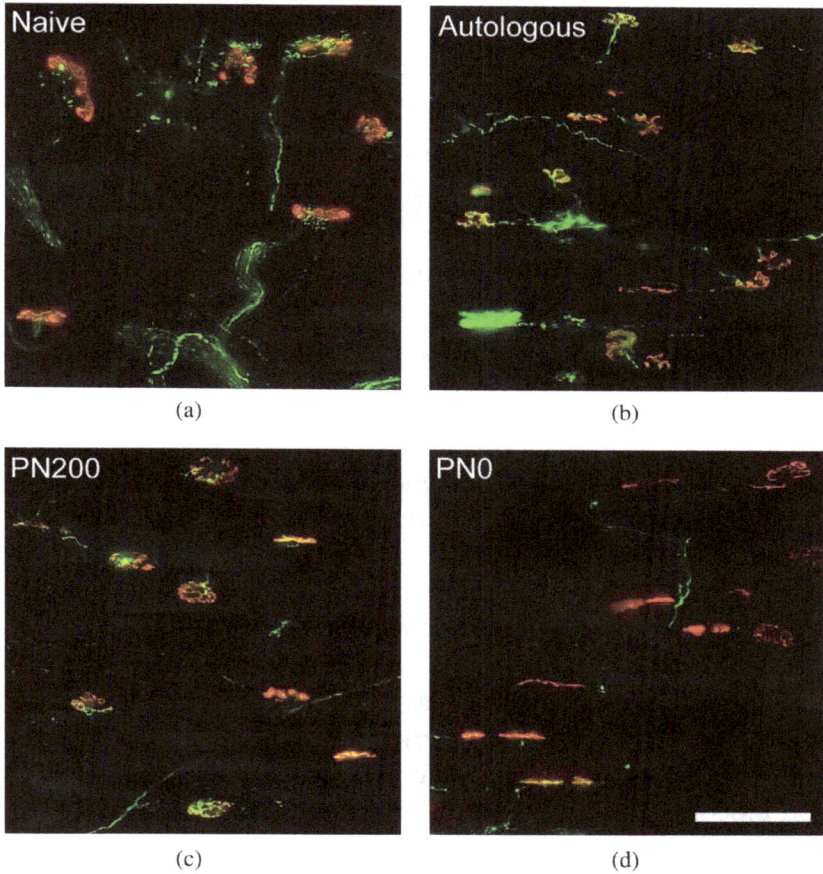

Fig. 8. Innervations of the gastrocnemius muscle at 12 weeks post surgery using double labelling with PGP 9.5 (an axonal marker; green) and α-bungarotoxin (a neuromuscular junction marker; red). The PN200 group (c) appeared to have comparable muscle innervations when compared to the naïve (a) and autologous (b) groups. However, the PN0 group (d) showed a poor muscle innervation. Scale bar = 100 u.m. (For interpretation of the references to colour in this figure legend, the reader is referred to the web version of this article.) The figure is adapted from (Huang *et al.*, 2011)

and MM blocks (Fig. 9) (Pawar and Edgar, 2012). Alginate hydrogel could be formed in the presence of divalent cations such as Ca^{2+} or Ba^{2+}, which are cross-linkers of alginate chains (Wang *et al.*, 2003). A Na-alginate solution can, under given conditions, undergo

(a)

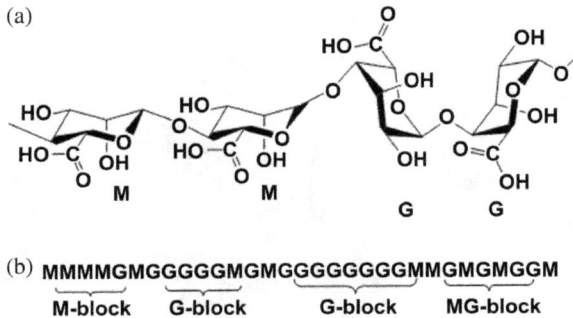

(b) MMMMGMGGGGGMGMGGGGGGGGMMGMGMGGM

M-block　　G-block　　　G-block　　MG-block

Fig. 9. Structure of alginate moleculars: a, chain conformation; b, block distribution. The figure is adapted from (Pawae and Edgar, 2012).

a sol-gel transition at pH below pKa for the uronides, which might be useful for the utilisation of alginates in drug delivery systems (Draget *et al.*, 2006). The free hydroxyl and carboxyl groups are rich in alginate and the inherent property can be improved by derivatization. The solubility and hydrophobicity, physicochemical and biological characteristics can be altered by chemical modification (Yang *et al.*, 2011). Alginate has been widely used as a bulking agent, drug delivery system, carrier for cell and model ECMs in tissue engineering due to its abundance in source, low price and good biocompatibility (Augst *et al.*, 2006). However, alginate also has some disadvantages, such as the non-uniform structure, poor mechanical property and difficulty in forming complex-shaped 3-D structures (Kuo and Ma, 2001).

3.5.2. *Category and characteristic*

Alginate has been used to repair the injury peripheral nerve (Pfister *et al.*, 2007; Mosahebi *et al.*, 2003; Kim *et al.*, 2006). Sufan *et al.* (2001) employed alginate in tubulation and nontubulation repairs of a 50 mm nerve gap in cat model. Three months post-operatively, axons were successfully elongated and regenerated and after 8 months, many regenerated myelinated axons were observed which suggested that alginate can promote the peripheral nerve regeneration (Sufan *et al.*, 2001). Matsuura *et al.* (2006) repaired a 2 mm nerve gap of Wistar rats using freeze-dried alginate sheet without sutures. After 12 weeks, the nerve regeneration was observed which demonstrated that alginate sponge has a potential to enhance nerve repair (Matsuura *et al.*, 2006). In Kataoka's study, freeze-dried alginate sponge cross-linked with covalent bond and collagen gel was implanted in the lesion of transected spinal cord (Kataoka *et al.*, 2004). In alginate-implanted rats, more regenerating axons were found to extend from the stump into the lesion compared with in collagen-implanted rats which suggested that alginate contributes to reducing the barrier composed of connective tissues and reactive astrocytic processes and could enhance nerve regeneration in peripheral nerves and spinal cords. However, a recent study found that filling the alginate gel into nerve conduits may impede the nerve regeneration process and the regenerated new tissue appears mainly on the outer surface of the implant (Marycz *et al.*, 2012). Therefore, it is required to remodel the alginate gel and prolong its time of hydration and gelation. Further study is needed to explore the influence of alginate's preparing conditions on the neuro regeneration process.

4. Conclusions and Future Perspective

Here we summarize the application of different kinds of natural materials in repairing the injured nerves, especially the peripheral nerves. The nerve conduits made from different natural biomaterials have different advantages. However, the disadvantages such as poor mechanical properties may limit their use in repairing the injured nerves. Many approaches such as improving the fabrication techniques and combining different biomaterials have been developed to improve the nerve conduits and to promote the functional recovery of injury nerves. Besides this, luminal fillers such as supporting cells and growth factors have been added into the hollow nerve conduits to promote the regeneration of nerves. Although some biomaterials have been approved by the US Food and Drug Administration (FDA) and Conformit Europe (CE) for the repair of peripheral nerve injuries, the efficiency of the nerve conduits made from natural biomaterials to enhance the peripheral nerve regeneration and functional recovery remains to be improved. Great efforts are still needed to be made to improve the nerve conduits to fulfill the requirements of the clinical application. In particular, the efficacy of natural nerve conduits in promoting nerve regeneration across long defect gaps and the clinical requirements for functional recovery should be considered when designing the nerve conduits.

References

Agrawal P, Strijkers GJ, Nicolay K. Chitosan-based systems for molecular imaging. *Adv Drug Delivery Rev* 2010;62:42–58.

Ahmed MR, Vairamuthu S, Shafiuzama M, Basha SH, Jayakumar R. Microwave irradiated collagen tubes as a better matrix for peripheral nerve regeneration. *Brain Res* 2005;1046:55–67.

Ahmed TA, Dare EV, Hincke M. Fibrin: A versatile scaffold for tissue engineering applications. *Tissue Eng Part B Rev* 2008;14:199–215.

Ahmed Z, Briden A, Hall S, Brown RA. Stabilisation of cables of fibronectin with micromolar concentrations of copper: *In vitro* cell substrate properties. *Biomaterials* 2004;25:803–812.

Ahmed Z, Idowu BD, Brown RA. Stabilization of fibronectin mats with micromolar concentrations of copper. *Biomaterials* 1999;20:201–209.

Allodi I, Udina E, Navarro X. Specificity of peripheral nerve regeneration: Interactions at the axon level. *Prog Neurobiol* 2012;98:16–37.

Alluin O, Wittmann C, Marqueste T, Chabas J-F, Garcia S, Lavaut M-N, et al. Functional recovery after peripheral nerve injury and implantation of a collagen guide. *Biomaterials* 2009;30:363–373.

Al-Majed AA, Neumann CM, Brushart TM, Gordon T. Brief electrical stimulation promotes the speed and accuracy of motor axonal regeneration. *J Neurosci* 2000;20:2602–2608.

Altman GH, Diaz F, Jakuba C, Calabro T, Horan RL, Chen J, et al. Silk-based biomaterials. *Biomaterials* 2003;24:401–416.

Aranaz I, Mengíbar M, Harris R, Paños I, Miralles B, Acosta N, et al. Functional characterization of chitin and chitosan. *Curr Chem Biol* 2009;3:203–230.

Archibald S, Shefner J, Krarup C, Madison R. Monkey median nerve repaired by nerve graft or collagen nerve guide tube. *J Neurosci* 1995;15:4109–4123.

Archibald SJ, Krarup C, Shefner J, Li ST, Madison RD. A collagen-based nerve guide conduit for peripheral nerve repair: An electrophysiological study of nerve regeneration in rodents and nonhuman primates. *J Comp Neurol* 1991;306: 685–696.

Artico M, Cervoni L, Nucci F, Giuffré R. Birthday of peripheral nervous system surgery: The contribution of Gabriele Ferrara (1543–1627). *Neurosurgery* 1996;39:380–383.

Asher RA, Morgenstern DA, Moon LD, Fawcett JW. Chondroitin sulphate proteoglycans: Inhibitory components of the glial scar. *Prog Brain Res* 2001;132:611–619.

Augst AD, Kong HJ, Mooney DJ. Alginate hydrogels as biomaterials. *Macromol Biosci* 2006;6:623–633.

Badylak SF. The extracellular matrix as a scaffold for tissue reconstruction. *Semin Cell Dev Biol* 2002;13:377–383.

Badylak SF. Xenogeneic extracellular matrix as a scaffold for tissue reconstruction. *Transplant Immunol* 2004;12:367–377.

Bailey S, Eichler M, Villadiego A, Rich K. The influence of fibronectin and laminin during Schwann cell migration and peripheral nerve regeneration through silicon chambers. *J Neurocytol* 1993;22:176–184.

Baron-Van Evercooren A, Gansmüller A, Gumpel M, Baumann N, Kleinman H. Schwann cell differentiation *in vitro*: Extracellular matrix deposition and interaction. *Dev Neurosci* 1986;8:182–196.

Beck K, Hunter I, Engel J. Structure and function of laminin: Anatomy of a multidomain glycoprotein. *FASEB J* 1990;4:148–160.

Bensaid W, Triffitt J, Blanchat C, Oudina K, Sedel L, Petite H. A biodegradable fibrin scaffold for mesenchymal stem cell transplantation. *Biomaterials* 2003;24:2497–2502.

Bertelli JA, Mira JC. Nerve repair using freezing and fibrin glue: Immediate histologic improvement of axonal coaptation. *Microsurgery* 1993;14:135–140.

Bigi A, Cojazzi G, Panzavolta S, Roveri N, Rubini K. Stabilization of gelatin films by crosslinking with genipin. *Biomaterials* 2002;23:4827–4832.

Bigi A, Cojazzi G, Panzavolta S, Rubini K, Roveri N. Mechanical and thermal properties of gelatin films at different degrees of glutaraldehyde crosslinking. *Biomaterials* 2001;22:763–768.

Bini E, Knight DP, Kaplan DL. Mapping domain structures in silks from insects and spiders related to protein assembly. *J Mol Biol* 2004;335:27–40.

Bovolenta P, Fernaud-Espinosa I. Nervous system proteoglycans as modulators of neurite outgrowth. *Prog Neurobiol* 2000;61:113–132.

Brown RA, Blunn GW, Ejim OS. Preparation of orientated fibrous mats from fibronectin: Composition and stability. *Biomaterials* 1994;15:457–464.

Brushart TM, Mathur V, Sood R, Koschorke G-M. Dispersion of regenerating axons across enclosed neural gaps. *J Hand Surg* 1995;20:557–564.

Bunge M, Bunge R, Kleitman N, Dean A. Role of peripheral nerve extracellular matrix in Schwann cell function and in neurite regeneration. *Dev Neurosci* 1989;11:348–360.

Cao J, Sun C, Zhao H, Xiao Z, Chen B, Gao J, *et al.* The use of laminin modified linear ordered collagen scaffolds loaded with laminin-binding ciliary neurotrophic factor for sciatic nerve regeneration in rats. *Biomaterials* 2011;32:3939–3948.

Chamberlain L, Yannas I, Arrizabalaga A, Hsu H, Norregaard T, Spector M. Early peripheral nerve healing in collagen and silicone tube implants: Myofibroblasts and the cellular response. *Biomaterials* 1998;19:1393–1403.

Chang JY, Ho TY, Lee HC, Lai YL, Lu MC, Yao CH, *et al.* Highly permeable genipin-cross-linked Gelatin conduits enhance peripheral nerve regeneration. *Artif Organs* 2009;33:1075–1085.

Chang JY, Lin JH, Yao CH, Chen JH, Lai TY, Chen YS. *In vivo* evaluation of a biodegradable EDC/NHS-cross-linked gelatin peripheral nerve guide conduit material. *Macromol Biosci* 2007;7:500–507.

Chang W-H, Chang Y, Lai P-H, Sung H-W. A genipin-crosslinked gelatin membrane as wound-dressing material: *In vitro* and *in vivo* studies. *J Biomater Sci Polym Ed* 2003;14:481–495.

Chaudhry V, Glass J, Griffin J. Wallerian degeneration in peripheral nerve disease. *Neurol Clin* 1992;10:613–627.

Chen G, Zhou P, Mei N, Chen X, Shao Z, Pan, *et al.* Silk fibroin modified porous poly(epsilon-caprolactone) scaffold for human fibroblast culture *in vitro*. *J Mater Sci Mater Med* 2004;15:671–677.

Chen MB, Zhang F, Lineaweaver WC. Luminal fillers in nerve conduits for peripheral nerve repair. *Ann Plast Surg* 2006;57:462–471.

Chen X, Yang Y, Yao J, Lin W, Li Y, Chen Y, *et al.* Bone marrow stromal cells-loaded chitosan conduits promote repair of complete transection injury in rat spinal cord. *J Mater Sci Mater Med* 2011;22:2347–2356.

Chen Y-S, Chang J-Y, Cheng C-Y, Tsai F-J, Yao C-H, Liu B-S. An *in vivo* evaluation of a biodegradable genipin-cross-linked gelatin peripheral nerve guide conduit material. *Biomaterials* 2005;26:3911–3918.

Cheng H, Huang Y-C, Chang P-T, Huang Y-Y. Laminin-incorporated nerve conduits made by plasma treatment for repairing spinal cord injury. *Biochem Biophys Res Commun* 2007;357:938–944.

Cheng M, Cao W, Gao Y, Gong Y, Zhao N, Zhang X. Studies on nerve cell affinity of biodegradable modified chitosan films. *J Biomater Sci Polym Ed* 2003;14:1155–1167.

Chung AE, Jaffe R, Freeman IL, Vergnes J-P, Braginski JE, Carlin B. Properties of a basement membrane-related glycoprotein synthesized in culture by a mouse embryonal carcinoma-derived cell line. *Cell* 1979;16:277–287.

Colognato H, Yurchenco PD. Form and function: The laminin family of heterotrimers. *Dev Dyn* 2000;218:213–234.

Cozzolino DJ, Cendron M, DeVore DP, Hoopes PJ. The biological behavior of autologous collagen-based extracellular matrix injected into the rabbit bladder wall. *Neurourol Urodyn* 1999;18:487–495.

Croisier F, Jérôme C. Chitosan-based biomaterials for tissue engineering. *Eur Polym J* 2013;49:780–792.

Cunha C, Panseri S, Antonini S. Emerging nanotechnology approaches in tissue engineering for peripheral nerve regeneration. *Nanomed Nanotechnol Biol Med* 2011;7:50–59.

De Olivera A. Peripheral nerve regeneration through the nerve tubulization technique. *Braz J Morphol Sci* 2004;21:225–231.

De Ruiter GC, Spinner RJ, Malessy MJ, Moore MJ, Sorenson EJ, Currier BL, *et al.* Accuracy of motor axon regeneration across autograft, single-lumen, and multichannel poly (lactic-co-glycolic acid) nerve tubes. *Neurosurgery* 2008;63: 144–155.

Deister C, Aljabari S, Schmidt CE. Effects of collagen 1, fibronectin, laminin and hyaluronic acid concentration in multi-component gels on neurite extension. *J Biomater Sci Polym Ed* 2007;18:983–997.

Deumens R, Bozkurt A, Meek MF, *et al.* Repairing injured peripheral nerves: bridging the gap. *Prog Neurobiol* 2010; 92:245–276.

Di Summa PG, Kalbermatten DF, Pralong E, Raffoul W, Kingham PJ, Terenghi G. Long-term *in vivo* regeneration of peripheral nerves through bioengineered nerve grafts. *Neuroscience* 2011;181:278–291.

Di Summa PG, Kingham PJ, Raffoul W, Wiberg M, Terenghi G, Kalbermatten DF. Adipose-derived stem cells enhance peripheral nerve regeneration. *J Plast Reconstr Aesthet Surg* 2010;63:1544–1552.

Dodane V, Vilivalam VD. Pharmaceutical applications of chitosan. *Pharm Sci Technol Today* 1998;1:246–253.

Doolittle RF. Fibrinogen and fibrin. *Annu Rev Biochem* 1984;53:195–229.

Draget KI, Skjak-Braek G, Stokke BT. Similarities and differences between alginic acid gels and ionically crosslinked alginate gels. *Food Hydrocolloids* 2006;20: 170–175.

Ehrbar M, Zeisberger SM, Raeber GP, Hubbell JA, Schnell C, Zisch AH. The role of actively released fibrin-conjugated VEGF for VEGF receptor 2 gene activation and the enhancement of angiogenesis. *Biomaterials* 2008;29:1720–1729.

Eysturskarð J, Haug IJ, Ulset A-S, Draget KI. Mechanical properties of mammalian and fish gelatins based on their weight average molecular weight and molecular weight distribution. *Food Hydrocolloids* 2009;23:2315–2321.

Farole A, Jamal BT. A bioabsorbable collagen nerve cuff (NeuraGen) for repair of lingual and inferior alveolar nerve injuries: A case series. *J Oral Maxillofac Surg* 2008;66:2058–2062.

Ferreira AM, Gentile P, Chiono V, Ciardelli G. Collagen for bone tissue regeneration. *Acta Biomater* 2012;8:3191–3200.

Fratzl P. Cellulose and collagen: From fibres to tissues. *Curr Opin Colloid Interface Sci* 2003;8:32–39.

Fu SY, Gordon T. The cellular and molecular basis of peripheral nerve regeneration. *Mol Neurobiol* 1997;14:67–116.

Gelse K, Pöschl E, Aigner T. Collagens — structure, function, and biosynthesis. *Adv Drug Delivery Rev* 2003;55:1531–1546.

Ghaznavi AM, Kokai LE, Lovett ML, Kaplan DL, Marra KG. Silk fibroin conduits: A cellular and functional assessment of peripheral nerve repair. *Ann Plast Surg* 2011;66:273–279.

Gómez-Guillén M, Giménez B, López-Caballero M, Montero M. Functional and bioactive properties of collagen and gelatin from alternative sources: A review. *Food Hydrocolloids* 2011;25:1813–1827.

Gordon T, Sulaiman O, Boyd JG. Experimental strategies to promote functional recovery after peripheral nerve injuries. *J Peripher Nerv Syst* 2003;8:236–250.

Gorodetsky R, Vexler A, Levdansky L, Marx G. Fibrin microbeads (FMB) as biodegradable carriers for culturing cells and for accelerating wound healing. *Biopolyme Methods Tissue Eng* 2004;238:11–24.

Grimpe B, Silver J. The extracellular matrix in axon regeneration. *Prog Brain Res* 2002;137:333–349.

Gu X, Ding F, Yang Y, Liu J. Construction of tissue engineered nerve grafts and their application in peripheral nerve regeneration. *Prog Neurobiol* 2011;93:204–230.

Hamilton SK, Hinkle ML, Nicolini J, Rambo LN, Rexwinkle AM, Rose SJ, *et al.* Misdirection of regenerating axons and functional recovery following sciatic nerve injury in rats. *J Comp Neurol* 2011;519:21–33.

Harding S, Afoke A, Brown R, MacLeod A, Shamlou P, Dunnill P. Engineering and cell attachment properties of human fibronectin-fibrinogen scaffolds for use in tissue engineered blood vessels. *Bioprocess Biosyst Eng* 2002;25:53–59.

Harish Prashanth K, Tharanathan R. Chitin/chitosan: Modifications and their unlimited application potential — an overview. *Trends Food Sci Tech* 2007;18:117–131.

Harley B, Spilker M, Wu J, Asano K, Hsu H-P, Spector M, *et al.* Optimal degradation rate for collagen chambers used for regeneration of peripheral nerves over long gaps. *Cells Tissues Organs* 2004;176:153–165.

Hsu S, Chan S H, Chiang C M, *et al.* Peripheral nerve regeneration using a microporous polylactic acid asymmetric conduit in a rabbit long-gap sciatic nerve transection model. *Biomaterials* 2011;32:3764–3775.

Hu X, Huang J, Ye Z, Xia L, Li M, Lv B, *et al.* A novel scaffold with longitudinally oriented microchannels promotes peripheral nerve regeneration. *Tissue Eng Part A* 2009;15:3297–3308.

Huang W, Begum R, Barber T, Ibba V, Tee N, Hussain M, *et al.* Regenerative potential of silk conduits in repair of peripheral nerve injury in adult rats. *Biomaterials* 2012;33:59–71.

Ilium L. Chitosan and its use as a pharmaceutical excipient. *Pharm Res* 1998;15:1326–1331.

Itoh S, Takakuda K, Kawabata S, Aso Y, Kasai K, Itoh H, *et al.* Evaluation of cross-linking procedures of collagen tubes used in peripheral nerve repair. *Biomaterials* 2002;23:4475–4481.

Jayakumar R, Prabaharan M, Sudheesh Kumar P, Nair S, Tamura H. Biomaterials based on chitin and chitosan in wound dressing applications. *Biotechnol Adv* 2011;29:322–337.

Jelinski LW. Establishing the relationship between structure and mechanical function in silks. *Curr Opin Solid State Mater Sci* 1998;3:237–245.

Jurga M, Dainiak MB, Sarnowska A, Jablonska A, Tripathi A, Plieva FM, *et al.* The performance of laminin-containing cryogel scaffolds in neural tissue regeneration. *Biomaterials* 2011;32:3423–3434.

Kadler KE, Baldock C, Bella J, Boot-Handford RP. Collagens at a glance. *J Cell Sci* 2007;120:1955–1958.

Kalbermatten DF, Pettersson J, Kingham PJ, Pierer G, Wiberg M, Terenghi G. New fibrin conduit for peripheral nerve repair. *J Reconstr Microsurg* 2009; 25:27–33.

Kaplan DL, Mello SM, Arcidiacono S, Fossey S, Senecal K, Muller W. Silk. In: McGrath K, Kaplan DL, editors. *Protein based materials.* Birkhauser, Boston, 1998;103–131.

Karim A, Bhat R. Fish gelatin: Properties, challenges, and prospects as an alternative to mammalian gelatins. *Food Hydrocolloids* 2009;23:563–576.

Kataoka K, Suzuki Y, Kitada M, Hashimoto T, Chou H, Bai H, *et al.* Alginate enhances elongation of early regenerating axons in spinal cord of young rats. *Tissue Eng* 2004;10:493–504.

Keilhoff G, Stang F, Wolf G, Fansa H. Bio-compatibility of type I/III collagen matrix for peripheral nerve reconstruction. *Biomaterials* 2003;24:2779–2787.

Khor E, Lim LY. Implantable applications of chitin and chitosan. *Biomaterials* 2003;24:2339–2349.

Kidd ME, Shin S, Shea LD. Fibrin hydrogels for lentiviral gene delivery *in vitro* and *in vivo. J Control Release* 2012;157:80–85.

Kim I-Y, Seo S-J, Moon H-S, Yoo M-K, Park I-Y, Kim B-C, *et al.* Chitosan and its derivatives for tissue engineering applications. *Biotechnol Adv* 2008; 26:1–21.

Kim SW, Bae HK, Nam HS, Chung DJ, Choung PH. Peripheral berve regeneration through nerve conduit composed of alginate-collagen-chitosan. *Macromol Res* 2006;14:94–100.

King VR, Henseler M, Brown RA, Priestley JV. Mats made from fibronectin support oriented growth of axons in the damaged spinal cord of the adult rat. *Exp Neurol* 2003;182:383–398.

Kitahara AK, Nishimura Y, Shimizu Y, Endo K. Facial nerve repair accomplished by the interposition of a collagen nerve guide. *J Neurosurg* 2000;93:113–120.

Kitahara AK, Suzuki Y, Qi P, Nishimura Y, Suzuki K, Kiyotani T, *et al*. Facial nerve repair using a collagen conduit in cats. *Scand J Plast Reconstr Surg and Hand Surg* 1999;33:187–193.

Kleinman HK, Sephel GC, Tashiro KI, Weeks BS, Burrous BA, Adler SH, *et al*. Laminin in neuronal development. *Ann N Y Acad Sci* 2006;580:302–310.

Koide A, Bailey CW, Huang X, Koide S. The fibronectin type III domain as a scaffold for novel binding proteins. *J Mol Biology* 1998;284:1141–1151.

Kundu B, Rajkhowa R, Kundu SC, Wang X. Silk fibroin biomaterials for tissue regeneration. *Adv Drug Delivery Rev* 2012;65:457–470.

Kuo CK, Ma PX. Ionically crosslinked alginate hydrogels as scaffolds for tissue engineering: part 1. Structure, gelation rate and mechanical properties. *Biomaterials* 2001;22:511–521.

Labrador RO, Butí M, Navarro X. Influence of collagen and laminin gels concentration on nerve regeneration after resection and tube repair. *Exp Neurol* 1998;149:243–252.

Lander AD, Fujii DK, Reichardt LF. Purification of a factor that promotes neurite outgrowth: Isolation of laminin and associated molecules. *J Cell Biol* 1985;101:898–913.

Lefcort F, Venstrom K, McDonald JA, Reichardt LF. Regulation of expression of fibronectin and its receptor, alpha 5 beta 1, during development and regeneration of peripheral nerve. *Development* 1992;116:767–782.

Li S, Harrison D, Carbonetto S, Fässler R, Smyth N, Edgar D, *et al*. Matrix assembly, regulation, and survival functions of laminin and its receptors in embryonic stem cell differentiation. *J Cell Biol* 2002;157:1279–1290.

Li S-T, Archibald SJ, Krarup C, Madison RD. Peripheral nerve repair with collagen conduits. *Clin Mater* 1992;9:195–200.

Lien S-M, Ko L-Y, Huang T-J. Effect of pore size on ECM secretion and cell growth in gelatin scaffold for articular cartilage tissue engineering. *Acta Biomater* 2009;5:670–679.

Lin Y-L, Jen J-C, Hsu S-h, Chiu I-M. Sciatic nerve repair by microgrooved nerve conduits made of chitosan-gold nanocomposites. *Surg Neurol* 2008;70: S9–S18.

Liu B-S, Yao C-H, Hsu S-H, Yeh T-S, Chen Y-S, Kao S-T. A novel use of genipin-fixed gelatin as extracellular matrix for peripheral nerve regeneration. *J Biomater Appl* 2004;19:21–34.

Liu BS. Fabrication and evaluation of a biodegradable proanthocyanidin-crosslinked gelatin conduit in peripheral nerve repair. *J Biomed Mater Res A* 2008;87: 1092–1102.

Lohmeyer JA, Siemers F, Machens H-G, Mailander P. The clinical use of artificial nerve conduits for digital nerve repair: A prospective cohort study and literature review. *J Reconstr Microsurg* 2009;25:55–61.

Longo FM, Hayman EG, Davis GE, Ruoslahti E, Engvall E, Manthorpe M, *et al.* Neurite-promoting factors and extracellular matrix components accumulating *in vivo* within nerve regeneration chambers. *Brain Res* 1984;309:105–117.

Lubec G, Holaubek J, Feidl C, Lubec B, Strouhal E. Use of silk in ancient Egypt. *Nature* 1993;362:25.

Lundborg G, Gelberman RH, Longo FM, Powell HC, Varon S. *In vivo* regeneration of cut nerves encased in silicone tubes: Growth across a six-millimeter gap. *J Neuropathol Exp Neurol* 1982;41:412–422.

Lundborg G. A 25-year perspective of peripheral nerve surgery: Evolving neuroscientific concepts and clinical significance. *J Hand Surg* 2000;25:391–414.

Madison R, da Silva CF, Dikkes P, Chiu T-H, Sidman RL. Increased rate of peripheral nerve regeneration using bioresorbable nerve guides and a laminin-containing gel. *Exp Neurol* 1985;88:767–772.

Madison RD, da Silva C, Dikkes P, Sidman RL, Chiu T-H. Peripheral nerve regeneration with entubulation repair: Comparison of biodegradeable nerve guides versus polyethylene tubes and the effects of a laminin-containing gel. *Exp Neurol* 1987;95:378–390.

Mao Y, Schwarzbauer JE. Fibronectin fibrillogenesis, a cell-mediated matrix assembly process. *Matrix Biol* 2005;24:389–399.

Martini R. Expression and functional roles of neural cell surface molecules and extracellular matrix components during development and regeneration of peripheral nerves. *J Neurocytol* 1994;23:1–28.

Marycz K, Tabakow P, Mierzwa J, Wozniak Z, Laska J, Szarek D, *et al.* Influence of alginates on tube nerve grafts of different elasticity — preliminary *in vivo* study. *J Biomater Nanobiotechnol* 2012;3:20–30.

Matsumoto K, Ohnishi K, Kiyotani T, Sekine T, Ueda H, Nakamura T, *et al.* Peripheral nerve regeneration across an 80-mm gap bridged by a polyglycolic acid (PGA)–collagen tube filled with laminin-coated collagen fibers: a histological and electrophysiological evaluation of regenerated nerves. *Brain Res* 2000;868:315–328.

Matsuura S, Obara T, Tsuchiya N, Suzuki Y, Habuchi T. Cavernous nerve regeneration by biodegradable alginate gel sponge sheet placement without sutures. *Urology* 2006;68:1366–1371.

McGrath A. Development of biosynthetic conduits for peripheral nerve repair. *Stockholm* 2012;1–55.

McPherson TB, Badylak SF. Characterization of fibronectin derived from porcine small intestinal submucosa. *Tissue Eng* 1998;4:75–83.

Meena C, Mengi S, Deshpande S. Biomedical and industrial applications of collagen. *Proc Indian Acad Sci* 1999;111:319–329.

Meinel L, Kaplan DL. Silk constructs for delivery of muskuloskeletal therapeutics. *Adv Drug Delivery Rev* 2012;64:1111–1122.

Mochizuki M, Kadoya Y, Wakabayashi Y, Kato K, Okazaki I, Yamada M, *et al.* Laminin-1 peptide-conjugated chitosan membranes as a novel approach for cell engineering. *FASEB J* 2003;17:875–877.

Morris JH, Hudson AR, Weddell G. A study of degeneration and regeneration in the divided rat sciatic nerve based on electron microscopy. *Zeitschrift Zellforschung Mikrosk Anat* 1972;124:103–130.

Mosahebi A, Wiberg M, Terenghi G. Addition of fibronectin to alginate matrix improves peripheral nerve regeneration in tissue-engineered conduits. *Tissue Eng* 2003;9:209–218.

Nazarov R, Jin H-J, Kaplan DL. Porous 3-D scaffolds from regenerated silk fibroin. *Biomacromolecules* 2004;5:718–726.

Ngo TTB, Waggoner PJ, Romero AA, Nelson KD, Eberhart RC, Smith GM. Poly (L-Lactide) microfilaments enhance peripheral nerve regeneration across extended nerve lesions. *J Neurosci Res* 2003;72:227–238.

Numata K, Kaplan DL. Silk-based delivery systems of bioactive molecules. *Adv Drug Delivery Rev* 2010;62:1497–1508.

Oliveira AL, Sun L, Kim HJ, Hu X, Rice W, Kluge J, *et al.* Aligned silk-based 3-D architectures for contact guidance in tissue engineering. *Acta Biomater* 2012;8:1530–1542.

Palazzi S, Vila-Torres J, Lorenzo JC. Fibrin glue is a sealant and not a nerve barrier. *J Reconstr Microsurg* 1995;11:135–139.

Pandit AS, Wilson DJ, Feldman DS, Thompson J. Fibrin scaffold as an effective vehicle for the delivery of acidic fibroblast growth factor (FGF-1). *J Biomater Appl* 2000;14:229–242.

Pankov R, Yamada KM. Fibronectin at a glance. *J Cell Sci* 2002;115:3861–3863.

Pannone PJ. Trends in biomaterials research. *Nova Publishers* 2007;143–190.

Pati F, Datta P, Adhikari B, *et al.* Collagen scaffolds derived from fresh water fish origin and their biocompatibility. *J Biomed Mater Res A* 2012;100: 1068–1079.

Patino MG, Neiders ME, Andreana S, Noble B, Cohen RE. Collagen: An overview. *Implant Dent* 2002;11:280–285.

Pawar SN, Edgar KJ. Alginate derivatization: A review of chemistry, properties and applications. *Biomaterials* 2012;33:3279–3305.

Pettersson J, Kalbermatten D, McGrath A, Novikova LN. Biodegradable fibrin conduit promotes long-term regeneration after peripheral nerve injury in adult rats. *J Plast Reconstr Aesthet Surg* 2010;63:1893–1899.

Pettersson J, McGrath A, Kalbermatten DF, Novikova LN, Wiberg M, Kingham PJ, *et al.* Muscle recovery after repair of short and long peripheral nerve gaps using fibrin conduits. *Neurosci Lett* 2011;500:41–46.

Pfister LA, Papaloïzos M, Merkle HP, Gander B. Hydrogel nerve conduits produced from alginate/chitosan complexes. *J Biomed Mater Res A* 2007;80:932–937.

Pfister LA, Papaloïzos M, Merkle HP, Gander B. Nerve conduits and growth factor delivery in peripheral nerve repair. *J Peripher Nerv Syst* 2007;12:65–82.

Phillips J, Bunting S, Ward Z, Hall S, Brown R. Filbronectin tubes as tissue engineering devices for peripheral nerve repair. *Proc IEEE-EMBS Special Topic Conf Mol Cell Tissue Eng* 2002;165–166.

Phillips JB, King VR, Ward Z, Porter RA, Priestley JV, Brown RA. Fluid shear in viscous fibronectin gels allows aggregation of fibrous materials for CNS tissue engineering. *Biomaterials* 2004;25:2769–2779.

Raisi A, Delirezh N, Heshmatian B, Amini K. Use of chitosan conduit for bridging small-gap peripheral nerve defect in sciatic nerve transection model of rat. *Iran J Vet Surg* 2010;5:89–100.

Ravi Kumar MN. A review of chitin and chitosan applications. *React Funct Polym* 2000;46:1–27.

Roy DC, Wilke-Mounts SJ, Hocking DC. Chimeric fibronectin matrix mimetic as a functional growth-and migration-promoting adhesive substrate. *Biomaterials* 2011;32:2077–2087.

Rutishauser U. Adhesion molecules of the nervous system. *Curr Opin Neurobiol* 1993;3:709–715.

Sanders FK. The repair of large gaps in the peripheral nerves. *Brain* 1942;65: 281–337.

Sarbon NM, Badii F, Howell NK. Preparation and characterization of chicken skin gelatin as an alternative to mammalian gelatin. *Food Hydrocolloids* 2013;30:143–151.

Scheibel T. Protein fibers as performance proteins: New technologies and applications. *Curr Opin Biotechnol* 2005;16:427–433.

Shen CC, Yang YC, Liu BS. Peripheral nerve repair of transplanted undifferentiated adipose tissue-derived stem cells in a biodegradable reinforced nerve conduit. *J Biomed Mater Res A* 2012;100A:48–63.

Shen H, Shen ZL, Zhang PH, Chen NL, Wang YC, Zhang ZF, *et al.* Ciliary neurotrophic factor-coated polylactic-polyglycolic acid chitosan nerve conduit promotes peripheral nerve regeneration in canine tibial nerve defect repair. *J Biomed Mater Res B* 2010;95:161–170.

Siemionow M, Bozkurt M, Zor F. Regeneration and repair of peripheral nerves with different biomaterials: review. *Microsurgery* 2010;30:574–588.

Silver FH, Wang M-C, Pins GD. Preparation and use of fibrin glue in surgery. *Biomaterials* 1995;16:891–903.

Spicer PP, Mikos AG. Fibrin glue as a drug delivery system. *J Control Release* 2010;148:49–55.

Standeven KF, Ariëns RA, Grant PJ. The molecular physiology and pathology of fibrin structure/function. *Blood Rev* 2005;19:275–288.

Stang F, Fansa H, Wolf G, Keilhoff G. Collagen nerve conduits-assessment of biocompatibility and axonal regeneration. *Biomed Mater Eng* 2005;15:3–12.

Stoll G, Griffin J, Li C, Trapp B. Wallerian degeneration in the peripheral nervous system: Participation of both Schwann cells and macrophages in myelin degradation. *J Neurocytol* 1989;18:671–683.

Sufan W, Suzuki Y, Tanihara M, Ohnishi K, Suzuki K, Endo K, *et al*. Sciatic nerve regeneration through alginate with tubulation or nontubulation repair in cat. *J Neurotrauma* 2001;18:329–338.

Sun W, Sun C, Lin H, Zhao H, Wang J, Ma H, *et al*. The effect of collagen-binding NGF-β on the promotion of sciatic nerve regeneration in a rat sciatic nerve crush injury model. *Biomaterials* 2009;30:4649–4656.

Sundback C, Hadlock T, Cheney M, Vacanti J. Manufacture of porous polymer nerve conduits by a novel low-pressure injection molding process. *Biomaterials* 2003;24:819–830.

Tan R, She Z, Wang M, Fang Z, Liu Y, Feng Q. Thermo-sensitive alginate-based injectable hydrogel for tissue engineering. *Carbohydr Polym* 2012;87:1515–1521.

Tate CC, Shear DA, Tate MC, Archer DR, Stein DG, LaPlaca MC. Laminin and fibronectin scaffolds enhance neural stem cell transplantation into the injured brain. *J Tissue Eng Regen Med* 2009;3:208–217.

Timpl R, Rohde H, Robey PG, Rennard SI, Foidart J-M, Martin GR. Laminin — a glycoprotein from basement membranes. *J Biol Chem* 1979;254:9933–9937.

Valentini R, Aebischer P, Winn S, Galletti P. Collagen-and laminin-containing gels impede peripheral nerve regeneration through semipermeable nerve guidance channels. *Exp Neurol* 1987;98:350–356.

VandeVord PJ, Matthew HW, DeSilva SP, Mayton L, Wu B, Wooley PH. Evaluation of the biocompatibility of a chitosan scaffold in mice. *J Biomed Mater Res* 2001;59:585–590.

Vollrath F, Knight D. Structure and function of the silk production pathway in the Spider *Nephila edulis*. *Int J Biol Macromol* 1999;24:243–249.

Waitayawinyu T, Parisi DM, Miller B, Luria S, Morton HJ, Chin SH, *et al*. A comparison of polyglycolic acid versus type 1 collagen bioabsorbable nerve conduits in a rat model: an alternative to autografting. *J Hand Surg* 2007;32:1521–1529.

Wan Y, Huang J, Zhang J, Yin D, Zheng Z, Liao C, *et al*. Investigation of mechanical properties and degradability of multi-channel chitosan–polycaprolactone/collagen conduits. *Polym Degrad Stabil* 2012;98:122–132.

Wang A, Ao Q, Wei Y, Gong K, Liu X, Zhao N, *et al*. Physical properties and biocompatibility of a porous chitosan-based fiber-reinforced conduit for nerve regeneration. *Biotechnol Lett* 2007;29:1697–1702.

Wang L, Shelton R, Cooper P, Lawson M, Triffitt J, Barralet J. Evaluation of sodium alginate for bone marrow cell tissue engineering. *Biomaterials* 2003;24:3475–3481.

Wang X, Hu W, Cao Y, Yao J, Wu J, Gu X. Dog sciatic nerve regeneration across a 30-mm defect bridged by a chitosan/PGA artificial nerve graft. *Brain* 2005;128:1897–1910.

Webber C, Zochodne D. The nerve regenerative microenvironment: early behavior and partnership of axons and Schwann cells. *Exp Neurol* 2010;223:51–59.

Wenk E, Merkle HP, Meinel L. Silk fibroin as a vehicle for drug delivery applications. *J Control Release* 2011;150:128–141.

Whitworth I, Brown R, Dore C, Green C, Terenghi G. Orientated mats of fibronectin as a conduit material for use in peripheral nerve repair. *J Hand Surg* 1995;20:429–436.

Winkler S, Wilson D, Kaplan DL. Controlling β-sheet assembly in genetically engineered silk by enzymatic phosphorylation/dephosphorylation. *Biochemistry* 2000;39:12739–12746.

Yamada KM. Cell surface interactions with extracellular materials. *Annu Rev Biochem* 1983;52:761–799.

Yamaguchi Y, Mann DM, Ruoslahti E. Negative regulation of transforming growth factor-β by the proteoglycan decorin. *Nature* 1990;346:281–284.

Yan X, Zhao Y, Wang W, Gu X, Yang Y. Biological safety assessment of the silk fibroin-based nerve guidance conduits *in vitro* and *in vivo*. *Adv Studies Biol* 2009;1:119–138.

Yan XL, Khor E, Lim LY. Chitosan-alginate films prepared with chitosans of different molecular weights. *J Biomed Mater Res* 2001;58:358–365.

Yang J-S, Xie Y-J, He W. Research progress on chemical modification of alginate: A review. *Carbohydr Polym* 2011;84:33–39.

Yang Y, Chen X, Ding F, Zhang P, Liu J, Gu X. Biocompatibility evaluation of silk fibroin with peripheral nerve tissues and cells *in vitro*. *Biomaterials* 2007;28:1643–1652.

Yang Y, Ding F, Wu J, Chen X, Liu J, Gu X. Development and Biocompatibility evaluation of silk fibroin-based nerve grafts for peripheral nerve regeneration. *7th Asian-Pacific Conf Med Biol Eng* 2008;19:4–8.

Yang Y, Gu X, Tan R, Hu W, Wang X, Zhang P, *et al*. Fabrication and properties of a porous chitin/chitosan conduit for nerve regeneration. *Biotechnol Lett* 2004;26:1793–1797.

Yang YC, Shen CC, Cheng HC, Liu BS. Sciatic nerve repair by reinforced nerve conduits made of gelatin–tricalcium phosphate composites. *J Biomed Mater Res A* 2011;96:288–300.

Yao CH, Liu BS, Hsu SH, Chen YS, Tsai CC. Biocompatibility and biodegradation of a bone composite containing tricalcium phosphate and genipin crosslinked gelatin. *J Biomed Mater Res A* 2004;69:709–717.

Yao L, de Ruiter GC, Wang H, Knight AM, Spinner RJ, Yaszemski MJ, *et al*. Controlling dispersion of axonal regeneration using a multichannel collagen nerve conduit. *Biomaterials* 2010;31:5789–5797.

Yin Q, Kemp GJ, Yu LG, Wagstaff SC, Frostick SP. Neurotrophin-4 delivered by fibrin glue promotes peripheral nerve regeneration. *Muscle Nerv* 2001;24:345–351.

Yoshii S, Ito S, Shima M, Taniguchi A, Akagi M. Functional restoration of rabbit spinal cord using collagen-filament scaffold. *J Tissue Eng Regen Med* 2009;3:19–25.

Yoshii S, Oka M, Shima M, Taniguchi A, Akagi M. 30 mm regeneration of rat sciatic nerve along collagen filaments. *Brain Res* 2002;949:202–208.

Yuan Y, Zhang P, Yang Y, Wang X, Gu X. The interaction of Schwann cells with chitosan membranes and fibers *in vitro*. *Biomaterials* 2004;25:4273–4278.

Yurchenco PD, Birk DE, Mecham RP. Extracellular matrix assembly and structure. *Biol Extra Mat* 1994; 315–340.

Zeng L, Huck S, Redl H, Schlag G. Fibrin sealant matrix supports outgrowth of peripheral sensory axons. *Scand J Plast Reconstr Surg Hand Surg* 1995;29:199–204.

Zhang X, Reagan MR, Kaplan DL. Electrospun silk biomaterial scaffolds for regenerative medicine. *Adv Drug Delivery Rev* 2009;61:988–1006.

Zheng L, Cui H-F. Enhancement of nerve regeneration along a chitosan conduit combined with bone marrow mesenchymal stem cells. *J Mater Sci Mater Med* 2012;23: 2291–2302.

Zhu Q, Xie M, Yang J, Chen Y, Liao K. Influence of chitosan and porosity on heat and mass transfer in chitosan-treated porous fibrous material. *Int J Heat Mass Transf* 2012;55:1997–2007.

Chapter 3

Native Polymer-Based 3D Substitutes for Cartilage Repair

Huitang Xia[*,†,a], *Yu Liu*[*,a], *Ran Tao*[*], *Chunlei Miao*[†], *Shengjian Tang*[†], *Biaobing Yang*[†], *Guangdong Zhou*[*,‡]

*Department of Plastic and Reconstructive Surgery
Shanghai Ninth People's Hospital, Shanghai Jiao Tong University
School of Medicine, Shanghai Key
Laboratory of Tissue Engineering
National Tissue Engineering Center of China,
Shanghai 200011, P. R. China
†Research Institute of Plastic Surgery,
Wei Fang Medical College
Wei Fang, Shan Dong 261041, P. R. China

1. Introduction

Cartilage degeneration is the major cause of disability and can be caused by trauma or common joint disorders such as osteoarthritis (OA). However, cartilage is a special tissue that lacks neural connections, vascularization, and a latent pool of stem cells/chondro-progenitors, which limits its self-reparative capacity through endogenous healing

‡Corresponding author. Department of Plastic and Reconstructive Surgery, Shanghai 9th People's Hospital; Address: 639 Zhi Zao Ju Road, Shanghai 200011, P. R. China; Tel./Fax: +86-21-53078128; E-mail: guangdongzhou@126.com
ª These authors contributed equally to this work.

mechanisms (Keeney *et al.*, 2011). Furthermore, the subsequent repair tissue for cartilage lesions is incompetent for the native tissue due to its inferior biochemical and mechanical properties. As a consequence, the repair of cartilage damage has been an enormous challenge in this field. Taking articular cartilage for example, a variety of treatments such as arthroscopic lavage, abrasion arthroplasty, subchondral drilling, and microfracture etc. have been attempted to reconstruct the function of injured cartilage. However, all had led to suboptimal healing and regeneration of cartilage (LaPorta *et al.*, 2012; Alford and Cole, 2005). Cartilage tissue engineering, with the development of tissue regenerative technology based on cells and biodegradable scaffolds, has been considered as the ideal treatment for cartilage damage (Kuo *et al.*, 2006; Thiede *et al.*, 2012; Alford and Cole, 2005).

A typical tissue engineering strategy can be separated into three components: cells, scaffolds, and tissue regeneration microenvironments. During the past score years, all these three aspects have obtained sufficient developments in cartilage tissue engineering (Berthiaume *et al.*, 2011). Firstly, besides autologous chondrocytes, many kinds of cells including multipotent stem cells derived from bone marrow, adipose, synovium, and so on, as well as pluripotent stem cells such as embryonic stem cells (ESCs) and induced pluripotent stem cells (iPSCs), have been explored as the cell sources for cartilage regeneration. Secondly, with the increasing understanding of the role of scaffolds in tissue engineering and the advancement of developmental biology and nanotechnology, recent researches on the scaffolds have mainly focused on the design and synthesis of functionalized scaffolds that can elicit desirable cell-material interactions to guide cell behavior and enhance tissue regeneration. Thirdly, many kinds of factors including biological (such as growth factors), physical (such as mechanical stress provided by bioreactor), and chemical (such as small molecule compounds) factors have been investigated for mimicking chondrogenic regenerative microenvironments and have played important roles in promoting proliferation of chondrogenic cells, inducing chondrogenesis, and enhancing cartilage extracelluar matrix (ECM) synthesis (Vinatier *et al.*, 2009; Danisovic *et al.*, 2012).

In spite of the above advancements, the clinical translation of cartilage regeneration technology only achieved poor results. More efforts still need to be performed in all these three aspects to further accelerate the process of clinical translation (Hunziker *et al.*, 2006). In terms of scaffolds, it has been proposed that the scaffold should actively participate in tissue neogenesis progress rather than just support a physical three-dimensional (3D) space for cartilage regeneration. To address this issue, the method to accurately mimic ECM microenvironments of native cartilage in components and structures obviously is an important direction in scaffold design for cartilage repair and regeneration. From this point, native biomaterials, which have special biological property in contrast to synthetic materials, is necessary for manufacturing an ideal cartilage scaffold (Raghunath *et al.*, 2007).

In this chapter, major characters and advances of native polymers as well as their roles in cartilage engineering are reviewed.

2. The Theory Basis of Native Polymer-Based 3D Substitutes for Cartilage Repair

The basic strategy of cartilage regeneration, similar to the others, involves cells, scaffolds, and regeneration microenvironments, alone or in combination (Alford and Cole, 2005; Thiede *et al.*, 2012). As the key component of tissue engineering, scaffolds are mainly responsible for mimicking ECM microenvironments to maintain a 3D space for cell survival and tissue formation. Native ECM, a dynamic structure, provides structural and anchoring support for cells and contributes to signalling, directing cell fate and function through cell-matrix interactions. Accordingly, an ideal scaffold should: support cell growth and maintenance; provide appropriate mechanical, chemical and biological characteristics mimicking native ECM; and facilitate effective nutrient transfer, gas exchange (i.e., O_2 and CO_2), metabolic waste removal, and signal transduction (Geckil *et al.*, 2010; Kim and Mooney, 1998; Lu and Spalazzi, 2009).

The polymers for scaffolds can be divided into native polymers and synthetic polymers according to different material resources.

Synthetic polymers are the most widely used scaffolds because of their high versatility, mechanical properties, reproducibility, and good workability. However, none of them can perfectly repair cartilage defects in animal models due to lack of biological cues of natural ECM. By means of mimicking the cartilage ECM, native polymers are playing an indispensable role in cartilage tissue engineering. Native polymers may mimic cartilage ECM in three aspects: molecular composition, natural structure, and biological function. In the following text, these three aspects are analyzed respectively (Ge *et al.*, 2012).

2.1. Native polymers mimic the molecular composition of cartilage ECM

The molecule composition of tissue is the basis of its structure and biological function. Therefore, the molecular composition of cartilage ECM is an important aspect that native polymer-based scaffolds need to mimic.

2.1.1. Molecular composition of cartilage ECM

ECM is a general name of the organic substances in tissue except cells. The composition of cartilage ECM is a complex mixture of structural and functional proteins, collagens, glycosaminoglycans, glycoproteins, and small molecules arranged in a unique, tissue specific three-dimensional architecture (Schenke-Layland, 2008; Singh and Schwarzbauer, 2012). There are three types of cartilage: hyaline cartilage, found in articulating joints, nose, trachea, and ribs; elastic cartilage, found in auricles, gullet, and epiglottis; and fibrocartilage, found in intervertebral disks and the meniscus (Verma, 2001).

The molecule composition of ECM in different types of cartilage is different. Fibrocartilage has a higher collagen content and lower proteoglycan (PG) content compared to hyaline cartilage, and elastic cartilage contains elastic fibers in addition to collagen and proteoglycans. Due to the urgency of clinical requirements in articular cartilage, more researches and progress are mainly focused on this. Therefore, this part will elaborate cartilage ECM components with articular cartilage as an example (Spiller *et al.*, 2011). Articular cartilage is a thick and highly organized tissue that lies in the articulating ends of diarthrodial joints, which is crucial for efficient functioning of the joint. This tissue has very exquisite organization at both the macro- and micro-scale that result in its unique biochemical and biomechanical properties. Articular cartilage is a multiphasic tissue consisting of < 5% chondrocytes, 60% – 85% interstitial fluid, and a solid ECM, composed of about 15% – 22% type II collagen, 4% – 7% proteoglycans, and other protein macromolecules (Ma and Langer, 1999). It can be regarded as a composite, organic solid matrix that is saturated with water and mobile ions. Within native cartilage, stiff and elastic cross-linked collagen fibrils/fiber bundles help cartilage to resist lateral expansion on axial compression by maintaining a rigid framework. The articular cartilage specific PG, aggrecan, with its highly sulfated glycosaminoglycan (GAG) chains attached to collagen fibrils, sequesters large amounts of ions and water via negative charges. Upon compression, some water molecules are forced out, causing reversible deformation of cartilage and temporarily increasing the contact area, while most water molecules remain under hydrostatic pressure at their original location through GAG, contributing to the compression stiffness and lubrication properties of cartilage (Mow *et al.*, 1984; Mow *et al.*, 1992).

2.1.2. *Native polymers mimic molecular composition of cartilage ECM*

Many different types of matrices have been tested *in vitro*, as well as in experimental animals and human patients, for their efficacy in facilitating or promoting articular cartilage repair. These matrices can be broadly categorized according to their chemical nature into protein-based polymers, carbohydrate-based ones, artificial materials, and combinations of these (Table 1) (Hunziker, 2002).

As mentioned above, the molecule composition is the basis of its structure and biological function. Obviously, only native polymer-based scaffolds, such as acellular cartilage matrix, collagen, and hyaluronic acid, can efficiently mimic the molecular composition of cartilage ECM. Although the applications of synthetic polymers (such

Table 1.   Chemical classes of matrix.

1. Protein-based polymers	Fibrin, collagen, gelatine
2. Carbohydrate-based polymers	Polylactic acid, polyglycolic acid, hyaluronan, alginate, agarose, chitosan
3. Artificial polymers	Dacron(polyethylene terephthalates), Tefilon (polytetrafluoroethylene), carbon fibers, polyesterurethane, polybutyric acid, polyethylmethacrylate, hydroxyapatite
4. Within/between classes	Crosslinkage, chemical modifications, geometrical modifications (to produce fibrillar forms or foams), matrix combination

as polylactic acid and polyglycolic acid) have obtained some satisfactory results in cartilage repair and regeneration, the synthetic polymers fail to avoid aseptic inflammation reaction caused by the degraded byproducts and fail to provide biological signals for ideal cartilage regeneration. With the development of researches, native polymer-based scaffolds with biological information participate in many fundamental biological processes beyond synthetic polymer capability and thus play an irreplaceable role in cartilage engineering and become an important consideration in cartilage scaffold design (Sadr *et al.*, 2012; Dahlin *et al.*, 2011).

It has been shown that chondrocyte phenotypes are under dual regulation by internal genetic factors and the external environment. ECM composition in the microenvironment has a crucial effect in directing progenitor cells to differentiate to chondrogenic cells (Pennesi *et al.*, 2011; Keung *et al.*, 2010). Studies reported that the incorporations of ECM components gelatin/chondroitin/HA in poly-(lactic-co-glycolic acid) could significantly augment the proliferation of MSCs and GAG synthesis. Various research groups have shown that incorporation of a biomimetic surface to the scaffold could dramatically alter the differentiation outcomes of MSCs (Patrascu *et al.*, 2013). Coating of the biomaterial with chondroitin sulfate (CS), a major GAG found in cartilage, resulted in enhanced formation of hyaline cartilage (Ouzzine *et al.*, 2012).

Though provision of some ECM components has been shown to improve the chondrogenic differentiation of MSC with enhancement in ECM deposition and formation of more hyaline cartilage, individual ECM components have different effects. For example, collagen type II has the tendency to induce hypertrophic chondrogenic phenotypes, while chondroitin sulfate helps to maintain pre-hypertrophic phenotypes.

2.2. *Native polymers mimic the structure of cartilage ECM*

2.2.1. *Structure of cartilage ECM*

Each type of cartilage has its structural characteristics. Of all kinds of cartilage, articular cartilage has the most complex structure. The solid matrix of articular cartilage has a highly specific ultrastructural arrangement consisting of different zones varying in depth from the articular surface, forming a hierarchical structure that is typically divided into three zones: superficial zone (surface to 10 – 20% of thickness), middle zone (20 – 70%), and deep zone (70 – 100%). Each zone has distinct ECM composition and organization, cell morphology, and metabolic activity (Coates and Fisher, 2010). Collagen fibrils in the superficial zone are oriented parallel to the articular surface and impart high tensile strength to withstand the tensile stress encountered under joint loads (Clark, 1991). The relatively small amount of PG attached onto the collagenous membrane can act as a barrier of high resistance against fluid flow when cartilage is compressed. In the middle zone, the larger collagen fibers are randomly oriented, while aggrecan content reaches its maximal level. In the deep zone, the collagen fibers form bundles that are oriented perpendicular and attached to calcified cartilage and subchondral bone. Collagen content per wet weight does not change significantly with depth, but depth-dependent increases in hydroxylysine and hydroxylysyl pyridinoline cross-links exist, and together with the presence of other minor collagen isoforms such as type IX and type XI, play a critical role in the regulation of fibril size, interfibril cross-linking, and interactions with cartilage proteoglycans, thereby contributing to the mechanical properties of articular cartilage (Bobick *et al.*, 2009; Bank *et al.*, 1998). The concentration of these

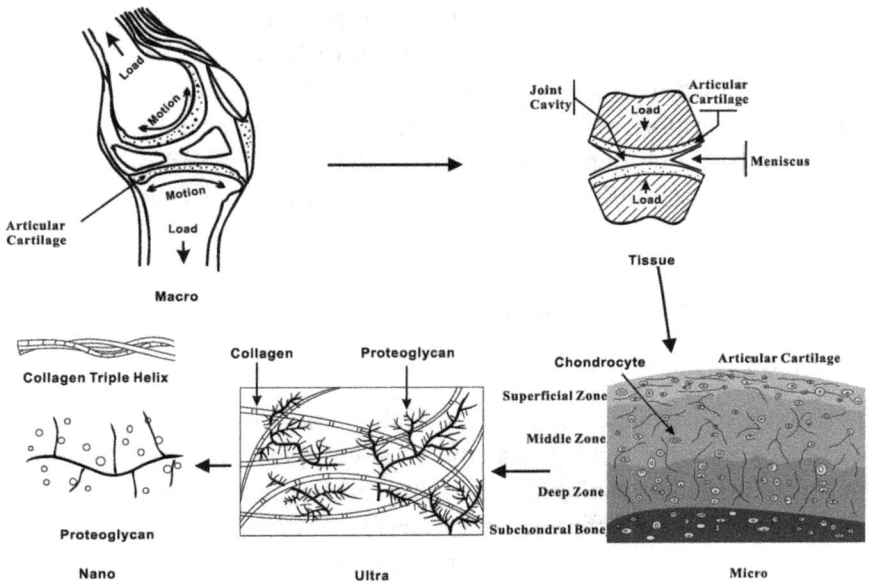

Fig. 1.   The structure of cartilage from macro to microscale (Ge Z *et al.*, 2012).

PG aggregates increases from the articular surface to the middle zone, which contributes to high swelling pressure and water content, and thus increases the compressive modulus of the tissue (Williamson *et al.*, 2003; Schinagl *et al.*, 1997). Chondrocytes at the different zones of articular cartilage have distinct morphologies and organization, and express zone-specific markers. Chondrocytes in superficial zone are flattened and clustered in a horizontal fashion at a relatively high density. Chondrocytes in the middle zone are more spherical and randomly oriented, while they are larger and organized in vertical columns in the deep zone (Klein *et al.*, 2009). Cartilage can be regarded as a biphasic material with complex mechanical properties such as anisotropy, nonlinearity, and viscoelasticity.

2.2.2.   *Native polymers mimic the structure of cartilage ECM*

The scaffold-based approach relies on biomaterials to provide structural and mechanical strength, and spatiotemporal organization for cartilage regeneration with biomimetic zones. The physical properties, together with the biological and chemical cues of the scaffolds can coax embedded cells to specific phenotypes (Nuernberger *et al.*, 2011). Different combination of materials and morphogenes can facilitate chondrogenesis or osteogenesis in continual but distinct portions of the scaffolds (Sherwood *et al.*, 2002), while small chemical groups can induce mesenchymal stem cells (MSCs) to differentiate into different lineages (Nuernberger *et al.*, 2011). Hydrogels with collagen gradient enhance cartilage regeneration by recruiting more MSCs, compared with hydrogels constituting of exactly the same materials but without collagen gradients (Mimura *et al.*, 2008). Agarose gels with depth-dependent mechanical properties were adapted to fabricate nonhomogenous layered cartilage. Nevertheless, differences in Young's Modulus of different layers became indistinct with extended *in vitro* culture (Roberts *et al.*, 2011).

The cell-based approach aims to seed chondrocytes or differentiate stem/progenitor cells to chondrocytes in different layers, relying on the cell-directed generation of zonal phenotypic specificity. Multilayered photopolymerizing hydrogel encapsulating chondrocytes harvested from superficial, middle, and deep zones, respectively, were cultured *in vitro* to make hierarchical cartilage layers similar to their native counterparts (Kim *et al.*, 2003). By incorporating a unique combination of ECM components (CS, HA, and metalloproteinase sensitive peptides) into polyethylene glycol (PEG) hydrogel, MSC can be differentiated in the presence of a distinct ratio of PG to collagen, forming tissue of different compressive modulus, corresponding to the mechanical strength of zonal cartilage (Suciati *et al.*, 2006). These results indicate the potential for a composite scaffold with specific biomaterial compositions and structures that can direct progenitor stem cells to give rise to zonal cartilage. Hybrid heterogenous scaffolds based on prototyped biomaterials and specific cell populations can be used to develop zonal cartilage. Cross-linkable hydrogels such as agarose, gelatin, or PEG, or synthetic biomaterials such as poly (e-caprolactone) are mixed with the cell suspension before being printed using a predesigned program to fabricate three-dimensional scaffolds (Fedorovich *et al.*, 2007). Biologically active molecules such as laminin, fibronectin, and glycosaminoglycan can be added to the bio-inks to promote zonal cartilage formation. Heterogeneous geometries and high cell viability can be achieved through these methods, but whether or not the cells and matrix deposition in the native orientation can facilitate biological properties and functions have to be investigated further.

2.3. *Native polymers mimic the function of cartilage ECM*

2.3.1. *Regulation of cell fate by native adhesion ligand*

Cell adhesion can be regulated by adsorbed protein surfaces by using a variety of naturally derived ECM molecules such as collagen and fibrin, or using these molecules to decorate synthetic polymers (Santos *et al.*, 2012). However, protein engineering allows us to isolate functional domains within large ECM molecules and incorporate them into otherwise inert substrates. Thus, epitopes that mediate cell adhesion can be mimicked using synthetic peptides. Among them, perhaps the most known are arginine-glycine-aspartic acid (RGD), derived from fibronectin, and tyrosine-isoleucine-glycine-serine-arginine (YIGSR), derived from laminin. PEG hydrogels can also be modified by novel polymerization mechanisms such as thiol-ene and thiol-acrylate chemistries (Anderson *et al.*, 2011), whereas other polymers like alginate are usually modified by means of carbodiimide chemistry. Not only the adhesion moieties themselves but also their density and spatial distribution on micrometer and nanometer scales influence cell fate (Chan and Mooney, 2008). By manipulating the way of cell adhesion, it is possible to induce major cellular processes such as migration, proliferation and differentiation (Mooney and Vandenburgh, 2008). With this idea, nanoscale patterns of RGD islands in hydrogels can be varied without altering the final ligand density. For instance, hydrogels with reduced island spacing were produced by uniformly distributing alginate chains containing a single ligand, whereas more increased island spacing was achieved by mixing unmodified chains and chains coupled with multiple peptides. This scaffold surface design testified that more closely spaced islands favored cell spreading, whereas more widely spaced islands supported differentiation (Comisar *et al.*, 2007).

Beyond these approaches, much attention has been lately paid to the patterning of adhesive moieties. Biomolecules can also be immobilized in micropatterned volumes within agarose gels with a multiphoton laser. When agarose or hyaluronic acid (HA) is covalently modified with a derivative of cysteine protected with a photocleavable group, the protection groups can be removed upon exposure with a laser beam (micrometric resolution). As a result, desired oligopeptides are covalently immobilized in patterned sites via Michael-type addition (Santos *et al.*, 2012). Using this procedure and by means of orthogonal physical binding pairs (barnasa–barstar and streptavidin–biotin), simultaneous patterning of multiple growth factors has been achieved to direct neural precursor cell differentiation. Two-photon laser scanning (TPLS) photolithography in PEG diacrylate (PEGDA) hydrogels can also be used to guide encapsulated dermal fibroblasts with precisely patterned RGD moieties.

2.3.2. *Regulation of chondrogenic differentiation and phenotypic maintenance*

Chondrogenic differentiation and phenotypic maintenance have been significant challenges for cartilage repair based on tissue engineering strategies. Functions of native polymers for both of these aspects will be introduced in the following part. Much effort has been made to derive the appropriate chondrogenic phenotype in differentiation development, including selection of proper materials, control of topography and mechanical properties of materials, regulating cell distribution through adjusting porosity and connection of the pores. Spherical morphologies of chondrocytes are closely related to their chondrogenic potential. On the other hand, dedifferentiated fibroblastic morphology is related to adhesion of chondrocytes via integrin and subsequent actin stress fibre arrangement. This dedifferentiation of chondrocytes can be reversed by seeding them in three-dimensional hydrogel or with cytochalasin D to artificially inhibit actin fiber arrangement. Proper cell aggregation could induce differentiation and appropriate phenotype of chondrocytes during both embryonic development and cartilage regeneration. Chondrogenic differentiation of stem cells is initiated by cell aggregation to undergo a condensation process with increased cell density and cell-cell interactions during embryonic development, which leads to distinct spherical cell morphology and expression of chondrogenic marker genes Sox9, Sox5, and Sox6 (Woods *et al.*, 2007). Correlation between stem cell differentiation lineage commitment and cell shape was reported. Cell shape change confers a switch between chondrogenic (round shape) and smooth muscle cell (spread and flattened) fates involving Rac1 signaling regulation of N-cadherin, underlying the tight coupling between lineage commitment and changes in cell shape and cell-cell adhesion that occur during morphogenesis (Gao *et al.*, 2010). Chondrocytes naturally aggregate into cell clusters in suspension, while surrounding pericellular matrix rich in collagen II and cartilage oligomeric protein (COMP) helps to maintain phenotype of chondrocytes. Direct cell-cell aggregation among chondrocytes might not be a normal occurrence in mature cartilage. Preaggregation of expanded chondrocytes before loading into a porous scaffold enhances the quality of regenerated tissues. It has been proposed that condensation and aggregation of chondrocytes during OA development protected them from apoptosis.

Cell aggregation induced by bioactive scaffolds based on native polymers, can be achieved through manipulation of the extracellular environment surrounding the cells. The designs of grooved topological structures that induce cell collision and chitosan-graft-lactose coated microcarriers have been shown to promote cell aggregation (Gigout *et al.*, 2008). Inclusion of biodegradable moieties in three-dimensional hydrogel can also enhance cell aggregation. The hydrophobicity of ECM plays an important role in dictating cellular behaviors. Spatial-design of hydrophobicity within the scaffold structure could be employed to promote cellular aggregation that enhances cartilage tissue formation. MSCs adopt specific morphology, migration pattern, and aggregation on these surfaces that results in specific lineage differentiation.

2.3.3. *Controlled delivery of biochemical factors*

Controlled delivery of biochemical factors as a powerful tool to provide a more biomimetic microenvironment for seeded cells has been an ongoing research topic in tissue regeneration, as it can significantly enhance tissue formation (Koria, 2012). Ideally, the delivery vehicles should have quantitative control of target-binding affinities and drug-carrying capacities, therapeutic unloading response, and intrinsic properties for labeled or nonlabeled tracing. With regards to cartilage tissue engineering, increasing attention has been paid to spatiotemporal delivery of morphogenetic effectors such as growth factors (Tabata, 2003). However, more comprehensive and in-depth understanding of the developmental and regenerative biology of articular cartilage, such as lineage differentiation and surface markers, is urgently needed before this aim can be achieved. Biochemical factors can be absorbed physically, cross-linked chemically, or embedded in degradable biomaterials, to provide a sustained bioactive factor release for cartilage regeneration (Richardson *et al.*, 2001). Micro/nano polymeric particles as potential carriers have been developed. Microparticles embedded with biochemical factors could be incorporated into the hydrogel as a beneficial carrier for sustained release of biochemical factors, especially in direct injectable systems to maintain the phenotype of chondrocytes (Bian *et al.*, 2011).

Specific targeting can be achieved through a combination of macromolecular therapeutics, or through local delivery such as transcutaneous delivery. The stimuli-responsive carrier can also be developed for the controlled release of bioactive factors and be used to enable on-demand controlled release profiles that may improve the cartilage tissue formation. Therefore, environmental stimulus, such as mechanical stimulus, temperature sensitivity, change of pH value, electrical signals, and optical signals, can be adopted to trigger the release of loaded biochemical factors (Stuart *et al.*, 2010; Lee *et al.*, 2000). Microparticles with sophisticated internal structures could potentially release biochemical factors in response to multiple stimuli.

2.3.4. *Regulation of mechanical properties*

Mechanical stimulation is not only essential for embryonic development of articular joint, but also critical for *in vitro* chondrogenesis as well as functional cartilage regeneration in adults through up-regulation of genes, activation of signaling pathways, and maturation of ECM.

Proper Young's Modulus (0.4 – 0.8MPa) is particularly important for neocartilage development (Mikic *et al.*, 2004; Salisbury *et al.*, 2010; Takahashi *et al.*, 1998). With the development of cartilage tissue engineering, the mechanical environment of cells and their microenvironments have aroused much interest, and many mechanoactive scaffolds have been developed. Mechanical compatibility of scaffolds with native cartilage is critical for functionality of regenerated cartilage at both the macroscopic and microscopic levels, at least for relatively large cartilage defects, as physiological loads cannot be effectively compensated by surrounding tissues.

It has been attempted widely to fabricate scaffolds with mechanical properties similar to articular cartilage. Cartilage tissue is regarded as biphasic, being predominantly composed of collagen fibers and PG. Based on this understanding, poly-(L-lactide-co-e-caprolactone) (PLCL) was adopted as a mimic to collagen fibers for providing a relative rigid frame structure, while chitosan was immobilized on PLCL as a substitute for GAG. Although the chitosan-PLCL scaffolds had a similar viscoelastic property to native cartilage, their Young's Modulus was much less than that of native cartilage. The inferiority of chitosan-PLCL scaffolds in mechanical properties could attribute to several drawbacks, such as inferior mechanical properties of PLCL, inappropriate PLCL/chitosan ratio, difference in charge density between chitosan and GAG, and lack of other functional molecules (fibronectin, hyuloronic acid, decorin, etc.). Integration of graphene and carbon nanotube could effectively enhance mechanical properties of polymeric scaffold. However, their nondegradation properties hinder clinical application to some extent (Fan *et al.*, 2010; Zhao *et al.*, 2010).

Mechanical properties of degradable biomaterials are easily altered by degradation, as degradation gradually erodes the integral structures. On the other hand, mechanical properties of cell-laden biomaterials may improve with ECM deposition and maturity. These dual effects on mechanical properties have to be considered in scaffold design (Wan *et al.*, 2011). The fundamental roles of bioactive scaffolds based on native polymers are not only to provide a temporary substrate on which transplanted cells can adhere but also to maintain mechanical integrity during the healing process, as well as deliver appropriate mechanical stimuli to the attached cells under normal physiological weight loading condition. Mechanical stiffness of tissue is not a static property. During development of tissues, their stiffness alters according to developmental changes in ECM. Understanding the effect of time-dependent stiffening on cell maturation may be a critical design parameter for future material-based cell therapies, especially the use of progenitor cells. Biomaterials with time-dependent elastic modulus or stiffness mimicking tissue development, which provide fine-tuned support for cell differentiation and tissue maturation, should be introduced in scaffold design for cartilage regeneration (Li *et al.*, 2012).

Mechanical signals are also involved in cell proliferation and differentiation through membrane and integrin activation/internalization, as well as cytoskeleton and cell rigidity changes. However, the mechanotransduction process by which mechanical signals are sensed, transmitted to biochemical pathways, and manifested as changes in cell behavior, is not clearly understood. It is believed that ion channels, integrins, and the intracellular cytoskeleton are involved (Discher *et al.*, 2009; Ramage *et al.*, 2009; Wescoe *et al.*, 2008).

Activation of ion channels in the plasma membranes of chondrocytes allows an influx of ions such as calcium into the cell that leads to the activation of intracellular signaling pathways. Transmembrane integrins physically anchor chondrocytes to ECM, forming transmembrane connections to the cytoskeleton, thus enabling signal transduction (Lee *et al.*, 2000; Knight *et al.*, 2006).

The response of chondrocytes to mechanical stimuli depends on the nature of the stimulus, and the presence of appropriate ECM composition to engage and to activate cell surface molecules. Understanding the role and mechanism of mechanical and electrochemical signals on chondrocyte function will enable future strategies to develop functional scaffolds.

2.3.5. *Integration of neocartilage with host tissues*

When transplanting the cell-scaffold constructs into cartilage defects, the integration of implants with host tissues, including native cartilage and subchondral bone, is crucial for both immediate functionality and long-term performance of the tissue graft. If the integration of neo-cartilage with native cartilage and subchondral bone is unstable, the neo cartilage can easily break away from the native cartilage and cause severe damage again.

Much attention has been focused on integration between neocartilage and native cartilage. As relatively low turnover of natural cartilage impedes the integration, bioglues or adhesives have been used with the aim to enhance integration. Two types of adhesives are broadly used — reactive and nonreactive adhesives, defined by whether or not the adhesive chemically reacts to the original materials. The strength of adhesion depends on many factors, including van der Waals forces between molecules, electro-staticforces, capillary forces, and chemical bonds between adhesive and substrate (Sun *et al.*, 2005; Singer *et al.*, 1998). Several nonreactive glues/ adhesives have been commercialized and used clinically, including derivatives of cyanoacrylates (Superglue), Bioglue (glutaraldehyde-albumin), and fibrin glue (Tisseal). Biomimetic topography has been developed by mimicking geckofeet with an aim to adhere in a dry environment without chemical reaction. However, the strength of the adhesive is usually too low to mechanically integrate neocartilage to surrounding cartilage and serves mainly to facilitate cell seeding. Reactive adhesives have been developed to enhance the integration. Tissue transglutaminase, naturally expressed in cartilage and other tissues, is a biocompatible and stronger adhesive than nonreactive adhesives (Capito *et al.*, 2008; Jürgensen *et al.*, 1997). Biopolymer chondroitin sulphate (CS), one of the major components of cartilage ECM, has been modified chemically as a novel bioadhesive, which not only integrates specific biomaterials with native cartilage, but also allows chondrocytes to migrate freely across the interface between the implant and native tissue. This in turn enhances mechanical stability of the hydrogel and tissue repair in cartilage defects in both the short term and long term (Zhang *et al.*, 2011).

As changes in mechanical and physiological properties of sub-chondral bone have been implicated in degradation of articular cartilage during development of osteoarthritis, it is reasonable to deduce that they are equally important for regeneration and functionality of cartilage. Integration with subchondral bone is of great importance for neocartilage to transduce physiological loads, exchange nutrients and metabolic waste, as well as shaping internal structures (Mente and Lewis, 1994; Pan *et al.*, 2009).

Many hybrid scaffolds have been made to regenerate cartilage and subchondral bone simultaneously. However, the integration of regenerated cartilage with subchondral bone has not been evaluated histologically and mechanically *in vivo*. Porous scaffolds with a specific design of intramedullary stem not only effectively regenerated articular cartilages, but also achieved firm integration with subchondral bone (Shao *et al.*, 2006; Zhou *et al.*, 2011).

3. The Main Native Polymers for Cartilage Regeneration

So far, there are many kinds of native polymers which are used for cartilage engineering. In the following part, we will respectively introduce these native polymers. Their properties, manufacturing procedures, and application prospective in the cartilage engineering setting will be emphasized.

3.1. *Acellular matrix*

Biological scaffolds derived from decellularized tissues and organs have been successfully used in both pre-clinical animal studies and in human trials. Removal of cells from a tissue or an organ leaves the complex mixture of structural and functional proteins that constitute ECM. Because of the removal of the xenogeneic or allogenic cellular antigens and preservation of most of the structural and functional proteins, biologic scaffolds derived from decellularized tissue and organs have great potential for tissue and even organ regeneration. Furthermore, many kinds of decellularized scaffolds, including human dermis, porcine SIS, porcine heart valves, and porcine urinary bladder, have received approval for use in humans (Gilbert, 2012; Gilbert *et al.*, 2006; Barnes *et al.*, 2011; Yang *et al.*, 2008).

Due to the distinctive merits such as great biocompatibility with chondro-inductive function, decollurised cartilage has aroused great attention as a potentially ideal scaffold for cartilage regeneration.

However, lack of standard manufacturing procedures as well as a consensus of evaluation criterions due to restricted knowledge concerning this kind of scaffolds has greatly limited decellularised cartilage for vast applications in cartilage engineering. Therefore, in the following part of this section, major characteristics of decellularised cartilage such as immunogenicity, tissue sytokine and host response will be addressed. Major methods for preparation of decellularised cartilage will also be emphasized.

3.1.1. *Properties of acellular matrix for cartilage regeneration*

The immune response to xenogeneic whole organ transplants has been well studied and has framed our understanding of the present limits of cell, tissue and organ transplantation. However, the mammalian immune response to acellular scaffolds derived from xenogeneic ECM has received much less attention. The consistency and uniformity of host response to autologous, allogenous and xenogeneic ECM scaffolds suggests that preservation of the constituent molecules has been significant. The uniformity of response may also suggest that ECM contains biologic mediators of the immune system that affect the host response to injury. Furthermore, the experimental results of tissue cytokine and humoral response to acellular matrix scaffolds support the application of them in the cartilage engineering. Several studies have examined the role of Th1 and Th2 lymphocytes in cell mediated immune responses to (cellular) xenografts. Activation of the Th1 pathway produces interleukin-(IL)2, interferon (IFN) gamma and tumour necrosis factor-(TNF) beta leading to macrophage activation, stimulation of complement fixing antibody isotypes (IgG2a and IgG2b in mice) and differentiation of CD8q cells to a cytotoxic phenotype. Activation of this pathway is associated with both allogeneic and xenogeneic transplant rejection. The Th2 lymphocyte response produces IL-4, IL-5, IL-6 and IL-10, cytokines that do not activate macrophages and lead to production of non-complement fixing antibody isotypes (IgG1 in mice). The Th2 pathway is associated with transplant acceptance (Badylak, 2004).

3.1.2. *Acellular matrix scaffolds for cartilage regeneration*

Any processing step intending to remove cells will alter the native three-dimensional architecture of the ECM. The most commonly utilized method for decellularization of tissues involves a combination of physical and chemical treatments. The physical treatments can include agitation or sonication, mechanical massage or pressure, or freezing and thawing. These methods disrupt the cell membrane, release cell contents, and facilitate subsequent rinsing and removal of the cell contents from the ECM. These physical treatments are generally insufficient to achieve complete decellularization and must be combined with a chemical treatment. Enzymatic treatments, such as trypsin, and chemical treatments, such as ionic solutions and detergents, disrupt cell membranes and the bonds responsible for intercellular and extracellular connections. Tissues are composed of both cellular materials and ECM arranged in variable degrees of compactness depending on the source of the tissue. The ECM must be adequately disrupted during the decellularization process to allow for adequate exposure of all cells to the chaotropic agents and to provide a path for cellular material to be removed from the tissue. The intent of most decellularization processes is to minimize the disruption and thus retain native mechanical properties and biological properties. The mechanisms of physical, enzymatic, and chemical decellularization for a variety of tissues are reviewed in the following sections and in Table 2 (Gilbert *et al.*, 2006).

It is unlikely that any combination of methods will remove 100% of all cell components from a tissue or organ. However, it seems apparent that methods which remove most or all of the visible cellular materials result in biologic scaffold materials that are safe for implantation. A number of naturally occurring ECM devices and related decellularization protocols have received regulatory approval for use in human patients, including human dermis, porcine SIS, porcine urinary bladder, and porcine heart valves. The growing list of biologic scaffolds used for tissue engineering/regenerative medicine applications makes the continued development of decellularization protocols a clinically relevant and important effort. A number of studies based on acellular matrix scaffolds prepared with some of above methods

Table 2.   Commonly used decellularization methods and chaotropic agents.

	Method	Mode of action	Effect on ECM
Physical	Snap freezing	Intracellular ice crystals disrupt cell membrane	ECM can be disrupted or fractured during rapid freezing
	Mechanical force	Pressure can burst cells and tissue removal eliminates cells	Mechanical force can cause damage to ECM
	Mechanical agitation	Can cause cell lysis, but more commonly used to facilitate chemical exposure and cellular material removed	Aggressive agitation or sonication can disrupt ECM as the cellular materials is removed
Chemical	Alkaline; acid	Solubilizes cytoplasmic components of cells; disrupts nucleic acids	Removes GAGs
	Triton X-100	Disrupts lipid–lipid and lipid–protein interactions, while leaving protein–protein interactions intact	Mixed results; efficiency dependent on tissue, removes GAGs
	CHAPS	Exhibit properties of non-ionic and ionic detergents	Efficient cell removal with ECM disruption similar to that of Triton X-100
	Sulfobetaine-10 and -16 (SB-10, SB-16)		Yielded cell removal and mild ECM disruption with Triton X-200
	Tri(n-butyl) phosphate	Organic solvent that disrupts protein–protein interactions	Variable cell removal; loss of collagen content, although effect on mechanical properties was minimal
	Hypotonic and hypertonic solutions	Cell lysis by osmotic shock	Efficient for cell lysis, but does not effectively remove the cellular remnants
	EDTA, EGTA	Chelating agents that bind divalent metallic ions, thereby disrupting cell adhesion to ECM	No isolated exposure, typically used with enzymatic methods (e.g., trypsin)

(*Continued*)

Table 2.   (*Continued*)

	Method	Mode of action	Effect on ECM
Enzymatic	Trypsin	Cleaves peptide bonds on the C-side of Arg and Lys	Prolonged exposure can disrupt ECM structure, removes laminin, fibronectin, elastin, and GAGs
	Endonucleases	Catalyze the hydrolysis of the interior bonds of ribonucleotide and deoxyribonucleotide chains	Difficult to remove from the tissue and could invoke an immune response
	Exonucleases	Catalyze the hydrolysis of the terminal bonds of ribonucleotide and deoxyribonucleotide chains	

have been made for cartilage repair. Qiang Yang *et al.* (2008) reports that BMSC-scaffold constructs successfully formed cartilage-like tissue in nude mice. Yang Q *et al.* (2011) repairs a osteochondral defect in a canine model, with acellular matrix scaffolds and bone marrow-derived mesenchymal stem cells (BMSCs).

Through the above, we have learnt the major properties of acellular matrix as a scaffold for cartilage engineering as well as development of the decellularization procedure. Although the great potentials of decellularised cartilage as an ideal scaffold for cartilage engineering have recently been realized and much research has been conducted, more thorough investigations are still needed to further improve the properties (such as mechanical strength and porosity) of decellularised cartilage so that its vast applications in cartilage engineering can be achieved and clinical translation of engineered catilage based on this type of scaffold can be realized.

3.2. *Collagen*

The family of collagens is mainly composed of fibrillar collagens, collagens with globular domains and other collagen structures (Kolácná *et al.,* 2007). Its rigid right-handed triple-helical rod structure, also known as tropocollagen, is comprised of three left-handed alpha helices of approximately 1000 amino acids in length. The collagen alpha helices contain extensive repeats of the tripeptide sequence Gly-X-Y, in which proline, lysine, and the post-translationally modified amino acids hydroxyproline and hydroxylysine are most commonly found in the X and Y residue positions. Glycine is necessarily present at every third position in the molecule to permit close packing of the alpha helices. Extensive hydrogen bonding between hydroxyprolines and glycines on adjacent polypeptide backbones confers stability to the structure. A total of 44 genes for 26 collagen types were described up to now; they are divided according to their supramolecular structure into nine groups; most collagens are associated either with different proteins of ECM or with a another collagen type. Three α-chains form the olypeptide structure of each collagen type: they are either identical (homotrimer) or only two chains are the same (two times α1) and the last chain is different (α2) = heterotrimer. Alpha-chains are characteristic for each collagen type: the chains are numbered according to the respective collagen type. Of the 19 types of collagen found in the human body, Types I, II and III are the most abundant and most fibrous. Type I collagen is the main fibrillar collagen of bone, tendon, and skin and provides tissues with tensile strength. Insoluble type I collagen has found ample usage in the biomedical field. Type II collagen is the principal collagenous component of cartilage, intervertebral discs and the vitreous body. Its mechanical function is to provide tensile strength and resists shearing forces. Self-aggregation and covalent cross-linking of one or more of these three collagen types facilitates higher-order packing into hexagonal parallel arrays or microfibrils which can be further assembled into sheets, bundles, and other configurations to form skin, bone and tendon (mainly Type I), cartilage (mainly Type II), and blood vessels (mainly Type III) (Geutjes *et al.,* 2006; Orgel *et al.,* 2011).

3.2.1. *Properties of collagen for cartilage engineering*

Collagen, as the basic structural element for most connective tissues, plays a prominent role in maintaining the biologic and structural integrity of ECM architecture and presents different morphologies in different tissues which perform different functions (Cen *et al.*, 2008). Collagen is the most widely utilized natural polymer for diverse biomedical applications on skin, nerve, cartilage and bladder. Pouya Mafi *et al.* listed the relative positive and negative aspects of biological protein-based collagen scaffolds in Table 3. And, several properties of

Table 3. List of studies commenting on the relative positive and negative aspects of biological protein-based collagen scaffolds.

Study	Aim	Scaffold type	Scaffold features
Jäger M, Krauspe R. 2007	To investigate the antigen pattern of cord blood stem cells cultivated onto a porous porcine collagen I/III scaffold.	Porous porcine collagen I/III	Porous, promotes differentiation of osteoblasts Acts as an acceptable blood progenitor cell carrier.
Itoh *et al.* 2001	To develop a 3D honeycomb carrier for cell culture.	Honeycomb Atelocollagen	Three-dimensional, elastic, hard, porous, supports differentiation and proliferation of cells
Masuoka *et al.* 2005	Culturing of chondrocytes to repair articular cartilage defects.	ACHMS scaffold: atelocollagen honeycomb-shaped sealed with a membrane	Elastic, mechanical strength, 3D, immunocompatible But lower rate of chondrocyte proliferation.
Sato *et al.* 2003	Using ACHMS-scaffold) for the culture of annulus fibrosus (AF) cells in tissue engineering procedures of intervertebral disc repair.	ACHMS: atelocollagen honeycomb-shaped sealed with a membrane	3D, porous, elastic, hard But slower cell proliferation.

(Continued)

Table 3.   (*Continued*)

Study	Aim	Scaffold type	Scaffold features
Cherubino P. *et al.* 2003	To present a new tissue-engineering technique for treatment of deep cartilage defects.	Collagen I/III	3D, porous, stimulates proliferation and differentiation of cells.
Ben-Yishay A *et al.* 1995	To evaluate the potential use of a prototype collagen-chondrocyte allograft in the repair of full thickness articular cartilage defects.	Type I collagen sponge	3D, porous, nonantigenic, as their telopeptides had been cleaved.
Middelkoop E *et al.* 1995	To evaluated dermal substitutes and obtain data on cell proliferation, the rate of degradation of the dermal equivalent, contractibility and de novo synthesis of collagen.	(1) Non-crosslinked collagen (2) Chemically crosslinked (3) Native collagen fibres	Rapid degradation of non-crosslinked by fibroblasts Chemically crosslinked showed cytotoxicity Native collagen fibres proved to be stable.
Nehrer S *et al.* 1997	Investigated the behavior of canine chondrocytes in two sponge like matrices in vitro.	Type I and Type II collagen matrices	Type I scaffold: uniform pore structure and wall thickness., maintained its porous structure, degraded more slowly than type II. Type II scaffold: thicker pore walls with pre-existing empty lacunae in a few areas.

(*Continued*)

Table 3. (*Continued*)

Study	Aim	Scaffold type	Scaffold features
Yates KE, Allemann F, Glowacki J. 2005	Assessed porous three dimensional collagen sponges for in vitro engineering of cartilage in both standard and serum-free culture conditions.	Ovine pepsin-digested acid soluble type I collagen	Porous 3D collagen. maintain chondrocyte viability, shape, and synthetic activity by providing an environment favorable for high-density chondrogenesis.
Rodkey WG, Steadman JR, Li ST 1999	A study of collagen meniscus implants to restore the injured meniscus.	Resorbable collagen scaffold (collagen meniscus implant)	Implantable, biocompatible, resorbable, supports new tissue regeneration and protects the chondral surfaces.
Stone KR *et al.*1992	Creating a regeneration template for the meniscal cartilage of the knee in order to induce complete meniscal regeneration, and to develop the technique for implanting the prosthetic appliance in vivo.	Copolymeric collagen-based scaffold	Biocompatible, may induce meniscus regeneration in mature dog and does not inhibit regeneration of meniscus in immature pig.

collagen as a scaffold for the cartilage repair will be introduced in the following part (Bastiaansen-Jenniskens *et al.*, 2008).

As a natural material, collagen has minimal potential for antigenicity after removal of telopeptides. Porous collagen sponges made from pepsin-digested (i.e., telopeptide-free) bovine skin collagen, were found to be biocompatible when implanted into subcutaneous pockets in normal rats, in contrast to sponges made in an identical manner from acid-soluble bovine skin collagen. Because sponges made from the telopeptide-containing collagens provoked an inflammatory reaction, albeit mild, they were not used in subsequent *in vitro* and *in vivo* studies to avoid concerns about compatibility with cells and regenerative processes (Emin *et al.*, 2008; Responte *et al.*, 2007).

Collagen can mediate the adhesion of cells onto scaffolds through the collagen-binding class of integrins (a1b1, a2b1, a10b1 and a11b1; see Hynes 2002). Each one of these four integrins binds multiple types of collagen (e.g. types I, II, III and IV) or other ECM molecules, at multiple ligands, with various affinities. Among the collagen-binding integrins, a1b1 and a2b1 are the most widely expressed and well studied. Indeed, integrins a1b1 and a2b1 have been shown to dominate binding of fibroblasts to collagen surfaces during the contraction process and binding of pancreatic cancer cells in CGSs. Studies of integrin-mediated cell adhesion in CGSs are under way in many laboratories (Yannas *et al.*, 2010).

Another distinctive characteristic of certain collagen-based scaffolds is their biological activity. Collagen-based scaffolds can promote the process of repair dependent on cell-binding activity. According to the contraction-blocking mechanism, extensive binding denies these cells the opportunity to induce contraction of the wound bed thus promoting the process of repair. Collagen-based scaffolds can achieve it by mediating binding of contractile cells on a scaffold surface. Structural features that control cell-scaffold binding include the chemical identity of the ligands, the pore size and the duration of the scaffold in undegraded form (Zippel *et al.*, 2010).

Collagen viscoelasticity and interstitial fluid pressurization play a crucial part in the mechanical property of articular cartilage. The collagen fibrillar matrix was assumed to be viscoelastic with a strain-dependent tensile modulus, and for axial tension, collagen viscoelasticity was found to account for most of the stress relaxation, while the effects of fluid pressurization on the tensile stress were negligible. The corresponding study by L.P. Li and his colleagues illustrates the essential role of both of them in the mechanical response of articular cartilage (Kiviranta *et al.*, 2006; Li *et al.*, 2005; Hardmeier *et al.*, 2010).

3.2.2. *Collagen scaffolds for cartilage regeneration*

Many efforts have been made for cartilage repair based on collagen scaffolds. A study by Ben-Yishay A and his colleagues was designed to evaluate the potential use of a prototype collagen-chondrocyte allograft in the repair of full-thickness articular cartilage defects in 1995. Masuoka K *et al.* indicated the effectiveness of implantation of allogenic chondrocytes cultured in atelocollagen honeycomb-shaped scaffold (ACHMS scaffold) in repairing articular cartilage defects in 2005. Kuroda R *et al.* reported that they repaired a full-thickness articular cartilage defect in the medial femoral condyle of an athlete with autologous bone-marrow stromal cells and a collagen scaffold. Besides articular cartilage repair, Sato M *et al.* studied the feasibility of tissue-engineered trachea based on an atelocollagen honeycomb-shaped scaffold, and Lin CH *et al.* demonstrated that it is a promising approach to construct tissue-engineered trachea with a composite scaffold comprising a poly(«-caprolactone) (PCL) stent and a type II collagen sponge.

There are two common forms of collagen-based scaffolds, Collagen sponges and hydrogels, in a particularly important position in their application to cartilage repair. Both of them will be respectively introduced:

(i) Collagen sponges: Collagen sponges made from 0.5 wt% solutions, lyophilized to maintain a porous surface, and cross-linked by ultraviolet irradiation have interconnected pores suitable for cellular ingrowth and histogenesis. They are highly porous (95%) with pore diameters in the range of 120–200 lm. The very porosity of collagen sponges can overcome one of the known limiting factors of traditional high-density 3D culture that cells in the centre of the mass do not survive because of poor exchange of nutrients and dissolved gases. Julie Glowacki *et al.* (2008) has evaluated the *in vivo* fate of engineered cartilage with porous collagen sponges. The positive results showed that collagen sponges can provide a framework for supporting chondrogenesis. Furthermore, the evidences showed that bioreactors that apply medium perfusion or hydrostatic fluid pressure enhance histogenesis by cells in porous collagen sponges (Glowacki and Mizuno, 2008).

(ii) Collagen hydrogels: One approach to combining the benefits inherent in both natural and synthetic polymers is to combine them in the form of hydrogels. Hydrogels are a class of materials that have gained widespread use as tissue engineering scaffolds due to their high water content and biocompatibility. Hydrogels can be designed from natural and/or synthetic polymers and have a wide range of physical and chemical properties. In addition, hydrogels can be molded for *ex vivo* tissue engineering and implantation, or alternatively injected to form a scaffold in situ. Methods of triggering gelation of precursor solutions include photo-polymerization, self-assembly, thermosensitivity and pH sensitivity. Their hydrophilicity makes them a natural choice for tissue engineering applications, given the high water content of most body tissues. Additionally, hydrogels can maintain the diffusion of nutrients and oxygen to growing cells through varying levels of porosity.

Sargeant TD *et al.* (2012) developed and evaluated in situ forming hydrogels composed of natural and synthetic polymers. Their study suggested that collagen and PEG hydrogels exhibit mechanical, physical and biological properties suitable for use as an injectable tissue scaffold for the treatment of a variety of simple and complex tissue defects (Sargeant *et al.*, 2012). Much research for cartilage repair based on collagen hydrogels has been done in recent years (Chen *et al.*, 2013; Rampichová *et al.*, 2013; Guenther *et al.*, 2013).

3.3. *Chitosan*

The history of chitosan dates back to the 19th century, when Rouget discussed the deacetylated form of chitosan in 1859 (Valérie and Vinod, 1998). Studies on chitosan have been intensified as biomaterials for tissue engineering applications during the past 25 years. Chitin, the source material for chitosan, is one of the most abundant organic materials, being second only to cellulose in the amount produced annually by biosynthesis. It is an important constituent of the exoskeleton in animals, especially in crustacean, molluscs and insects. It is also the principal fibrillar polymer in the cell wall of certain fungi (Eugene and Lee, 2003). Chitosan is a linear polysaccharide, composed of glucosamine and N-acetyl glucosamine units linked by β (1-4) glycosidic bonds. The content of glucosamine is known as the degree of deacetylation (DD). Depending on the source and preparation procedure, its molecular weight may range from 300 to over 1000 kD with a DD from 30% to 95%. In its crystalline form, chitosan is normally insoluble in aqueous solution above pH 7; however, in dilute acids (pHb 6.0), the protonated free amino groups on glucosamine facilitate solubility of the molecule. Generally, chitosan has three types of reactive functional groups, an amino group as well as both primary and secondary hydroxyl groups at the C(2), C(3), and C(6) positions, respectively. These groups allow modification of chitosan like graft copolymerization for specific applications, which can produce various useful scaffolds for tissue engineering applications. The chemical nature of chitosan in turn provides many possibilities for covalent and ionic modifications which allow extensive adjustment of mechanical and biological properties (Kim *et al.*, 2008; Suh and Matthew, 2000).

3.3.1. *Properties of chitosan for cartilage regneration*

Chitosan is a cationic polymer and has been widely used in food, cosmetic, biomedical, and pharmaceutical applications. In particular, chitin and chitosan have been extensively used in many biomedical applications which include cartilage engineering due to their properties. In the following part, several significant properties of Chitosan-based scaffolds for cartilage regneration will be introduced.

Firstly, chitosan-based scaffolds are biocompatible and viable for the tissue engineering, including cartilage engineering. A number of researchers have examined the host tissue response to various chitosan-based implants and proved the biocompatibility of chitosan-based scaffolds. Furthermore, chitosan is involved in many biological processes, such as cell proliferation and migration, integration of the implanted material with the host tissue, which is crucial for tissue engineering including cartilage engineering. Suh and Mattew (2000) reported that chitosan and its fragments on immune cells may stimulate the induction local cell proliferation and ultimately integration of the implanted material with the host tissue. Okamoto *et al.* (1995) reported that chitosan influenced all stages of wound repair in experimental animal models. In the inflammatory phase, chitosan has unique hemostatic properties that are independent of the normal clotting cascades. *In vivo*, these polymers can also stimulate the proliferation of fibroblasts and modulate the migration behaviour of neutrophils and macrophages modifying subsequent repair processes such as fibroplasias and re-epithelialization (Kim *et al.*, 2008).

Secondly, the porous structure and the cationic nature of chitosan also are important properties for cartilage engineering. Chitosan can be formed as interconnected-porous structures by freezing and lyophilizing of chitosan solution or by processes such as an "internal bubbling process (IBP)" where $CaCO_3$ is added to chitosan solutions to generate chitosan–$CaCO_3$ gels in specific shapes by using suitable molds. The porous structure of chitosan is crucial for the architecture of engineered cartilage corresponding to the complex macro- and micro- constructure of native cartilage, as well as for controlling cellular colonization rates and organization within an engineered tissue. With regard to the cationic nature of chitosan, we focus on the property that allows for pH-dependent electrostatic interactions with anionic glycosaminoglycans (GAG) and proteoglycans which are distributed widely throughout the body and other negatively charged species. This property is one of the important elements for tissue engineering applications because numbers of cytokines/growth factors are known to be bound and modulated by GAG including heparin and heparan sulfate. A scaffold incorporating a chitosan–GAG complex may serve means of retaining and concentrating desirable factors secreted by colonizing cells. Moreover, Nishikawa *et al.* (2000) reported that chitosan, structurally resembling GAG which consists of long-chain, unbranched, and repeating disaccharide units, is regarded to play a key role in modulating cell morphology, differentiation, and function (Chen *et al.*, 1996).

Thirdly, the degradation and the mechanical properties of chitosan-based scaffolds need to be introduced. The degradability of a scaffold plays a crucial role on the long-term performance of tissue-engineered cell/material construct because it affects many cellular process, including cell growth, tissue regeneration, and host response. The degradation rate of chitosan-based scaffolds needs to be regulated for the cartilage engineering though the combination with other polymers. With regard to the mechanical properties of chitosan-based scaffolds, they are dependent on the pore sizes and pore orientations. Chen and Hwa (1996) reported the effect of the molecular weight of used chitosans and their crystallinity on the mechanical property of chitosan membrane. The results implied that the lower tensile strength of the membrane was the result of less crystallinity in the chitosan membrane prepared from low molecular weight of chitosan.

Finally, the last mentioned property of chitosan is that it confers considerable antibacterial activity against a broad spectrum of bacteria. Aimin *et al.* (1999) has shown that chitosan can reduce the infection rate of experimentally induced osteomyelitis by Staphylococcus aureus in rabbits. The cationic nature of chitosan by amino group is related to anions on the bacterial cell wall. The interaction between positively charged chitosan and negatively charged microbial cell wall leads to the leakage of intracellular constituents. Above all, due to this antibacterial property, chitosan has been blended with other polymers.

3.3.2. *Chitosan-based scaffolds for cartilage regeneration*

In this part, we emphasize on three aspects: the modification of chitosan, the combination of chitosan with other materials, chitosan nanofibers. These three aspects of chitosan largely determine the efficient application of chitosan in the cartilage engineering.

3.3.2.1. Modification of chitosan

The early use of chitosan was mainly restricted to unmodified forms in tissue engineering field. Recently, there has been a growing interest in modification of chitosan to improve its solubility, introduce desired properties, and widen the field of its potential applications by choosing various types of side chains (Tseng *et al.*, 2013). Three different approaches have been investigated to introduce beneficial bioactivity to chitosan by using the reactive hydroxyl and amine groups of chitosan. The first approach is based on ionic complexation between positively charged chitosan and negatively charged bioactive agent of interest. For example, negatively charged heparin molecule can be introduced to chitosan systems by ionically interacting with positively charged chitosan molecules. The second approach is based on physical adsorption or entrapment of biomolecules on chemically modified chitosan surface or within chitosan substrates. This strategy sometimes involves chemical modification of chitosan to alter its physicochemical properties through introduction of certain structural moieties to chitosan polymer chains. For instance, the modified chitosan in combination with heparin hydrogels were used to immobilize fibroblast growth factor-2 (FGF-2) to attain neovascularization and fibrous tissue formation in a mouse model. Chitosan can also be further modified with multiple functionalities such as lactose and aside via sequential condensation reaction at the chitosan amine sites. This resulting polymer can be photo-crosslinked and avoids the use of other toxic cross-linking agents during scaffold fabrication and biomolecules incorporation, which shows great potential in incorporating growth factors under mild conditions for a variety of biomedical applications such as for wound healing and controlled angiogenesis. The third approach is to prepare a reactive chitosan intermediate, for example thiolated chitosan, carboxylated chitosan, or maleimide activated chitosan through the amine groups. These active chitosan intermediates can further react with sulphydryl or amino groups of some peptide sequences or whole protein molecules to form covalent immobilization. Other chemical modification methods of chitosan through its hydroxyl groups for biological purposes have also been reported. For example, carboxylation can also be achieved through the hydroxyl groups on chitosan followed by imide bonding formation with peptides. In addition, photochemical reactions can occur between the hydroxyl groups on chitosan and azido derivative of peptides to form covalent bonding. All the aforementioned modification methods take advantage of the amine and hydroxyl functionalities of chitosan, to develop biologically active chitosans with cell recognition and specificities for various tissue engineering and regenerative medicine applications.

3.3.2.2. Combination of chitosan with other materials

The combination of chitosan with other materials appears to be a common theme in various reports. Blending with other polymers is widely investigated. Blends with synthetic and natural polymers can imbibe the wide range of physicochemical properties and processing techniques of synthetic polymers as well as the biocompatibility and biological interactions of natural polymers. Huang *et al.* (2005) blended chitosan with gelatin to improve the biological activity since (i) gelatin contains Arg-Gly-Asp (RGD)-like sequence that promotes cell adhesion and migration, and (ii) forms a polyelectrolyte complex. Addition of gelatin affected the stiffness of 2D and 3D scaffolds, facilitated the degradation rate and maintained the dimension in the presence of lysozyme. Sarasam and Madihally (2005) reported the effect of blending chitosan with poly(ε-caprolactone) (PCL). As previously mentioned, these blending membranes improved mechanical properties as well as cellular support. The γ-poly (glutamic acid) (γ-PGA), a hydrophilic and biodegradable polymer, was also used to modify chitosan matrices and the γ-PGA/chitosan composite matrix was found to enhance hydrophilicity and serum proteins adsorption, and to increase the maximum strength through addition of γ-PGA in tissue engineering applications. Chung *et al.* (2002a, b) prepared galactosylated chitosan-based scaffolds by combining with alginate to improve mechanical properties and biocompatibility. The scaffolds exhibited the usual pore configurations, and the pore sizes were dependent on the freezing pre-treatments, the molecular weight of chitosan and amount of galactosylated chitosan.

3.3.2.3. Chitosan nanofibers

Recently, much attention has been focused on making polymeric nanofibers by electrospinning process as a unique technique because it can produce chitosan nanofibers with diameter in the range from several micrometers down to tens of nanometers, depending on polymer and processing conditions. Electrospinning applies high voltages to a capillary droplet of polymer solution or a melt to overcome liquid surface tension and thus enables the formation of much finer fibers than conventional fiber spinning methods. These nanofibers that mimic the structure and function of natural ECM are of great interest in tissue engineering as scaffolding materials to restore, maintain or improve the function of human tissue, because they have several useful properties such as high specific surface area and high porosity (Li and Hsieh, 2006). The recent attempts have been made to prepare chitosan-based nanofibrous structures by electrospinning, with varying degrees of success. Min *et al.* (2004) produced chitin and chitosan nanofibers with an average diameter of 110 nm and their diameters ranged from 40 to 640 nm by the SEM image analysis. Bhattarai *et al.* (2005) further concluded that these chitosan-based nanofibers promoted the adhesion of chondrocyte and osteoblast cells and maintained characteristic cell morphology.

Chitosan is an attractive candidate biomaterial that shows great potential in the area of tissue engineering and regenerative medicine due to its biocompatibility, controlled biodegradability and functionality. With the advances in the areas of controlled delivery and bioactive biomaterials, biologically active chitosan systems have been recently developed and investigated. These systems, either by incorporating various growth factors or by physically or chemically attaching biomolecules or peptide sequences, exhibit specific bioactivity attributed to the loaded or attached factors. Although much progress has been achieved to apply bioactive chitosan systems in tissue engineering, a few limitations of the current systems need to be overcome in order to develop more clinically meaningful chitosan substrates for various tissue regeneration. Most of the current chitosan growth factor delivery systems use either spongy scaffolds with adsorbed growth factors, or individual growth factor encapsulated micro- or nanospheres lacking overall structural integrity. The former systems show poor control over the growth factor release and the mechanical properties are generally low. The latter systems, although have a better spatial distribution of growth factors and better release kinetics, can only be used as a filling material without providing any structural support and function. Therefore, it is promising to develop bioactive chitosan systems with controlled delivery capacity that also shows tissue compatible mechanical properties. On the other hand, modification of chitosan by introducing various recognition molecules or domains represents an intriguing research direction. With the understanding of the immobilization processes and cellular responses to the modified two dimensional films, it is essential to develop three dimensional structures exhibiting recognition and specificity for tissue engineering applications. In addition, considering that chitosan is biodegradable *in vivo* and the biomolecule modified surface will lose the bioactivity after being degraded, chitosan bulk modification with particular biomolecules will be of interest and importance (Jiang *et al.*, 2008).

3.4. *Hyaluronic acid*

Hyaluronan (sodiumhyaluronate, hyaluronic acid, HA), a common component of synovial fluid (SF) and ECM, is a linear high molar mass, natural polysaccharide composed of alternating (1 fi 4)-b linked D-glucuronic and (1 fi 3)-b linked N-acetyl-D-glucosamine residues. HA belongs to a group of substances known as glycosaminoglycans (GAGs), being structurally the most simple, the only one not covalently associated with a core protein, not synthesized in Golgi apparatus, and the only non-sulfated one (Bastow *et al.*, 2008). The molar mass of HA can reach as high as 107 Da. Such high molar mass and its associated unique viscoelastic and rheological properties predispose HA to play important physiological roles in living organisms and make it an attractive biomaterial for various medical applications (Kogan *et al.*, 2007). Although HA is almost omnipresent (albeitin relatively small amounts) in the human body and in other vertebrates, the highest amounts of HA are found in the ECM of soft connective tissues. Besides vertebrates, HA is also present in the capsules of some bacteria, but is absent in fungi, plants, and insects. The largest content by far of HA is found in rooster combs.

In cartilage, despite its relatively low content, HA functions as an important structural element of the matrix, forming an aggregation center for aggrecan, a large chondroitin sulfate proteoglycan that retains its macromolecular assembly in the matrix due to specific HA–protein interactions. These aggregates have enormous molar mass of up to 100 MDa and are embedded within a collagenous framework.

3.4.1. *Properties of hyaluronic acid for cartilage regeneration*

Hyaluronan, has historically been considered a readily available and desirable scaffold material for tissue engineering applications. As a structural molecule, HA provides tissues with many of their physio-mechanical properties, and in tissues, HA plays a central role in many normal biologic processes through its interaction with cell receptors and other ECM molecules. From both of these aspects, the part that HA plays in the cartilage engineering will be introduced.

HA have biophysical properties that distinguish it from other components of the ECM. In solution as a long-chain biopolymer, it behaves as a stiffened random coil with a large hydrated volume, such that each molecule interacts with its neighbours to create viscoelastic solutions. The distinctive structure of HA and its plasticity have special significance for engineered cartilage with complex macro- and micro- structure.

The other distinctive physical property of HA is its visco-elasticity in the hydrated state. HA is the major hydrodynamic non-protein component of joint synovial fluid (SF) and many tissues and organs, and its unique visco-elastic properties confer remarkable shock absorbing and lubricating abilities to fluids and tissues. With regard to fluids, its macromolecular size and hydrophilicity serve to retain fluid in the joint cavity during articulation. HA restricts the entry of large plasma proteins and cells into fluids, but facilitates solute exchange between the synovial capillaryies and cartilage and other joint tissues. For tissues, HA can form a pericellular coat around cells, interact with proinflammatory mediators, and bind to cell receptors where it modulates cell proliferation, migration, and gene expression. Furthermore, the anomalous viscosity of HA solutions may provide an ideal biological lubricant, at least by reducing the workload during rapid movements. In general, the visco-elasticity of HA can promote the regeneration of cartilage especially articular cartilage through the above mechanisms (Fraser *et al.*, 1997).

In the other aspect, HA plays a central role in many normal biologic processes through its interaction with cell receptors and other ECM molecules. Firstly, HA-Chondrocyte CD44 interactions assist the application in the cartilage engineering through several mechanisms. Chondrocytes express the standard isoforms of CD44, CD44s. CD44-HA interactions link chondrocytes with their matrix and these interactions are essential for maintaining normal cartilage homeostasis. Treatment of cartilage explants with antisense oligonucleotides that inhibit CD44 expression causes substantial loss of aggrecan, and treatment with HA oligosaccharides that compete with endogenous HA for binding to CD44 have both anabolic and catabolic effects. These studies suggest that chondrocyte CD44-HA interactions modulate cartilage metabolism, and that disrupting these interactions promotes matrix remodelling. Other studies have shown that CD44-HA interactions also modify chondrocyte survival and apoptosis (Ishida *et al.*, 1997).

Secondly, HA can mediate epithelial-mesenchymal interactions and then promote cartilage formation. Normal developmental processes, organogenesis and growth depend on interactive signaling between cells and tissues. Epithelial-mesenchymal interactions are critical mechanisms employed during embryogenesis that promote cartilage formation from the primary mesenchyme in parallel with the formation of the neural crest from the ectoderm. A considerable number of experiments revealed that HA plays a crucial role in the transition of epithelia to mesenchyme during embryogenesis, and that the EMT (Epithelial Mesenchymal Transition) program is activated and regulated mainly by HA (Astachov *et al.*, 2011).

Thirdly, the carboxyl groups of HA are fully ionized at extracellular pH and osmotic activity is very high in relation to HA molar mass. As a consequence, there are effects on the distribution and movement of water influencing water homeostasis. Secondary hydrogen bonds form along the axis of the polysaccharide, creating a twist in the chains, imparting some stiffness, and generating hydrophobic patches that permit association with other HA chains. Nonspecific interaction with cellular membranes and other lipid structures may then occur (Scott, 1992). The stiffness of the HA polymers promotes an extended random-coil configuration, ensuring the occupancy of enormous molecular domains. Alone or in conjunction with collagen fibers and other macromolecular elements of the extracellular matrix, this ultimately reduces the mobility of HA itself and determines its permeability to other substances, whether transported by diffusion or driven by hydrodynamic bulk flow (Volpi *et al.*, 2009).

3.4.2. *Hyaluronic acid-based scaffolds for cartilage regeneration*

A comprehensive overview of the different sources from which HA can be isolated for biological and biotechnological purposes and the contribution of potential impurities was published recently (Shiedlin *et al.*, 2004).

(i) Biosynthesis of HA in Vertebrates and Bacteria: HA is synthesized at the inner plasma membrane from "activated" UDP-D-glucuronic acid and UDP-N-acetyl-Dglucosamine. Both are connected with each other at the reducing end of the growing chain. This process is catalyzed by a membrane-located enzyme, the HA-synthases (Weigel *et al.*, 1997). The involved enzymes are very similar, i.e. the microbial HA-synthase shows a marked homology with the HA synthase of vertebrates. This is often explained by a lateral gene-exchange from the animal host to the bacterium.

(ii) Isolation of HA from Biological Sources: HA can be isolated from tissues when bovine eyeballs, rooster comb, umbilical cord and connective tissues such as skin or cartilage are particularly rich in HA. However, HA prepared through this method often contains residual proteins that may easily lead to immune reactions when used for medical purposes. Furthermore, obtaining very pure HA from these animal tissue sources is rather difficult. As a consequence, the use of animal sources for sampling HA is in continuous decline.

(iii) HA Manufacturing by Biotechnology: There are some pathogenic bacteria with a so-called HA capsule and the HA secreted by microorganisms such as certain attenuated strains of Streptococcus zooepidemicus, S.equi, etc. is currently offered by many companies, with production of up to several tons per year. In this method, the HA capsule initially needs to be prepared and be bound to the bacteria. And then, the capsule will float in solution after being removed from the cells. Using the fermentation process, molar masses of HA of about 0.5 – 2.5 MDa can be obtained and the yield of HA correlates closely with the time of fermentation (Widner *et al.,* 2005).

(iv) Purification of Microbial HA: Subsequent to cultivation, the highly viscous culture media are diluted with water and the bacteria removed by filtration or centrifugation. The remaining HA-containing solution needs to be subjected to several purification and concentrating steps, After which, the HA is normally precipitated from the aqueous solution with solvents (miscible with water) such as ethanol, acetone or isopropanol. During the complete process, great care must be taken to avoid degradation of the HA. In particular, the dissolved HA macromolecules must avoid getting in contact with metals, or be subjected to shear stress that might result in diminution of the molar mass.

Besides such researches for the sources of HA, many efforts also have been made for cartilage engineering based on HA scaffolds (Moss *et al.,* 2011; Chang *et al.,* 2006). In the process of cartilage repair with HA based scaffolds, two aspects, chemical modifications and HA hydrogel, play a particularly important role. Both of them will be introduced in the following part. The simplicity of HA makes it easy for chemical modifications, which can promote the interaction of HA with cell receptors and other ECM molecules and thus cartilage repair (Vasita *et al.,* 2008; Reitinger *et al.,* 2013).

The chemical modification of HA has focused on its two principal functional groups, hydroxyl groups and carboxyl groups. These groups have been utilized in a number of chemistries including: divinylsulfone, esterification, and carbodiimide mediated reactions. These techniques are used to cross-link HA to form hydrogels with dynamic physical properties, depending on the extent and type of modification (Darr and Calabro, 2009).

HA hydrogel is the most common form of HA based scaffolds. HA hydrogels support chondrocyte matrix deposition and chondrogenic differentiation of mesenchymal stem cells (MSCs). Moreover, in direct comparison to PEG hydrogels, HA hydrogels enable more robust MSC chondrogenesis and cartilaginous matrix formation both *in vitro* and *in vivo*. HA is also easily functionalized into formats that are both photopolymerizable and/or hydrolytically degradable, and can include MMP sensitive peptides and RGD sequences for cell-mediated degradation and cellular adhesion. With its natural pro-chondrogenic properties and facile tunability, HA hydrogels will be a particularly important scaffold for cartilage regeneration (Chang *et al.*, 2003; Kim *et al.*, 2011).

Above all, it is no doubt that HA is a particularly important native polymer for cartilage engineering. However, there are still many limitations of HA-based scaffolds. More studies such as biological mechanism of HA-based scaffolds for cartilage repair, the optimized modifications of HA, the combination of HA with other materials which possess complementary advantages should be done.

3.5. *Other native polymers for cartilage regeneration*

3.5.1. *Silk*

Silks are naturally occurring protein polymers produced by a wide variety of insects and spiders. Silk in its natural form is composed of a filament core protein, silk fibroin, and a glue-like coating consisting of a family of sericin proteins. The most widely studied silks are cocoon silk from the silkworm Bombyx mori and dragline silk from the spider Nephila clavipes. Structurally, silk fibroins from these species are characterized as natural block copolymers, composed of hydrophobic blocks with highly preserved repetitive sequence consisting of short side-chain amino acids such as glycine and alanine, and hydrophilic blocks with more complex sequences that consist of larger side-chain amino acids as well as charged amino acids (Chao *et al.*, 2010; Zhang *et al.*, 2009).

Although silk has been used clinically as sutures for centuries, only recently has it been exploited as a scaffold biomaterial for cell culture and tissue engineering *in vitro* and *in vivo* (Seth *et al.*, 2013; Mandal *et al.*, 2011). Blends of SF and other natural polymers such as chitosan have been evaluated for use in cartilage TE. *In vitro* culture tests revealed that the novel genipin-cross-linked CS/SF sponges promoted the adhesion, proliferation and matrix production of chondrocyte-like cells. Similarly, Bhardwaj *et al.* found that the blended scaffolds composed of SF/CS supported attachment, growth and chondrogenic phenotype of bovine chondrocytes for two weeks of *in vitro* culture and biomechanical studies revealed that the static and dynamic stiffness at high frequencies was higher in cell-seeded constructs than non-seeded controls (Wang *et al.*, 2006). Baek *et al.* and Jin *et al.* treated nanofibrous SF mesh with microwave-induced argon plasma treatment which significantly improved the hydrophilicity and the cytocompatibility with human articular chondrocytes. Similarly, Cheon *et al.* found that the argon plasma treatment increased the hydrophilicity of SF/keratose e-spun scaffold and induced deeper and more cylindrical pores than nontreated scaffolds. Also, the attachment and proliferation of neonatal human knee articular chondrocytes on treated SF/keratose scaffolds increased significantly, followed by increased glycosaminoglycan (GAG) synthesis. Hence, the microwave-induced argon plasma treatment could be explored as a means of improving chondrogenic cell growth and cartilage-specific ECM formation (Kasoju and Bora, 2012).

Despite the progress made, SF based clinical product is still a dream, probably due to significant knowledge gaps. We feel that efforts aiming at integrating the principles of developmental biology and tissue biology with materials science and engineering, and following bioinspired and biomimetic engineering strategies, could possibly speed up the development of a full-fledged SF based tissue engineered product. Nevertheless, we envision a tremendous growth in the investigations, both qualitatively and quantitatively, on SF based tissue engineering research and possibilities of clinical success in the near future.

3.5.2. *Fibrin*

Normal fibrinogen is synthesized by the liver and secreted into the blood and lymph systems. It consists of two sets of three chains (α, β and γ). In addition, a variant of fibrinogen with an extended α chain,

α E, whose function is not yet clear, appears as a small percentage of the total adult human fibrinogen. Fibrinogen circulates in the blood vessels at concentrations of 2 – 5 mg/ml and it forms fibrin clots upon its activation. Its major role is to block blood flow after an injury and reduce hemorrhage upon injury (Wu *et al.*, 2012).

The biodegradable, non-toxic and non-immunogenic properties of fibrin glue itself, with no stabilizing agents, suggest it as the basic material for producing bio-engineered matrices. *In vitro* responses of cultured cells to fibrin and its components have been evaluated. For example, a fibrin clot recruited cultured human fibroblasts and endothelial cells from the surrounding area. Fibrinogen only slightly increased fibroblast proliferation while thrombin enhanced their proliferation by a factor of 1.5 – 1.8. A cell attachment assay showed the haptotactic response of normal and transformed cells to fibrin (ogen) and thrombin (Gorodetsky, 2008).

The use of FG was advocated in tissue engineering for its 3-D characteristics, providing appropriate cells with contact with the culture environment through their surfaces. Besides, FG combines some important advantages such as high seeding efficiency, uniform cell distribution and adhesion capabilities. FG is non-cytotoxic, and is a naturally porous physiologic scaffold that can stimulate cell adhesion and growth. Furthermore, when produced from the patient's own blood, it could be used as an autologous scaffold without the potential risk of foreign body reaction or infection (Ahmed *et al.*, 2008).

The disadvantage of fibrin sealants as the sole component of cell-binding matrices for tissue regeneration is the fast rate of fibrinolysis once introduced into the tissues. A few approaches that tried to design fibrin matrices were adopted to overcome this problem. Such solutions included the formation of soft beads from alginate and fibrinogen mixtures as cell carriers, so that the alginate stabilized the fibrin matrix to form beads onto which cells could be loaded. Others designed composite scaffolds made of whole fibrinogen or separated chains, bound covalently to hydrogels. These can form a more durable and controlled pore size meshwork for culturing cells in 3D conditions for implantation.

3.5.3. *Alginate*

Alginate is derived from brown marine algae. This family of poly-anionic co-polymers comprise of 1,4-linked β-D-mannuronic and α-L-guluronic residues, which are soluble in aqueous solutions and can be cross-linked with bivalent cations such as Ba^{2+} and Ca^{2+} to form a stable gel (Awad *et al.*, 2004). Alginate has been intensively studied *in vivo* for use in cartilage engineering due to its excellent bio-inductive capacity on chondrocytes. Alginate implants have been shown to be capable of supporting chondrocyte viability and production of a cartilage-like ECM as early as four weeks after implantation. In contrast to growth on chitosan scaffolds, chondrocytes stop proliferating and tend to more readily produce type II collagen and assume a differentiated phenotype with rounded morphology, when grown on an alginate substrate (Chia *et al.*, 2005). Cell aggregation and differentiation on alginate matrices have been attributed to the high hydrophilicity that minimizes cell and protein adhesion and allows cell–cell interactions to predominate and rapidly establish functional tissue. The rapid switch from a proliferative to a differentiated phenotype could be problematic if a threshold population of cells has not yet been established, but this potential disadvantage can be largely offset by the efficient, rapid, and high-density seeding that is possible when using a dry alginate scaffold with >90% porosity. The efficient seeding properties of alginate are facilitated by pore tortuosity which entraps the cells and highly hydrophilic matrix that wicks up the cells when dry, resulting in homogenous distribution throughout the scaffold. The property of high-density cell seeding is desirable for rapid regeneration of ECM and secretion of tissue-regenerating growth factors. Unfortunately, alginate implants fall short of matching the mechanical properties of native cartilage tissue in mechanically demanding requirements. Therefore, its application will likely be useful to the creation of complex 3D shapes, for use in reconstructing facial structures (ears, nose bridge, etc.) where auricular cartilage is important for maintain proper dimensions, rather than mechanical integrity (Chia *et al.*, 2004).

3.5.4. *Agarose*

Agarose is a linear polysacharride consisting of basic repeating units of agarobiose. It is derived from Asian seaweeds and, like alginate, can encapsulate cells in its hydrogel form and support the chondrogenic phenotype. Agarose is solid at body temperature, but requires higher temperatures to acquire its liquid state. Concerns that such temperatures may affect cell viability have been overcome by the development of a low-melting-point agarose. Agarose is used at low concentrations of 1 – 3% (w/v) to optimize nutrient and waste exchange, as the stiffer gels formed at higher concentrations exhibit lower hydraulic permeability. A recent study comparing CC seeded in 2% and 3% agarose showed that, at 28 days of culture in 20% (v/v) foetal bovine serum, 2% agarose constructs were stiffer and had a higher collagen content (Ng *et al.*, 2005). This result demonstrates the importance of determining the ideal scaffold to support the production of constituent proportions of ECM and subsequently the mechanical properties of the developing neocartilage. Agarose has been useful for investigating the effects of mechanical loading on cartilage formation (Kelly *et al.*, 2004).

Rahfoth *et al.* (1998) in Germany performed allograft transplants of CC in agarose gel into osteochondral defects in the knees of rabbits. They reported no graft versus-host rejections or foreign-body immune responses and, after 18 months, 47% of grafts developed a morphologically stable hyaline cartilage.

Alginate and agarose gels have been investigated for their use as vehicles to support synthesis and organization of articular cartilage ECM components (Tripathi and Kumar, 2011). Alginate gels (as opposed to collagen gels) were shown to support the differentiation phenotype of chondrocytes. Although cell numbers in alginate gels exhibited an initial loss, a finding consistent with the report that rabbit articular chondrocytes de-differentiated in monolayer cultures will re-express a cartilage phenotype when maintained in anchorage-independent culture in agarose gels. Implantation of cell-laden agarose gels holds some promise as an approach to enhance deposition and organization of osteochondral components. In another study, agarose gels containing allograft chondrocytes were implanted into rabbit full-thickness defects, and exhibited formation of new subchondral bone.

The newly synthesized components appeared to integrate with host articular cartilage when compared with the controls.

4. Cartilage Regeneration and Repair Based on Native Polymers

In the past score years, cartilage regeneration and repairs have achieved a sufficient development in both *in vivo* and *in vitro*. Some techniques have been used for patients, although the current status is still imperfect. In this part, the related advances and challenges in cartilage regeneration and repair based on native polymers will be described in details.

4.1. *Cartilage regeneration based on native polymers*

Due to their excellent bio-characteristics such as similarities with the extracellular matrix, high chemical versatility, typically good biological performance and inherent cellular interaction, native polymers are considered as one of the most attractive options to be used in cartilage tissue engineering. The present section intends to introduce major types of native polymers such as collagen, gelatin, fibrin, and alginate which are being widely used for cartilage regeneration research. Examples of these native polymers with applications in *in vitro* studies or in preliminary researches in cartilage regeneration settings will also be summarized in tables.

4.1.1. *Collagen*

Collagen is always regarded as an ideal scaffold for cartilage regeneration because it is the major protein component of native cartilage matrix. In its native environment, collagen interacts with chondrocytes and transduces essential signals for the regulation of cell anchorage, migration, proliferation, differentiation, and survival. A total of twenty-seven types of collagens have been identified to date, but collagen type I is the most abundant and the most investigated for biomedical applications. Table 4 intends to summarize some relevant applications reported as research works.

Table 4. Collagen-based scaffolds for cartilage engineering.

Scaffold	Seed cell type	Regeneration model	References
Collagen sponge	Chondrocytes	Nude mice	Fujisato T *et al.*, 1996
Collagen gel	Bone marrow stromal cells	Mouse	Xu XL *et al.*, 2005
Collagen sponge	Chondrocytes	Sheep	Dorotka R *et al.*, 2005
Collagen sponge and hydrogel	Intervertebral disc cells	*In vitro*	Gruber HE *et al.*, 2006 Gruber HE *et al.*, 2004
Collagen membrane	Chondrocytes	Rabbit	De Franceschi L *et al.*, 2005

4.1.2. *Gelatin*

Gelatin is a natural polymer that is derived from collagen, and is commonly used for cartilage regeneration because of its appropriate biodegradability and excellent biocompatibility in physiological environments. Moreover, gelatin has relatively low antigenicity because of being denatured in contrast to collagen which is known to have antigenicity due to its animal origin. Due to its easy processability and gelation properties, gelatin has been manufactured in a range of shapes including sponges and injectable hydrogels, but definitively the most used carriers are gelatin microspheres which are normally incorporated in a second scaffold such as a hydrogel or a synthetic polymer with stronger mechanical properties such as PCL. Examples of gelatin with application in cartilage regeneration research have been summarized in Table 5.

4.1.3. *Fibrin*

Fibrin is a protein matrix produced from fibrinogen, which can be autologously harvested from the patient, providing an immunocompatible carrier for delivery of seed cells in tissue engineering research. Fibrin and fibrinogen have a well-established application in cartilage regeneration due to their innate ability to induce improved cellular interaction and subsequent scaffold remodelling compared to synthetic scaffolds. Fibrin as support for cell delivery has been applied to the cartilage regeneration field as presented in Table 6.

Table 5. Gelatin-based scaffolds for cartilage regeneration.

Scaffold	Seed cell type	Regeneration model	References
Gelatin microspheres encapsulated in a hydrogel injectable matrix	Chondrocytes	*In vitro*	Holland TA *et al.*, 2003 Holland TA *et al.*, 2004
Porous gelatin disks (Surgifoam®)	Adipose-derived stem cells	*In vitro*	Awad HA *et al.*, 2004
Macroporous gelatin microcarriers beads (CultiSpher G®)	Chondrocytes	Nude mice	Malda J *et al.*, 2003
Porous gelatin sponge (Gelfoam®)	Mesenchymal stem cells	Rabbit	Ponticiello MS *et al.*, 2000
Gelatin and chemically modified hyaluronic acid injectable hydrogel	Bone marrow derived stromal cells	Rabbit	Liu Y *et al.*, 2006

Table 6. Fibrin-based scaffolds for cartilage engineering.

Scaffold	Seed cell type	Regeneration model	References
Fibrin-collagen gel	Embryonic chondrogenic cells	*In vitro*	Perka C *et al.*, 2000
Fibrin gel	Chondrocytes	*In vitro*	Hunter CJ *et al.*, 2004 Schmoekel H *et al.*, 2004
Fibrin gel in a PGA non-woven mesh	Chondrocytes	*In vitro*	Ameer GA *et al.*, 2002
Porous fibrin gel	Chondrocytes	*In vitro*	Perka C *et al.*, 2000
Fibrin glue combined with plotted PCL	Bone marrow derived stromal cells	Rabbit	Shao XX *et al.*, 2006

4.1.4. *Alginate*

Alginate is one of the most studied and applied native polymers in tissue engineering field. They are abundant in nature and are found as structural components of marine brown algae and as capsular polysaccharides in some soil bacteria. Due to its biocharacteristics and the mild gelation process conditions, alginate templates are by far one of the most widely applied native polymers for cartilage engineering applications, as Table 7 demonstrates.

Table 7. Alginate-based scaffolds for cartilage engineering (Malafaya *et al.*, 2007).

Scaffold	Seed cell type	Regeneration model	References
Alginate hydrogel	Bone marrow stromal cells	Mouse	Xu XL *et al.*, 2005
Alginate beads	Chondrocytes	*In vitro*	Gründer T *et al.*, 2004 Masuda K *et al.*, 2006 Gaissmaier C *et al.*, 2005
Alginate beads	Chondrocytes	Nude mice	Kaps C *et al.*, 2005
Alginate beads	Adipose-derived stem cells	Mude mice	Erickson GR *et al.*, 2002
Alginate hydrogel and film	Periostal cells	*In vitro*	Stevens MM *et al.*, 2004 Stevens MM *et al.*, 2004
Alginate hydrogel	Chondrocytes	*In vitro*	Heywood HK *et al.*, 2006
Alginate beads	—	Rabbit knee	Mierisch CM *et al.*, 2002
Alginate hydorgel	Synovium-derived progenitor cells	*In vitro*	Park Y *et al.*, 2005
Alginate freeze-dried sponges and combined with hyaluronic acid	Chondrocytes	Rat	Dausse Y *et al.*, 2005
Alginate gel alone and in PGA-PLA pads	Rib chondro-progenitor cells	Rabbit	Cohen SB *et al.*, 2003
Alginate-chitosan microcapsules	Bone marrow stromal cells	Nude mice	Pound JC *et al.*, 2006

4.1.5. *Remarks and future directions*

Native polymers have received considerable interest for tissue engineering and regenerative medicine. This section introduces the properties of major types of native polymers that render them attractive for cartilage regeneration. We also review research works where these properties have been described, and summarize them in comprehensive tables for each type of polymer.

Although native polymers present some drawbacks namely the difficulties in controlling the variability from batch to batch, mechanical properties or limited processability, their advantages clearly surplus the drawbacks. Their degradability, biocompatibility, low cost and availability, similarity with the extracellular matrix and intrinsic cellular interaction makes them attractive candidates for biomedical applications, in particular for cartilage engineering applications as described in this section.

The aforementioned drawbacks are obviously limiting the widespread use of native polymers, mainly in clinical purposes. To try to overcome this disadvantage and for rational design of scaffolds for cartilage engineering, it is essential to study the effect of individual components. Therefore, a better controlled development in methods for production, purification, or in material properties such as molecular weight, mechanical behavior or degradation rate is essential to widespread the use of these class of polymers.

4.2. *Cartilage repair based on native polymers*

4.2.1. *Articular cartilage repair based on native polymers*

Damage or loss of articular cartilage as a consequence of congenital anomaly, degenerative joint disease or injury leads to progressive debilitation, which has a negative impact on the quality of life of affected individuals in all age groups. Articular cartilage repair is always a clinical challenge with the addition of its limited regenerative ability of chondrocytes *in vivo* (Kessler and Grande, 2008). There are many strategies, such as arthroscopic lavage, abrasion arthroplasty, subchondral drilling, and microfracture etc. for articular cartilage regeneration. Although these strategies can mostly promote pain relief and enhanced joint functions, they fail to restore the native structure and biomechanical properties of articular cartilage. Therefore, investigators have focused on reconstructing cartilage using tissue engineering strategies (Ahmed and Hincke, 2010). In this part, the application status, the major challenges, and prospect of native polymers for articular cartilage regeneration will be discussed.

The most common natural materials for articular cartilage regeneration are hyaluronan and collagen based matrices because of their similar molecule components to natural articular cartilage (Safran *et al.*, 2008). Hyalograft C (a derivative of HA) is a hyluronan based scaffold which has been used clinically in the treatment of articular cartilage lesions. Nettles *et al.* (2004) demonstrates in a rabbit osteochondral defect model that photocross-linkable HA-MA promoted the retention of the chondrocytic phenotype and cartilage matrix synthesis for encapsulated chondrocytes *in vitro* and accelerated healing in an *in vivo* osteochondral defect model. In a recent clinical research, a tissue-engineered graft composed of autologous chondrocytes and HYAFF 11(another derivative of HA) scaffolds was used for artilcular cartilage defect repair of 141 patients with satisfactory results. These positive clinical results indicate that HA-based graft is a safe and effective therapeutic option for the treatment of articular cartilage lesions (Marcacci *et al.*, 2005). For collagen, it has been proven that collagen hydrogels along with chondrocytes into rabbit chondral defects led to satisfactory hyaline cartilage regeneration as indicated by histological and biomechanical assessment. Furthermore, combining collagen I with GAGs and chitosan or fibrin have been shown to be helpful for maintaining chondrocyte proliferation, morphology, and expression of specific markers.

Besides hyaluronan and collagen, other native polymers are also involved in articular cartilage regeneration. For example, alginate beads could significantly increase the accumulation of cartilage specific ECM such as GAGs and collagen II, and that the seeding of chondrocytes on chitosan-coated coverslips better maintained the spherical chondrocyte morphology and promoted expression of collagen II and aggrecan. All these findings predict the enormous potential of native polymers for articular cartilage regeneration.

Despite the above advances in articular cartilage repair based on native polymers, the current native polymers still have their inherent limitations for achieving desirable articular cartilage regeneration. In general, native polymer-based scaffolds lack the mechanical strength and bioactive factors to meet the functional requirements of articular defect repair. In addition, the scaffolds based on native polymers still cannot accurately mimic the unique zonal structure of articular cartilage. Therefore, much more endeavours should be devoted to improve the design of native polymer-based scaffolds. At least three approaches may be tried in future studies. Firstly, combinations of more than one matrix type, the formation of copolymers, and cross-linkage with various substrates may be very important approaches for enhancing the mechanical strength and improving biodegradability. Secondly, chemical modifications are also important approaches for native polymers in the articular cartilage engineering. The functionalization of a matrix by incorporating either agents to promote cell adhesion or growth factors — whereby it serves as a delivery system — is also a very promising approach to improving its performance in articular cartilage repair. Geometrical modifications too, elicited by the production of sponge- or foam-like matrices, have been shown to enhance the differentiation potential and metabolic activity of chondroprogenitor cells and mature chondrocytes, respectively. Finally, some new techniques, such as nanotechnology and 3D printing technology, could be adopted to accurately control 3D structure of the scaffold at micro- and macro-levels for mimicking the unique zonal structure of articular cartilage (Mow *et al.*, 1984; Nukavarapu and Dorcemus, 2012; Chiang and Jiang, 2009; Kerker and Leo, 2008).

4.2.2. *Tracheal cartilage repair based on native polymers*

Trachea, affected by pressure difference, levels of kinds of hormone, bacteria in the lumen, function of smooth muscle, is an awfully complicated air duct in the living body. Thus, as an important branch of tissue engineering, the regeneration and reconstruction of trachea is one of the most troublesome fields of all time. With the wide application of native polymer in cartilage engineering, some developments have been achieved in trachea reconstruction.

Due to the complex structural characteristic of trachea, decellular trachea matrix is an obvious ideal candidate polymer for tracheal reconstruction. One of the most impressive studies is that Macchiarini P and his colleagues repaired a patient's left main bronchus with allogenic decellular trachea matrix as well as bronchus-derived epithelial cells and bone marrow-derived mesenchymal stem cells obtained from the recipient with end-stage bronchomalacia. To alleviate the immunogenicity from allogenic trachea, both cells and MHC antigens were removed from the donor trachea. After the surgery, the patient's left lung ventilated well immediately. Moreover, in the following three months, all tests showed satisfactory results and the quality of patient's life was improved. Since then, nine other patients received trachea transplantation using the similar approach.

The preliminary success of the above clinical trial mainly attributes to the advantage of acellular trachea scaffold, such as low immunogenicity, similar structure to native trachea, excellent cell affinity, and biocompatibility. Besides this, the authors considered that the acellular trachea matrix also preserved several angiogenic cytokines, such as basic fibroblast growth factor, transforming growth factor β, which would be very helpful when revascularization occurs.

However, the long-term outcomes of these cases are unpredictable. All of the cases have had to face the risk of hemorrhea, and even one of them died as a result. The recipient in the very first case had to receive several stent inserted over the past five years because of narrowed or collapsed airway. Meanwhile, additional lab experiment was performed to observe what happened to decellularized trachea over time. The outcomes indicated that the trachea became significantly weaker and lost some of the microstructure after a year in saline medium, which may be beneficial for adherence of new cells and generation of new tissue according to the analysis of authors (Vogel, 2013). On the contrary, it can be deduced that weaker and microstructure-lost trachea scaffold may be one of the important reasons that cause stenosis and collapse of air duct after implantation.

Except for the decellularized trachea, other native polymers, such as HA, gelatine, and collagens, have also been studied in trachea repair or construction of tubular cartilage (Walles *et al.*, 2004; Kobayashi *et al.*, 2010). Despite the terrific cell affinity, biocompatibility, and minimal immunogenicity, some inherent drawbacks which include inferior mechanical properties, difficulty in shape control, and fast degradation rate greatly restrict their research and application in trachea engineering. To address these problems, some new strategies need to be introduced. Composite scaffolds based on natural and synthetic substances may be an important direction, since the synthetic polymers have complementary advantages in mechanical properties, shape control, and biodegradability (Omori *et al.*, 2008; Tada *et al.*, 2008; Schagemann *et al.*, 2010).

4.3. *Challenges in cartilage repair*

4.3.1. *Tissue integration*

The poor integration of repair tissue with native cartilage is a problem that has been recognized by several authors and is one that must be resolved if we are to achieve enduring healing results and biomechanical competence. Various measures have been employed to promote this process by establishing good contact between implants and native articular cartilage, including the use of collagen cross-linkers, biological glues (e.g., tissue transglutaminase) and other adhesives. Another possible approach that has been shown to have a beneficial effect is the brief enzymatic degradation of proteoglycans along defect surfaces, these molecules being known to impede both cell and matrix adhesion (Zhang *et al.*, 2005).

4.3.2. *The scale of cartilage defect repair*

Investigators pay too little attention to the dimensional aspects of articular cartilage repair. What volume needs to be filled with repair tissue? Can the biologics of the engineered construct cope with this? Such simple questions — despite being of the utmost importance — are rarely addressed in reported tissue regeneration studies. Even just a casual glance at the respective histologies of knee-joint articular cartilage in humans and experimental animals commonly used for the testing of repair concepts reveals very obvious differences, not only in structural organization and cellularity but also in height and volume. In the human tibial plateau or medial femoral condyle, a shallow full thickness defect would span a height of approximately 2 – 4 mm, frequently cover an area of about 3 – 4 cm² and thus embrace a tissue volume of 0.9 – 1.0 cm³ (i.e., ml), which represents a vast void to be filled with engineered tissue. Indeed, routine tissue culturing methodologies cannot cope with this scale of production, which requires the use of special technologies, such as bioreactor systems. Albeit so, it is still absolutely essential that investigators give due consideration to defect design in their chosen animal model, so as to be able to simulate the human situation as closely as possible (Solchaga *et al.,* 2001).

4.3.3. *Defect design in animal models*

A survey of the literature pertaining to tissue engineering aspects of articular cartilage repair reveals a depressing consistency in the degree to which investigators disregard defect dimensions relative to the make-up of the tissue surrounds and fail to consider their pertinence to the human condition they ultimately wish to treat (Ahern *et al.,* 2009). The first and the most fundamental question to be asked is: do I wish to treat a partial-thickness defect, which has no access to the blood vascular spaces, or a full-thickness one, in which case the influence of blood-borne cell populations and signalling substances must be brought under control?

The creation of partial-thickness defects in most commonly used experimental animals poses great technical difficulties owing to the small scale of the dimensions involved, and these are not easily surmounted on a reproducible basis. It is probably for this reason that most investigators have chosen to work with full-thickness defect models. But by so doing, they have, on one hand, won a head's start for their treatment principle, in that any repair response elicited will be promoted or boosted by the spontaneously-generated one. On the other hand, they have thereby rendered an interpretation of their findings more difficult, particularly if they have failed to set up systematically the control experiments required for the drawing of unequivocal conclusions. The volumes of such full-thickness defects are, of course, more akin to those of partial-thickness ones in human patients, but then appropriate measures must be taken to render the microenvironment alike too. One means of achieving this is to create a so-called 'virtual' partial-thickness defect, which would involve treating the floor and walls of a full-thickness one in such a manner as to render it impermeable to blood-borne cells and signalling substances emanating from the subchondral bone-tissue spaces.

Not only defect design but also study design needs to be better considered in the future, so as to yield more clear-cut evidence on which to base clinical decisions regarding the application of a new methodology in human patients.

4.4. *Future directions in native polymer-based scaffolds and cartilage regeneration*

Interdisciplinary efforts made for the development of tissue engineering scaffolds have made current scaffolding systems versatile. Better understanding in cell biology and biomaterial science such as cell-cell, cell-ECM, cell-biomaterial, and biomaterial-protein interactions, hierarchical arrangement of cell in tissue, and effects of various signalling molecules and their combinations on cell behavior, have contributed significantly to the design of scaffolds for improving cartilage regeneration. Identification of functional domains of various ECM proteins has provided biomimetic approaches that involve precision immobilization of peptides (functional domains) and proteins on the surface of polymeric scaffolds. The ability to control non-specific adsorption of proteins and enable oriented immobilization without denaturing protein conformation is desirable for two reasons (i) improving cell attachment, and (ii) developing controlled delivery system for therapeutic proteins.

Surface modification, bioactive regulation, and nanotechnology will still be the main directions in native polymer-based scaffolds. Future studies, both *in vitro* and *in vivo*, on scaffolding systems that possess nanoscale surface features will bring more clarity in the area of biomaterial surface engineering. It is hoped that this clarity will instigate the development of next generation scaffolding systems with nanoengineered surfaces that elicit improved cell behavior and can act as a delivery system for therapeutic proteins. Similarly, fusion protein-based protein delivery systems that can potentially circumvent the limitation associated with the conventional approaches have demonstrated potential to be developed as the protein delivery systems of the future. Therefore, a combination of physical modification (nano-engineering) and protein immobilization would enable the successful design of scaffolding systems for cartilage regenerative applications.

Recent advances in stem cells have provided hope for repairing large cartilage defect by offering large number of cells with chondrogenic potential. However, how complex signals interact with each other to regulate stem cell differentiation remains poorly understood, and high throughput strategies may provide a valuable tool for overcoming this hurdle and identifying an optimal 3D niche. For successful cartilage regeneration, it is crucial to appreciate the structural heterogeneity in the native cartilage. Novel strategies that facilitate the recreation of such zonal organization will likely lead to functional superior cartilage tissues. As compared to the direct injection of autologous chondrocytes in current practice, future therapies will likely utilize more complex strategies involving 3D scaffolds with spatial control of biological and mechanical signals to promote for zonal specific cartilage regeneration. Another important factor to consider is to evaluate novel strategies in appropriate animal models for monitoring cell fate, inflammatory response, long-term functional stability, etc. When designing the animal experiment, it is important to take into account the variation of cartilage thickness and different loading distribution in cartilage across species. Another important but often overlooked factor is the diseased animal model. The majority of tissue engineered cartilage grafts will be used for repairing degenerated cartilage such as osteoarthritis. *In vitro* and *in vivo* models need to be developed to better understand how diseased microenvironment influences the engineered cartilage tissues, which will provide valuable information in predicting therapeutic outcomes.

References

Ahern BJ, Parvizi J, Boston R, Schaer TP. Preclinical animal models in single site cartilage defect testing: A systematic review. *Osteoarthritis Cartilage* 2009;17(6):705–713.

Ahmed TA, Dare EV, Hincke M. Fibrin: A versatile scaffold for tissue engineering applications. *Tissue Eng Part B Rev* 2008;14(2):199–215.

Ahmed TA, Hincke MT. Strategies for articular cartilage lesion repair and functional restoration. *Tissue Eng Part B Rev* 2010;16(3):305–329.

Aimin C, Chunlin H, Juliang B, Tinyin Z, Zhichao D. Antibiotic loaded chitosan bar. An *in vitro, in vivo* study of a possible treatment for osteomyelitis. *Clin Orthop Relat Res* 1999;(366):239–247.

Alford JW, Cole BJ. Cartilage restoration, part 1: Basic science, historical perspective, patient evaluation, and treatment options. *Am J Sports Med* 2005 Feb;33(2):295–306.

Alford JW, Cole BJ. Cartilage restoration, part 2: Techniques, outcomes, and future directions. *Am J Sports Med* 2005;33(3):443–460.

Ameer GA, Mahmood TA, Langer R. A biodegradable composite scaffold for cell transplantation. *J Orthop Res* 2002;20(1):16–19.

Anderson SB, Lin CC, Kuntzler DV, Anseth KS. The performance of human mesenchymal stem cells encapsulated in cell-degradable polymer-peptide hydrogels. *Biomaterials* 2011;32(14):3564–3574.

Astachov L, Vago R, Aviv M, Nevo Z. Hyaluronan and mesenchymal stem cells: From germ layer to cartilage and bone. *Front Biosci* 2011;16:261–276.

Awad HA, Wickham MQ, Leddy HA, Gimble JM, Guilak F. Chondrogenic differentiation of adipose-derived adult stem cells in agarose, alginate, and gelatin scaffolds. *Biomaterials* 2004;25(16):3211–3222.

Badylak SF. Xenogeneic extracellular matrix as a scaffold for tissue reconstruction. *Transpl Immunol* 2004;12(3–4):367–377.

Bank RA, Bayliss MT, Lafeber FP, Maroudas A, Tekoppele JM. Ageing and zonal variation in post-translational modification of collagen in normal human articular cartilage. The age-related increase in non-enzymatic glycation affects biomechanical properties of cartilage. *Biochem J* 1998;330(Pt 1):345–351.

Barnes CA, Brison J, Michel R, Brown BN, Castner DG, Badylak SF, Ratner BD. The surface molecular functionality of decellularized extracellular matrices. *Biomaterials* 2011;32(1):137–143.

Bastiaansen-Jenniskens YM, Koevoet W, de Bart AC, van der Linden JC, Zuurmond AM, Weinans H, Verhaar JA, van Osch GJ, Degroot J. Contribution of collagen network features to functional properties of engineered cartilage. *Osteoarthritis Cartilage* 2008;16(3):359–366.

Bastow ER, Byers S, Golub SB, Clarkin CE, Pitsillides AA, Fosang AJ. Hyaluronan synthesis and degradation in cartilage and bone. *Cell Mol Life Sci* 2008;65(3):395–413.

Berthiaume F, Maguire TJ, Yarmush ML. Tissue engineering and regenerative medicine: History, progress, and challenges. *Annu Rev Chem Biomol Eng.* 2011;2:403–430.

Bhattarai N, Edmondson D, Veiseh O, Matsen FA, Zhang M. Electrospun chitosan-based nanofibers and their cellular compatibility. *Biomaterials* 2005;26(31):6176–6184.

Bian L, Zhai DY, Tous E, Rai R, Mauck RL, Burdick JA. Enhanced MSC chondrogenesis following delivery of TGF-β3 from alginate microspheres within hyaluronic acid hydrogels *in vitro* and *in vivo*. *Biomaterials* 2011;32(27):6425–6434.

Bobick BE, Chen FH, Le AM, Tuan RS. Regulation of the chondrogenic phenotype in culture. *Birth Defects Res C Embryo Today* 2009;87(4):351–371.

Capito RM, Azevedo HS, Velichko YS, Mata A, Stupp SI. Self-assembly of large and small molecules into hierarchically ordered sacs and membranes. *Science* 2008;319(5871):1812–1816.

Cen L, Liu W, Cui L, Zhang W, Cao Y. Collagen tissue engineering: Development of novel biomaterials and applications. *Pediatr Res* 2008;63(5):492–496.

Chan G, Mooney DJ. New materials for tissue engineering: towards greater control over the biological response. *Trends Biotechnol* 2008;26(7):382–392.

Chang CH, Kuo TF, Lin CC, Chou CH, Chen KH, Lin FH, Liu HC. Tissue engineering-based cartilage repair with allogenous chondrocytes and gelatin-chondroitin-hyaluronan tri-copolymer scaffold: A porcine model assessed at 18, 24, and 36 weeks. *Biomaterials* 2006;27(9):1876–1888.

Chang CH, Liu HC, Lin CC, Chou CH, Lin FH. Gelatin-chondroitin-hyaluronan tri-copolymer scaffold for cartilage tissue engineering. *Biomaterials* 2003;24(26):4853–4858.

Chao PH, Yodmuang S, Wang X, Sun L, Kaplan DL, Vunjak-Novakovic G. Silk hydrogel for cartilage tissue engineering. *J Biomed Mater Res B Appl Biomater* 2010;95(1):84–90.

Chen RH, Hwa HD. Effect of molecular weight of chitosan with the same degree of deacetylation on the thermal, mechanical, and permeability properties of the prepared membrane. *Carbohydr Polym* 1996;29:353–358.

Chen X, Zhang F, He X, Xu Y, Yang Z, Chen L, Zhou S, Yang Y, Zhou Z, Sheng W, Zeng Y. Chondrogenic differentiation of umbilical cord-derived mesenchymal stem cells in type I collagen-hydrogel for cartilage engineering. *Injury* 2013;44(4):540–549.

Chia SH, Homicz MR, Schumacher BL, Thonar EJ, Masuda K, Sah RL, Watson D. Characterization of human nasal septal chondrocytes cultured in alginate. *J Am Coll Surg* 2005;200(5):691–704.

Chia SH, Schumacher BL, Klein TJ, Thonar EJ, Masuda K, Sah RL, Watson D. Tissue-engineered human nasal septal cartilage using the alginate-recovered-chondrocyte method. *Laryngoscope* 2004;114(1):38–45.

Chiang H, Jiang CC. Repair of articular cartilage defects: Review and perspectives. *J Formos Med Assoc* 2009;108(2):87–101.

Chung TW, Lu YF, Wang SS, Lin YS, Chu SH. Growth of human endothelial cells on photochemically grafted Gly-Arg-Gly-Asp (GRGD) chitosans. *Biomaterials* 2002;23(24):4803–4809.

Clark JM. Variation of collagen fiber alignment in a joint surface: A scanning electron microscope study of the tibial plateau in dog, rabbit, and man. *J Orthop Res* 1991;9(2):246–257.

Coates EE, Fisher JP. Phenotypic variations in chondrocyte subpopulations and their response to *in vitro* culture and external stimuli. *Ann Biomed Eng* 2010;38(11):3371–3388.

Cohen SB, Meirisch CM, Wilson HA, Diduch DR. The use of absorbable co-polymer pads with alginate and cells for articular cartilage repair in rabbits. *Biomaterials* 2003;24(15):2653–2660.

Comisar WA, Kazmers NH, Mooney DJ, Linderman JJ. Engineering RGD nanopatterned hydrogels to control preosteoblast behavior: A combined computational and experimental approach. *Biomaterials* 2007;28(30):4409–4417.

Dahlin RL, Kasper FK, Mikos AG. Polymeric nanofibers in tissue engineering. *Tissue Eng Part B Rev* 2011;17(5):349–364.

Danisovic L, Varga I, Zamborsky R, Böhmer D. The tissue engineering of articular cartilage: Cells, scaffolds and stimulating factors. *Exp Biol Med (Maywood)* 2012;237(1):10–17.

Darr A, Calabro A. Synthesis and characterization of tyramine-based hyaluronan hydrogels. *J Mater Sci Mater Med* 2009;20(1):33–44.

Dausse Y, Grossin L, Miralles G, Pelletier S, Mainard D, Hubert P, Baptiste D, Gillet P, Dellacherie E, Netter P, Payan E. Cartilage repair using new polysaccharidic biomaterials: Macroscopic, histological and biochemical approaches in a rat model of cartilage defect. *Osteoarthritis Cartilage* 2003;11(1):16–28.

De Franceschi L, Grigolo B, Roseti L, Facchini A, Fini M, Giavaresi G, Tschon M, Giardino R. Transplantation of chondrocytes seeded on collagen-based scaffold in cartilage defects in rabbits. *J Biomed Mater Res A* 2005;75(3):612–622.

Discher DE, Mooney DJ, Zandstra PW. Growth factors, matrices, and forces combine and control stem cells. *Science* 2009;324(5935):1673–1677.

Dorotka R, Bindreiter U, Macfelda K, Windberger U, Nehrer S. Marrow stimulation and chondrocyte transplantation using a collagen matrix for cartilage repair. *Osteoarthritis Cartilage* 2005;13(8):655–664.

Emin N, Koç A, Durkut S, Elçin AE, Elçin YM. Engineering of rat articular cartilage on porous sponges: Effects of tgf-beta 1 and microgravity bioreactor culture. *Artif Cells Blood Substit Immobil Biotechnol* 2008;36(2):123–137.

Erickson GR, Gimble JM, Franklin DM, Rice HE, Awad H, Guilak F. Chondrogenic potential of adipose tissue-derived stromal cells *in vitro* and *in vivo*. *Biochem Biophys Res Commun* 2002;290(2):763–769.

Fan H, Wang L, Zhao K, Li N, Shi Z, Ge Z, Jin Z. Fabrication, mechanical properties, and biocompatibility of graphene-reinforced chitosan composites. *Biomacromolecules* 2010;11(9):2345–2351.

Fedorovich NE, Alblas J, de Wijn JR, Hennink WE, Verbout AJ, Dhert WJ. Hydrogels as extracellular matrices for skeletal tissue engineering: State-of-the-art and novel application in organ printing. *Tissue Eng* 2007;13(8):1905–1925.

Fraser JR, Laurent TC, Laurent UB. Hyaluronan: Its nature, distribution, functions and turnover. *J Intern Med* 1997;242(1):27–33.

Fujisato T, Sajiki T, Liu Q, Ikada Y. Effect of basic fibroblast growth factor on cartilage regeneration in chondrocyte-seeded collagen sponge scaffold. *Biomaterials* 1996;17(2):155–162.

Gaissmaier C, Fritz J, Krackhardt T, Flesch I, Aicher WK, Ashammakhi N. Effect of human platelet supernatant on proliferation and matrix synthesis of human articular chondrocytes in monolayer and three-dimensional alginate cultures. *Biomaterials* 2005;26(14):1953–1960.

Gao L, McBeath R, Chen CS. Stem cell shape regulates a chondrogenic versus myogenic fate through Rac1 and N-cadherin. *Stem Cells.* 2010;28(3):564–572.

Ge Z, Li C, Heng BC, Cao G, Yang Z. Functional biomaterials for cartilage regeneration. *J Biomed Mater Res A* 2012;100(9):2526–2536.

Geckil H, Xu F, Zhang X, Moon S, Demirci U. Engineering hydrogels as extracellular matrix mimics. *Nanomedicine (Lond)* 2010;5(3):469–484.

Geutjes PJ, Daamen WF, Buma P, Feitz WF, Faraj KA, van Kuppevelt TH. From molecules to matrix: Construction and evaluation of molecularly defined bioscaffolds. *Adv Exp Med Biol* 2006;585:279–295.

Gigout A, Jolicoeur M, Nelea M, Raynal N, Farndale R, Buschmann MD. Chondrocyte aggregation in suspension culture is GFOGER-GPP- and beta1 integrin-dependent. *J Biol Chem* 2008;283(46):31522–31530.

Gilbert TW, Sellaro TL, Badylak SF. Decellularization of tissues and organs. *Biomaterials* 2006;27(19):3675–3683.

Gilbert TW. Strategies for tissue and organ decellularization. *J Cell Biochem* 2012;113(7):2217–2222.

Glowacki J, Mizuno S. Collagen scaffolds for tissue engineering. *Biopolymers* 2008;89(5):338–344.

Gorodetsky R. The use of fibrin based matrices and fibrin microbeads (FMB) for cell based tissue regeneration. *Expert Opin Biol Ther* 2008;8(12):1831–1846.

Gruber HE, Hoelscher GL, Leslie K, Ingram JA, Hanley EN Jr. Three-dimensional culture of human disc cells within agarose or a collagen sponge: Assessment of proteoglycan production. *Biomaterials* 2006;27(3):371–376.

Gruber HE, Leslie K, Ingram J, Norton HJ, Hanley EN. Cell-based tissue engineering for the intervertebral disc: In vitro studies of human disc cell gene expression and matrix production within selected cell carriers. *Spine J* 2004;4(1):44–55.

Gründer T, Gaissmaier C, Fritz J, Stoop R, Hortschansky P, Mollenhauer J, Aicher WK. Bone morphogenetic protein (BMP)-2 enhances the expression of type II collagen and aggrecan in chondrocytes embedded in alginate beads. *Osteoarthritis Cartilage* 2004;12(7):559–567.

Guenther D, Oks A, Ettinger M, Liodakis E, Petri M, Krettek C, Jagodzinski M, Haasper C. Enhanced migration of human bone marrow stromal cells in modified collagen hydrogels. *Int Orthop* 2013 May 5.

Hardmeier R, Redl H, Marlovits S. Effects of mechanical loading on collagen propeptides processing in cartilage repair. *J Tissue Eng Regen Med* 2010;4(1):1–11.

Heywood HK, Bader DL, Lee DA. Glucose concentration and medium volume influence cell viability and glycosaminoglycan synthesis in chondrocyte-seeded alginate constructs. *Tissue Eng* 2006;12(12):3487–3496.

Holland TA, Tabata Y, Mikos AG. In vitro release of transforming growth factor-beta 1 from gelatin microparticles encapsulated in biodegradable, injectable oligo(poly(ethylene glycol) fumarate) hydrogels. *J Control Release* 2003;91(3):299–313.

Holland TA, Tessmar JK, Tabata Y, Mikos AG. Transforming growth factor-beta 1 release from oligo(poly(ethylene glycol) fumarate) hydrogels in conditions that model the cartilage wound healing environment. *J Control Release* 2004;94(1):101–114.

Hunter CJ, Mouw JK, Levenston ME. Dynamic compression of chondrocyte-seeded fibrin gels: effects on matrix accumulation and mechanical stiffness. *Osteoarthritis Cartilage* 2004;12(2):117–130.

Hunziker E, Spector M, Libera J, Gertzman A, Woo SL, Ratcliffe A, Lysaght M, Coury A, Kaplan D, Vunjak-Novakovic G. Translation from research to applications. *Tissue Eng* 2006;12(12):3341–3364.

Hunziker EB. Articular cartilage repair: Basic science and clinical progress. A review of the current status and prospects. *Osteoarthritis Cartilage* 2002;10(6):432–463.

Ishida O, Tanaka Y, Morimoto I, Takigawa M, Eto S. Chondrocytes are regulated by cellular adhesion through CD44 and hyaluronic acid pathway. *J Bone Miner Res* 1997;12(10):1657–1663.

Jiang T, Kumbar SG, Nair LS, Laurencin CT. Biologically active chitosan systems for tissue engineering and regenerative medicine. *Curr Top Med Chem* 2008;8(4):354–364.

Jürgensen K, Aeschlimann D, Cavin V, Genge M, Hunziker EB. A new biological glue for cartilage-cartilage interfaces: Tissue transglutaminase. *J Bone Joint Surg Am* 1997;79(2):185–193.

Kaps C, Bramlage C, Smolian H, Haisch A, Ungethüm U, Burmester GR, Sittinger M, Gross G, Häupl T. Bone morphogenetic proteins promote cartilage differentiation and protect engineered artificial cartilage from fibroblast invasion and destruction. *Arthritis Rheum* 2002;46(1):149–162.

Kasoju N, Bora U. Silk fibroin in tissue engineering. *Adv Healthc Mater* 2012;1(4):393–412.

Keeney M, Lai JH, Yang F. Recent progress in cartilage tissue engineering. *Curr Opin Biotechnol* 2011;22(5):734–740.

Kelly TA, Wang CC, Mauck RL, Ateshian GA, Hung CT. Role of cell-associated matrix in the development of free-swelling and dynamically loaded chondrocyte-seeded agarose gels. *Biorheology* 2004;41(3–4):223–237.

Kerker JT, Leo AJ, Sgaglione NA.Cartilage repair: Synthetics and scaffolds: Basic science, surgical techniques, and clinical outcomes. *Sports Med Arthrosc* 2008;16(4):208–216.

Kessler MW, Grande DA. Tissue engineering and cartilage. *Organogenesis* 2008;4(1):28–32.

Keung AJ, Healy KE, Kumar S, Schaffer DV. Biophysics and dynamics of natural and engineered stem cell microenvironments. *Wiley Interdiscip Rev Syst Biol Med* 2010;2(1):49–64.

Kim BS, Mooney DJ. Development of biocompatible synthetic extracellular matrices for tissue engineering. *Trends Biotechnol* 1998;16(5):224–230.

Kim IL, Mauck RL, Burdick JA. Hydrogel design for cartilage tissue engineering: A case study with hyaluronic acid. *Biomaterials* 2011;32(34):8771–8782.

Kim IY, Seo SJ, Moon HS, Yoo MK, Park IY, Kim BC, Cho CS. Chitosan and its derivatives for tissue engineering applications. *Biotechnol Adv* 2008;26(1):1–21.

Kim TK, Sharma B, Williams CG, Ruffner MA, Malik A, McFarland EG, Elisseeff JH. Experimental model for cartilage tissue engineering to regenerate the zonal organization of articular cartilage. *Osteoarthritis Cartilage* 2003 Sep;11(9):653–664.

Kiviranta P, Rieppo J, Korhonen RK, Julkunen P, Töyräs J, Jurvelin JS. Collagen network primarily controls Poisson's ratio of bovine articular cartilage in compression. *J Orthop Res* 2006;24(4):690–699.

Klein TJ, Malda J, Sah RL, Hutmacher DW. Tissue engineering of articular cartilage with biomimetic zones. *Tissue Eng Part B Rev* 2009;15(2):143–157.

Knight MM, Toyoda T, Lee DA, Bader DL. Mechanical compression and hydrostatic pressure induce reversible changes in actin cytoskeletal organisation in chondrocytes in agarose. *J Biomech* 2006;39(8):1547–1551.

Kobayashi K, Suzuki T, Nomoto Y, Tada Y, Miyake M, Hazama A, Wada I, Nakamura T, Omori K. A tissue-engineered trachea derived from a framed collagen scaffold, gingival fibroblasts and adipose-derived stem cells. *Biomaterials* 2010;31(18):4855–4863.

Kogan G, Soltés L, Stern R, Gemeiner P. Hyaluronic acid: A natural biopolymer with a broad range of biomedical and industrial applications. *Biotechnol Lett* 2007;29(1):17–25.

Kolácná L, Bakesová J, Varga F, Kostáková E, Plánka L, Necas A, Lukás D, Amler E, Pelouch V. Biochemical and biophysical aspects of collagen nanostructure in the extracellular matrix. *Physiol Res* 2007;56 (Suppl 1):S51–S60.

Koria P. Delivery of growth factors for tissue regeneration and wound healing. *BioDrugs* 2012;26(3):163–175.

Kuo CK, Li WJ, Mauck RL, Tuan RS. Cartilage tissue engineering: Its potential and uses. *Curr Opin Rheumatol* 2006 Jan;18(1):64–73.

LaPorta TF, Richter A, Sgaglione NA, Grande DA. Clinical relevance of scaffolds for cartilage engineering. *Orthop Clin North Am* 2012;43(2):245–254.

Lee HS, Millward-Sadler SJ, Wright MO, Nuki G, Salter DM. Integrin and mechanosensitive ion channel-dependent tyrosine phosphorylation of focal adhesion proteins and beta-catenin in human articular chondrocytes after mechanical stimulation. *J Bone Miner Res* 2000 Aug;15(8):1501–1509.

Lee KY, Peters MC, Anderson KW, Mooney DJ. Controlled growth factor release from synthetic extracellular matrices. *Nature* 2000;408(6815):998–1000.

Li C, Wang L, Yang Z, Kim G, Chen H, Ge Z. A viscoelastic chitosan-modified three-dimensional porous poly(L-lactide-co-ε-caprolactone) scaffold for cartilage tissue engineering. *J Biomater Sci Polym Ed* 2012;23(1–4):405–424.

Li L, Hsieh YL. Chitosan bicomponent nanofibers and nanoporous fibers. *Carbohydr Res* 2006;341(3):374–381.

Li LP, Herzog W, Korhonen RK, Jurvelin JS. The role of viscoelasticity of collagen fibers in articular cartilage: Axial tension versus compression. *Med Eng Phys* 2005;27(1):51–57.

Liu Y, Shu XZ, Prestwich GD. Osteochondral defect repair with autologous bone marrow-derived mesenchymal stem cells in an injectable, in situ, cross-linked synthetic extracellular matrix. *Tissue Eng* 2006;12(12):3405–3416.

Lu HH, Spalazzi JP. Biomimetic stratified scaffold design for ligament-to-bone interface tissue engineering. *Comb Chem High Throughput Screen* 2009;12(6): 589–597.

Ma PX, Langer R. Morphology and mechanical function of long-term in vitro engineered cartilage. *J Biomed Mater Res* 1999;44(2):217–221.

Malafaya PB, Silva GA, Reis RL. Natural-origin polymers as carriers and scaffolds for biomolecules and cell delivery in tissue engineering applications. *Adv Drug Deliv Rev* 2007;59(4–5):207–233.

Malda J, Kreijveld E, Temenoff JS, van Blitterswijk CA, Riesle J. Expansion of human nasal chondrocytes on macroporous microcarriers enhances redifferentiation. *Biomaterials* 2003;24(28):5153–5161.

Mandal BB, Park SH, Gil ES, Kaplan DL. Multilayered silk scaffolds for meniscus tissue engineering. *Biomaterials* 2011;32(2):639–651.

Marcacci M, Berruto M, Brocchetta D, Delcogliano A, Ghinelli D, Gobbi A, Kon E, Pederzini L, Rosa D, Sacchetti GL, Stefani G, Zanasi S. Articular cartilage engineering with Hyalograft C: 3-year clinical results. *Clin Orthop Relat Res* 2005;(435):96–105.

Masuda K, Pfister BE, Sah RL, Thonar EJ. Osteogenic protein-1 promotes the formation of tissue-engineered cartilage using the alginate-recovered-chondrocyte method. *Osteoarthritis Cartilage* 2006;14(4):384–391.

Mente PL, Lewis JL. Elastic modulus of calcified cartilage is an order of magnitude less than that of subchondral bone. *J Orthop Res* 1994;12(5):637–647.

Mierisch CM, Cohen SB, Jordan LC, Robertson PG, Balian G, Diduch DR. Transforming growth factor-beta in calcium alginate beads for the treatment of articular cartilage defects in the rabbit. *Arthroscopy* 2002;18(8):892–900.

Mikic B, Isenstein AL, Chhabra A. Mechanical modulation of cartilage structure and function during embryogenesis in the chick. *Ann Biomed Eng* 2004;32(1):18–25.

Mimura T, Imai S, Kubo M, Isoya E, Ando K, Okumura N, Matsusue Y. A novel exogenous concentration-gradient collagen scaffold augments full-thickness articular cartilage repair. *Osteoarthritis Cartilage* 2008;16(9):1083–1091.

Min BM, Lee SW, Lim JN, You Y, Lee TS, Kang PH, *et al.* Chitin and chitosan nanofibers: Electrospinning of chitin and deacetylation of chitin nanofibers. *Polymer* 2004;45:7137–7142.

Mooney DJ, Vandenburgh H. Cell delivery mechanisms for tissue repair. *Cell Stem Cell.* 2008;2(3):205–213.

Moss IL, Gordon L, Woodhouse KA, Whyne CM, Yee AJ. A novel thiol-modified hyaluronan and elastin-like polypetide composite material for tissue engineering of the nucleus pulposus of the intervertebral disc. *Spine (Phila Pa 1976).* 2011;36(13):1022–1029.

Mow VC, Holmes MH, Lai WM. Fluid transport and mechanical properties of articular cartilage: A review. *J Biomech* 1984;17(5):377–394.

Mow VC, Ratcliffe A, Poole AR.. Cartilage and diarthrodial joints as paradigms for hierarchical materials and structures. *Biomaterials* 1992;13(2):67–97.

Nettles DL, Vail TP, Morgan MT, Grinstaff MW, Setton LA. Photocrosslinkable hyaluronan as a scaffold for articular cartilage repair. *Ann Biomed Eng* 2004;32(3):391–397.

Ng KW, Wang CC, Mauck RL, Kelly TA, Chahine NO, Costa KD, Ateshian GA, Hung CT. A layered agarose approach to fabricate depth-dependent inhomogeneity in chondrocyte-seeded constructs. *J Orthop Res* 2005;23(1):134–141.

Nishikawa H, Ueno A, Nishikawa S, Kido J, Ohishi M, Inoue H, Nagata T. Sulfated glycosaminoglycan synthesis and its regulation by transforming growth factor-beta in rat clonal dental pulp cells. *J Endod* 2000;26(3):169–171.

Nuernberger S, Cyran N, Albrecht C, Redl H, Vécsei V, Marlovits S. The influence of scaffold architecture on chondrocyte distribution and behavior in matrix-associated chondrocyte transplantation grafts. *Biomaterials* 2011; 32(4):1032–1040.

Nukavarapu SP, Dorcemus DL. Osteochondral tissue engineering: Current strategies and challenges. *Biotechnol Adv* 2012 Nov 19.

Okamoto Y, Shibazaki K, Minami S, Matsuhashi A, Tanioka S, Shigemasa Y. Evaluation of chitin and chitosan on open would healing in dogs. *J Vet Med Sci* 1995;57(5):851–854.

Omori K, Tada Y, Suzuki T, Nomoto Y, Matsuzuka T, Kobayashi K, Nakamura T, Kanemaru S, Yamashita M, Asato R. Clinical application of in situ tissue engineering using a scaffolding technique for reconstruction of the larynx and trachea. *Ann Otol Rhinol Laryngol* 2008;117(9):673–678.

Orgel JP, San Antonio JD, Antipova O. Molecular and structural mapping of collagen fibril interactions. *Connect Tissue Res* 2011;52(1):2–17.

Ouzzine M, Venkatesan N, Fournel-Gigleux S. Proteoglycans and cartilage repair. *Methods Mol Biol* 2012;836:339–355.

Pan J, Zhou X, Li W, Novotny JE, Doty SB, Wang L. In situ measurement of transport between subchondral bone and articular cartilage. *J Orthop Res* 2009;27(10):1347–1352.

Park Y, Sugimoto M, Watrin A, Chiquet M, Hunziker EB. BMP-2 induces the expression of chondrocyte-specific genes in bovine synovium-derived progenitor cells cultured in three-dimensional alginate hydrogel. *Osteoarthritis Cartilage* 2005;13(6):527–536.

Patrascu JM, Krüger JP, Böss HG, Ketzmar AK, Freymann U, Sittinger M, Notter M, Endres M, Kaps C. Polyglycolic acid-hyaluronan scaffolds loaded with bone marrow-derived mesenchymal stem cells show chondrogenic differentiation in vitro and cartilage repair in the rabbit model. *J Biomed Mater Res B Appl Biomater* 2013.

Pennesi G, Scaglione S, Giannoni P, Quarto R.. Regulatory influence of scaffolds on cell behavior: How cells decode biomaterials. *Curr Pharm Biotechnol* 2011;12(2):151–159.

Perka C, Schultz O, Lindenhayn K, Spitzer RS, Muschik M, Sittinger M, Burmester GR. Joint cartilage repair with transplantation of embryonic chondrocytes embedded in collagen-fibrin matrices. *Clin Exp Rheumatol* 2000;18(1):13–22.

Perka C, Spitzer RS, Lindenhayn K, Sittinger M, Schultz O. Matrix-mixed culture: new methodology for chondrocyte culture and preparation of cartilage transplants. *J Biomed Mater Res* 2000;49(3):305–311.

Ponticiello MS, Schinagl RM, Kadiyala S, Barry FP. Gelatin-based resorbable sponge as a carrier matrix for human mesenchymal stem cells in cartilage regeneration therapy. *J Biomed Mater Res* 2000;52(2):246–255.

Pound JC, Green DW, Chaudhuri JB, Mann S, Roach HI, Oreffo RO. Strategies to promote chondrogenesis and osteogenesis from human bone marrow cells and articular chondrocytes encapsulated in polysaccharide templates. *Tissue Eng* 2006;12(10):2789–2799.

Raghunath J, Rollo J, Sales KM, Butler PE, Seifalian AM. Biomaterials and scaffold design: key to tissue-engineering cartilage. *Biotechnol Appl Biochem* 2007; 46(Pt 2):73–84.

Rahfoth B, Weisser J, Sternkopf F, Aigner T, von der Mark K, Bräuer R. Transplantation of allograft chondrocytes embedded in agarose gel into cartilage defects of rabbits. *Osteoarthritis Cartilage* 1998;6(1):50–65.

Ramage L, Nuki G, Salter DM. Signalling cascades in mechanotransduction: Cell-matrix interactions and mechanical loading. *Scand J Med Sci Sports* 2009;19(4):457–469.

Rampichová M, Buzgo M, Křížková B, Prosecká E, Pouzar M, Štrajtová L. Injectable hydrogel functionalised with thrombocyte-rich solution and microparticles for accelerated cartilage regeneration. *Acta Chir Orthop Traumatol Cech* 2013;80(1):82–88.

Reitinger S, Lepperdinger G. Hyaluronan, a ready choice to fuel regeneration: A mini-review. *Gerontology* 2013;59(1):71–76.

Responte DJ, Natoli RM, Athanasiou KA. Collagens of articular cartilage: Structure, function, and importance in tissue engineering. *Crit Rev Biomed Eng* 2007;35(5):363–411.

Richardson TP, Peters MC, Ennett AB, Mooney DJ. Polymeric system for dual growth factor delivery. *Nat Biotechnol* 2001;19(11):1029–1034.

Roberts JJ, Earnshaw A, Ferguson VL, Bryant SJ. Comparative study of the viscoelastic mechanical behavior of agarose and poly(ethylene glycol) hydrogels. *J Biomed Mater Res B Appl Biomater* 2011;99(1):158–169.

Sadr N, Pippenger BE, Scherberich A, Wendt D, Mantero S, Martin I, Papadimitropoulos A. Enhancing the biological performance of synthetic polymeric materials by decoration with engineered, decellularized extracellular matrix. *Biomaterials* 2012;33(20):5085–5093.

Safran MR, Kim H, Zaffagnini S. The use of scaffolds in the management of articular cartilage injury. *J Am Acad Orthop Surg* 2008;16(6):306–311.

Salisbury Palomares KT, Gerstenfeld LC, Wigner NA, Lenburg ME, Einhorn TA, Morgan EF. Transcriptional profiling and biochemical analysis of mechanically induced cartilaginous tissues in a rat model. *Arthritis Rheum* 2010;62(4):1108–1118.

Santos E, Hernández RM, Pedraz JL, Orive G. Novel advances in the design of three-dimensional bio-scaffolds to control cell fate: Translation from 2D to 3D. *Trends Biotechnol* 2012;30(6):331–341.

Sarasam A, Madihally SV. Characterization of chitosan-polycaprolactone blends for tissue engineering applications. *Biomaterials* 2005;26(27):5500–5508.

Sargeant TD, Desai AP, Banerjee S, Agawu A, Stopek JB. An in situ forming collagen-PEG hydrogel for tissue regeneration. *Acta Biomater* 2012;8(1):124–132.

Schagemann JC, Chung HW, Mrosek EH, Stone JJ, Fitzsimmons JS, O'Driscoll SW, Reinholz GG. Poly-epsilon-caprolactone/gel hybrid scaffolds for cartilage tissue engineering. *J Biomed Mater Res A* 2010;93(2):454–463.

Schenke-Layland K. Non-invasive multiphoton imaging of extracellular matrix structures. *J Biophotonics* 2008;1(6):451–462.

Schinagl RM, Gurskis D, Chen AC, Sah RL. Depth-dependent confined compression modulus of full-thickness bovine articular cartilage. *J Orthop Res* 1997; 15(4):499–506.

Schmoekel H, Schense JC, Weber FE, Grätz KW, Gnägi D, Müller R, Hubbell JA. Bone healing in the rat and dog with nonglycosylated BMP-2 demonstrating low solubility in fibrin matrices. *J Orthop Res* 2004;22(2):376–381.

Scott JE. Supramolecular organization of extracellular matrix glycosaminoglycans, in vitro and in the tissues. *FASEB J* 1992;6(9):2639–2645.

Seth A, Chung YG, Gil ES, Tu D, Franck D, Di Vizio D, Adam RM, Kaplan DL, Estrada CR Jr, Mauney JR.The performance of silk scaffolds in a rat model of augmentation cystoplasty. *Biomaterials* 2013 Jul;34(20):4758–4765.

Shao X, Goh JC, Hutmacher DW, Lee EH, Zigang G. Repair of large articular osteochondral defects using hybrid scaffolds and bone marrow-derived mesenchymal stem cells in a rabbit model. *Tissue Eng* 2006;12(6):1539–1551.

Shao XX, Hutmacher DW, Ho ST, Goh JC, Lee EH. Evaluation of a hybrid scaffold/cell construct in repair of high-load-bearing osteochondral defects in rabbits. *Biomaterials* 2006;27(7):1071–1080.

Sherwood JK, Riley SL, Palazzolo R, Brown SC, Monkhouse DC, Coates M, Griffith LG, Landeen LK, Ratcliffe A. A three-dimensional osteochondral composite scaffold for articular cartilage repair. *Biomaterials* 2002;23(24): 4739–4751.

Shiedlin A, Bigelow R, Christopher W, Arbabi S, Yang L, Maier RV, Wainwright N, Childs A, Miller RJ. Evaluation of hyaluronan from different sources: Streptococcus zooepidemicus, rooster comb, bovine vitreous, and human umbilical cord. *Biomacromolecules* 2004;5(6):2122–2127.

Singer AJ, Hollander JE, Valentine SM, Turque TW, McCuskey CF, Quinn JV. Prospective, randomized, controlled trial of tissue adhesive (2-octylcyanoacrylate) vs standard wound closure techniques for laceration repair. Stony Brook Octylcyanoacrylate Study Group. *Acad Emerg Med* 1998;5(2):94–99.

Singh P, Schwarzbauer JE. Fibronectin and stem cell differentiation - lessons from chondrogenesis. *J Cell Sci* 2012;125(Pt 16):3703–3712.

Solchaga LA, Goldberg VM, Caplan AI. Cartilage regeneration using principles of tissue engineering. *Clin Orthop Relat Res* 2001;(391 Suppl):S161–170.

Spiller KL, Maher SA, Lowman AM. Hydrogels for the repair of articular cartilage defects. *Tissue Eng Part B Rev* 2011;17(4):281–299.

Stevens MM, Marini RP, Martin I, Langer R, Prasad Shastri V. FGF-2 enhances TGF-beta1-induced periosteal chondrogenesis. *J Orthop Res* 2004;22(5):1114–1119.

Stevens MM, Qanadilo HF, Langer R, Prasad Shastri V. A rapid-curing alginate gel system: Utility in periosteum-derived cartilage tissue engineering. *Biomaterials* 2004;25(5):887–894.

Stuart MA, Huck WT, Genzer J, Müller M, Ober C, Stamm M, Sukhorukov GB, Szleifer I, Tsukruk VV, Urban M, Winnik F, Zauscher S, Luzinov I, Minko S. Emerging applications of stimuli-responsive polymer materials. *Nat Mater.* 2010;9(2):101–113.

Suciati T, Howard D, Barry J, Everitt NM, Shakesheff KM, Rose FR. Zonal release of proteins within tissue engineering scaffolds. *J Mater Sci Mater Med* 2006;17(11):1049–1056.

Suh JK, Matthew HW. Application of chitosan-based polysaccharide biomaterials in cartilage tissue engineering: A review. *Biomaterials* 2000;21(24):2589–2598.

Sun W, Neuzil P, Kustandi TS, Oh S, Samper VD. The nature of the gecko lizard adhesive force. *Biophys J* 2005;89(2):L14–17.

Tabata Y. Tissue regeneration based on growth factor release. *Tissue Eng* 2003;9 (Suppl 1):S5–S15.

Tada Y, Suzuki T, Takezawa T, Nomoto Y, Kobayashi K, Nakamura T, Omori K. Regeneration of tracheal epithelium utilizing a novel bipotential collagen scaffold. *Ann Otol Rhinol Laryngol* 2008;117(5):359–365.

Takahashi I, Nuckolls GH, Takahashi K, Tanaka O, Semba I, Dashner R, Shum L, Slavkin HC. Compressive force promotes sox9, type II collagen and aggrecan and inhibits IL-1beta expression resulting in chondrogenesis in mouse embryonic limb bud mesenchymal cells. *J Cell Sci* 1998;111 (Pt 14):2067–2076.

Thiede RM, Lu Y, Markel MD. A review of the treatment methods for cartilage defects. *Vet Comp Orthop Traumatol* 2012;25(4):263–272.

Tripathi A, Kumar A.Multi-featured macroporous agarose-alginate cryogel: Synthesis and characterization for bioengineering applications. *Macromol Biosci* 2011;11(1):22–35.

Tseng HJ, Tsou TL, Wang HJ, Hsu SH. Characterization of chitosan-gelatin scaffolds for dermal tissue engineering. *J Tissue Eng Regen Med* 2013;7(1):20–31.

Vasita R, Shanmugam I K, Katt DS. Improved biomaterials for tissue engineering applications: Surface modification of polymers. *Curr Top Med Chem* 2008;8(4):341–353.

Verma, G.P. Cartilage and Bone. Fundamentals of Histology. New Dehli: New Age International Limited, 2001.

Vinatier C, Mrugala D, Jorgensen C, Guicheux J, Noël D. Cartilage engineering: A crucial combination of cells, biomaterials and biofactors. *Trends Biotechnol* 2009 May;27(5):307–314.

Vogel G. Trachea transplants test the limits. *Science* 2013;340(6130):266–268.

Volpi N, Schiller J, Stern R, Soltés L. Role, metabolism, chemical modifications and applications of hyaluronan. *Curr Med Chem* 2009;16(14):1718–1745.

Walles T, Giere B, Hofmann M, Schanz J, Hofmann F, Mertsching H, Macchiarini P. Experimental generation of a tissue-engineered functional and vascularized trachea. *J Thorac Cardiovasc Surg* 2004;128(6):900–906.

Wan LQ, Jiang J, Miller DE, Guo XE, Mow VC, Lu HH. Matrix deposition modulates the viscoelastic shear properties of hydrogel-based cartilage grafts. *Tissue Eng Part A* 2011;17(7–8):1111–1122.

Wang Y, Kim HJ, Vunjak-Novakovic G, Kaplan DL. Stem cell-based tissue engineering with silk biomaterials. *Biomaterials* 2006;27(36):6064–6082.

Weigel PH, Hascall VC, Tammi M. Hyaluronan synthases. *J Biol Chem* 1997;272(22):13997–14000.

Wescoe KE, Schugar RC, Chu CR, Deasy BM. The role of the biochemical and biophysical environment in chondrogenic stem cell differentiation assays and cartilage tissue engineering. *Cell Biochem Biophys* 2008;52(2):85–102.

Widner B, Behr R, Von Dollen S, Tang M, Heu T, Sloma A, Sternberg D, Deangelis PL, Weigel PH, Brown S. Hyaluronic acid production in Bacillus subtilis. *Appl Environ Microbiol* 2005;71(7):3747–3752.

Williamson AK, Chen AC, Masuda K, Thonar EJ, Sah RL. Tensile mechanical properties of bovine articular cartilage: Variations with growth and relationships to collagen network components. *J Orthop Res* 2003;21(5):872–880.

Woods A, Wang G, Beier F. Regulation of chondrocyte differentiation by the actin cytoskeleton and adhesive interactions. *J Cell Physiol* 2007;213(1):1–8.

Wu X, Ren J, Li J. Fibrin glue as the cell-delivery vehicle for mesenchymal stromal cells in regenerative medicine. *Cytotherapy* 2012;14(5):555–562.

Xu XL, Lou J, Tang T, Ng KW, Zhang J, Yu C, Dai K. Evaluation of different scaffolds for BMP-2 genetic orthopedic tissue engineering. *J Biomed Mater Res B Appl Biomater* 2005;75(2):289–303.

Yang Q, Peng J, Guo Q, Huang J, Zhang L, Yao J, Yang F, Wang S, Xu W, Wang A, Lu S. A cartilage ECM-derived 3-D porous acellular matrix scaffold for in vivo cartilage tissue engineering with PKH26-labeled chondrogenic bone marrow-derived mesenchymal stem cells. *Biomaterials* 2008;29(15):2378–2387.

Yang Q, Peng J, Lu SB, Guo QY, Zhao B, Zhang L, Wang AY, Xu WJ, Xia Q, Ma XL, Hu YC, Xu BS. Evaluation of an extracellular matrix-derived acellular biphasic scaffold/cell construct in the repair of a large articular high-load-bearing osteochondral defect in a canine model. *Chin Med J (Engl)* 2011;124(23):3930–3938.

Yannas IV, Tzeranis DS, Harley BA, So PT. Biologically active collagen-based scaffolds: Advances in processing and characterization. *Philos Trans A Math Phys Eng Sci* 2010;368(1917):2123–2139.

Zhang C, Sangaj N, Hwang Y, Phadke A, Chang CW, Varghese S. Oligo(trimethylene carbonate)-poly(ethylene glycol)-oligo(trimethylene carbonate) triblock-based hydrogels for cartilage tissue engineering. *Acta Biomater* 2011; 7(9):3362–3369.

Zhang X, Reagan MR, Kaplan DL. Electrospun silk biomaterial scaffolds for regenerative medicine. *Adv Drug Deliv Rev* 2009;61(12):988–1006.

Zhang Z, McCaffery JM, Spencer RG, Francomano CA. Growth and integration of neocartilage with native cartilage *in vitro*. *J Orthop Res* 2005;23(2):433–439.

Zhao Q, Yin J, Feng X, Shi Z, Ge Z, Jin Z. A biocompatible chitosan composite containing phosphotungstic acid modified single-walled carbon nanotubes. *J Nanosci Nanotechnol* 2010;10(11):7126–7129.

Zhou J, Xu C, Wu G, Cao X, Zhang L, Zhai Z, Zheng Z, Chen X, Wang Y. In vitro generation of osteochondral differentiation of human marrow mesenchymal stem cells in novel collagen-hydroxyapatite layered scaffolds. *Acta Biomater* 2011;7(11):3999–4006.

Zippel N, Schulze M, Tobiasch E. Biomaterials and mesenchymal stem cells for regenerative medicine. *Recent Pat Biotechnol* 2010;4(1):1–22.

Chapter 4

Native Polymer-based 3D Substitutes for Bone Repair

Yan Huang PH.D.[1], *Kerong Dai M.D.*[1,2],
Xiaoling Zhang PH.D.[1,2,*]

[1]*Key Laboratory of Stem Cell Biology, Institute of Health Sciences,
Shanghai Institutes for Biological Sciences (SIBS), Chinese
Academy of Sciences (CAS) and Shanghai Jiao Tong
University School of Medicine (SJTUSM), P. R. China.*
[2]*Shanghai Key Laboratory of Orthopaedic Implant, Department
of Orthopaedic Surgery, Shanghai Ninth People's Hospital,
Shanghai Jiao Tong University School of Medicine (SJTUSM),
P. R. China.*

1. Introduction

The bone is a rigid and remarkable organ, which plays key roles in critical functions in human physiology, such as protection, movement and support of other organs of the body, mineral storage, homeostasis, blood pH regulation, blood production, progenitor cell (mesenchymal and hemopoietic) housing, etc. (Porter *et al.*, 2009). The abnormal bone shape or segmental bone loss associated with traumatic injuries, cancer treatment, selective surgery and congenital abnormalities, have, in many cases, presented insurmountable challenges to the treatment for bone repair. Significant bone defects or post-traumatic

---

*Corresponding author.

complications may require bone grafting in order to fill the defect. Surgeries are considered as the most popular and efficient treatment of bone defects. These surgeries include the Ilizarov method, bone transport and bone graft transplant, which fill spaces and provide support, and may enhance the biological repair of the defect.

An ideal bone graft substitute is often described with good biological properties of osteoinductivity, osteoconductivity and osteogenicity. The bone graft scaffold with such properties involves the stimulation of osteoprogenitor cells to differentiate into osteoblasts, and allows the colonization and new bone ingrowth as a result of its three-dimensional structure. It will not only serve as a support for currently existing osteoblasts but will also trigger the formation of new osteoblasts, theoretically promoting faster integration of the graft. These illustrious characteristics of the ideal graft is mainly determined not only by the porosity properties of the scaffold but also by its chemical and physical properties of the substrate that promote attachment, proliferation, migration, and phenotypic expression of bone cells leading to formation of new bones in direct apposition to the biomaterial (Burchardt, 1983; Giannoudis *et al.*, 2005; LeGeros, 2002).

Previous clinical bone graft materials are autogenous bone, allograft bone and specially treated xenograft bone. However, these grafting bones had numerous drawbacks, such as limited donor bone supply, donor-site morbidity and pain, anatomical and structural complications, high cost, issues of processing, sterilisation and storage, and most importantly potential immunogenic response by the host to the foreign tissue and disease transmission. These drawbacks have posed many problems and inhibited bone restoration (Finkemeier, 2002). With the advancement in tissue engineering technology, the traditional treatment modalities for bone defects have been changed. One strategy to overcome these problems is to develop the bone grafting substitutes based on biomaterials.

Because of their similar composition to human skeleton, bone substitutes made of calcium phosphate ceramics and glasses have been widely accepted as bone implants. However, these substitutes have only osteoconductive and not osteoinductive properties which still limits their wide applications. In scaffold-based bone repair strategies, a key component is the scaffold that serves as a template for cell interactions and for the formation of the extracellular matrix (ECM) providing structural support to the newly formed tissue. A temporary 3D scaffold mimicking the physiological functions of the ECM is critical to preserve the ability of cells to differentiate into their native phenotypes and to constitute a structural template to fill the tissue lesion (Leong *et al.*, 2003; Karageorgiou and Kaplan, 2005). Obviously, the inorganic ceramics or glasses cannot meet these requirements, and, as such, the polymer-based substitutes might be a better choice.

In recent years, due to its biocompatible and biodegradable behavior, natural polymer-based composites have been focused with increasing attention than the synthetic ones for bone tissue engineering applications. The native polymers are the natural-origin biopolymers including proteins (soy, collagen, fibrin gels, silk), polysaccharides (starch, alginate, chitin/chitosan, hylauronic acid derivatives, celluloses) and some polyesters (polyhydroxyalkanoates, polyhydroxybutyrates) (Rezwan *et al.*, 2006), which often possess highly organized structures and may contain an extracellular substance ligand that is necessary to bind with cell receptors and guide cells to grow at various stages of development (Cheung *et al.*, 2007). Thus, by first comparing with other synthetic ones, native polymer-based substitutes will offer a wider range of advantages for bone tissue engineering application in, for example activating biological signaling, improving cell adhesion and responsive degradation and increasing bone remodeling. Second, environmental considerations made the use of many natural materials attractive because of their biodegradability, low toxicity and low disposal costs. Finally, the use of natural materials is economical because of the high price of synthetic polymer matrices and the low prices of such natural products. Therefore, we here focus on and review some extensively applied natural polymer scaffolds, hoping to help highlight theoretical principles and their potential in creating more suitable substitutes for bone repair.

2. Proteins

The employment of proteins in repair may attempt to mimic the functions of extracellular matrix (ECM), which is the optimized milieu that nature has been developing to maintain homeostasis and to direct tissue development. The biomolecules, such as proteins, preserve a potential space for new tissue development after cell seeding, which would help to overcome one of the main drawbacks in the use of synthetic materials — lacking cell recognition signals (Mano *et al.*, 2007).

2.1. *Collagen*

Collagen is a group of naturally occurring proteins found in animals, especially in the flesh and connective tissues of vertebrates (Muller, 2003). It is the main component of connective tissue, making up about 25% to 35% of the whole-body protein content, with a high content of glycine (Gly) (near 33%) and of imino acids (near 20%) (Di Lullo *et al.*, 2002). In the form of elongated fibrils, collagen is found abundantly in fibrous tissues and many tissues (such as ligament, cornea, skin, bone, cartilage, tendons, blood vessels, teeth, the gut, and intervertebral disc), where it provides the principal structural and mechanical support (Lee *et al.*, 2001). More than 20 genetically distinct forms have been identified, and its characteristics such as high mechanical strength, ability of cross-linking, good biocompatibility, biodegradability and low antigenicity make collagen a valuable material for bone tissue engineering applications (Weinburg and Bell, 1986).

Bone tissues are mainly constructed from type I with a small quantity of type V collagen, forming a framework that anchors hydroxyapatite (HA) crystals. Both organic and mineral phases compose a closely interwoven and highly complex but ordered composite. As the major structural component of bone extracellular matrix, type I collagen had been proven to induce a multipotential mesenchymal cell line to produce an extracellular matrix capable of osteoinductive activity *in vivo* and of stimulating alkaline phosphatase activity *in vitro* (Shi *et al.*,

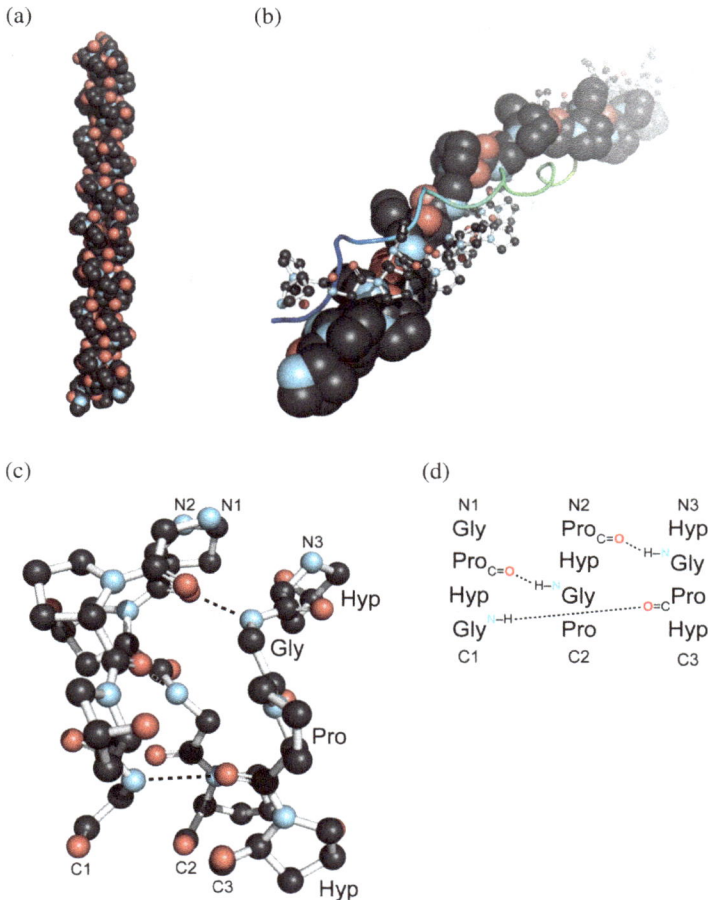

Fig. 1. Overview of the collagen triple helix. (a) First high-resolution crystal structure of a collagen triple helix (b) View down the axis of a triple helix with the three strands depicted in space-filling, ball-and-stick, and ribbon representation. (c) Ball-and-stick image of a segment of collagen triple helix. (d) Stagger of the three strands in the segment in panel c. (Shoulders and Raines, 2009).

1996). Additionally, collagen I could modulate the Ca^{2+} — and cAMP-signaling pathways in osteoblasts by altering PKC activity, which may influence bone remodeling processes (Green *et al.*, 1995). Mizuno M. found bone marrow stromal cells differentiated into osteoblasts on type I collagen matrix *in vitro*, but types II, III, and V collagens did not possess this activity, which implied that type I collagen matrix offered a suitable environment for the induction of osteoblastic differentiation *in vitro* and osteogenesis *in vivo* (Mizuno *et al.*, 1997).

Collagen sponges have also been reported to promote cell and tissue attachment and growth (O'Brien *et al.*, 2005) and to enhance bone formation by promoting the differentiation of osteoblasts (Takahashi *et al.*, 2005). A study performed by seeding MSCs into collagen gels followed by implanting osteochondral defects in rabbits led to both bone and hyaline cartilage formation (Wakitani *et al.*, 1994). Rocha LB *et al.* (2002) implanted a collagen matrix, with a sponge-like structure and heterogeneous pore size, into the surgically created bone defects in rat tibias, and then found a preliminary evaluation of the osteoconductiveness of the matrices, which suggested that the anionic collagen matrices as promising alternatives for bone defects treatment. Tampieri *et al.* (2008) developed the composite osteochondral scaffolds organized in different integrated layers consisting of a lower layer of the developed biomineralized collagen, an upper layer of hyaluronic acid-charged collagen, and an intermediate layer of the same nature, which are biomimetic of an osteochondral tissue were shown to have the capacity to differentially support cartilage and bone tissue generation.

However, the low biomechanical stiffness and high degradation rate of collagen, which always rapidly led to the loss of mechanical properties, made use of collagen as bone scaffolds difficult (Angele *et al.*, 2004). Many attempts have been made to overcome this problem, for example, by adding mineral crystals or by combining with either inorganic bioceramics, natural (Daamen *et al.*, 2003) or synthetic polymers (Rao *et al.*, 1994), or by applying various cross-linking methods (Glowacki and Mizuno, 2008).

From a biological perspective, polymers and bioceramics have been combined to fabricate biomimetic scaffolds for bone tissue engineering, because the native bone is in similar combination of naturally occurring polymers and biological apatite. Moreover, the composite scaffolds are ideal candidates because they can gradually degrade while new tissue is formed *in vivo*. The inorganic/ceramic materials, as mentioned above, have good osteoconductivity and have been studied for mineralized tissue engineering, but show drawbacks such as poor processability into highly porous structures and brittleness. In contrast, polymers offer great design flexibility because their composition and structure can be tailored to specific needs. Additionally, a series of processing techniques, such as solvent casting (Murphy *et al.*, 2002; Yoon and Park, 2001), phase inversion (Karp *et al.*, 2003), fiber bonding (Kim and Mooney, 1998), melt-based technologies (Malafaya *et al.*, 2003), high-pressure-based methods (Shea *et al.*, 2000), freeze-drying (Mao *et al.*, 2003) and rapid prototyping techniques (Taboas *et al.*, 2003) have been developed with the aim of producing scaffolds over the years. Therefore, today, lots of porous collagen scaffolds with ceramic particles are being formulated and studied by a number of scientists and engineers for bone tissue engineering purposes (Takahashi *et al.*, 2005; Shibata *et al.*, 2005).

It was reported that matrix made by cross-linked collagen fibers with modified HA implanted in cranial defects in rats demonstrated good biocompatibility and improved osteoconductivity with respect to the application of the two materials alone (Liu *et al.*, 1999). A three-dimensional porous biomimetic hydroxyapatite/collagen composites had been developed recently, which well supported the adhesion, proliferation, viability and differentiation of MG63 osteoblast-like cells (Thomas *et al.*, 2007). Electrospun nanofibrous biocomposite scaffolds of type I collagen and Hap nanoparticles were developed by Thomas *et al.* (2007). The incorporation of inorganic nanoparticles resulted in significant increase of tensile strength and modulus. Moreover, the nanocomposite matrices showed *in vitro* bioactivity inducing rapid formation of bone-like apatite minerals on their surfaces when incubated in simulated body fluid (SBF), and the osteoblastic cells showed favorable growth on the nanocomposite scaffolds and their alkaline phosphatase (ALP) activity was significantly higher than that on the pure collagen scaffolds (Kim *et al.*, 2006).

In addition, to improve the bioactivity of natural polymers, bioactive glass was also introduced in different types of scaffolds. It was proven that a bioglass particle–collagen hydrogel composite exhibited *in vitro* osteoconductivity properties, whereas the individual components alone did not (Pohunkova and Adam, 1995). This synergistic effect was attributed to the ability of the protein to bind calcium ions, which can further associate with silicic acid to form a bioactive layer. Andrade *et al.* (2007) produced collagen fibers coated with a bioactive glass, which showed *in vitro* bioactivity, improving the calcium, the phosphate precipitation on the collagen surface when immersed into SBF solution and the collagen secretion by osteoblasts which is an important factor for bone healing. Furthermore, the bioactive glassfiber–collagen nanocomposite exhibited an active induction of apatite minerals in contact with SBF, and human osteoblastic cells grew favorably and expressed significantly higher ALP levels than those on collagen alone, which both showed an excellent bioactivity of the composites *in vitro* (Andrade *et al.*, 2007).

Furthermore, as mentioned above, since collagen has a rapid adsorption rate and possesses weak mechanical strength, some other polymer additions including synthetic ones are often incorporated for enhancing the mechanical properties of the material constructs. Besides, polymers by themselves lack cell recognition signals (Kim and Mooney, 1998), and the addition of collagen provides the necessary binding sites for cell-material interactions (Teo and Ramakrishna, 2006). Liao *et al.* (2008) selected two types of polymers, namely collagen and poly (lactic-co-glycolic acid) (PLGA), natural and synthetic of its kind respectively, to electrospin into nanofibrous scaffolds and then observed that the formation of bone-like apatite into collagen was relatively abundant and significantly more uniform than PLGA, which showed another selection for the material of bone scaffolding system.

2.2. Silk

As fibrous proteins, silk are spun by a variety of species, including hymenoptera (bees, wasps, and ants), silverfish, mayflies, thrips, leafhoppers, beetles, lacewings, fleas, flies, midges and various arachnids such as spiders (Sutherland *et al.*, 2010). Silk proteins are usually produced within specialized glands after biosynthesis in epithelial cells, followed by secretion into the lumen of these glands where the proteins are stored prior to spinning into fibers. The wide composition, structure and properties of silks depend on their specific source. Each of these different silks has a different amino acid composition and exhibits mechanical properties tailored to their specific functions, including cocoons to protect eggs or larvae, lines for prey capture, lifeline support (dragline), and capture nets able to withstand high impacts and trap insects. The most notable characterized silks are from the domesticated silkworm and from spiders, which have been studied extensively. The protein fiber of silk is composed mainly of fibroin, with a predominance of alanine, glycine and serine, which is characterized by a highly repetitive primary sequence that leads to significant homogeneity in secondary structure with collagens (Altman *et al.*, 2003). Compared with other globular proteins, silk fibers exhibit enhanced environmental stability and are insoluble in many organic and aqueous solvents, due to their extensive hydrogen bonding, the hydrophobic nature and the significant crystallinity. Besides this, silks are also flexible and able to resist tensile and compressive forces with good mechanical properties similar to some biological tissues. Moreover, they can be processed in aqueous solutions into gels, sponges, powder and membranes, and easily modified with surface decorations.

Since the primary silk-like material used in biomedical applications as sutures (Vepari and Kaplan, 2007), silk fibers have proven to be effective in many clinical applications and also as attractive biomaterials for bone scaffolds because of their outstanding characters described above. Meinel L.'s group found that the RGD-silk scaffolds are particularly suitable for autologous bone tissue engineering, presumably because of their stable macroporous structure, tailorable mechanical properties matching those of native bone, and slow degradation (Meinel *et al.*, 2004). Sofia S. *et al.* (2001) reported on the covalent decoration of silk films involved in the induction of bone formation. Wang X. *et al.* (2005) developed a completely aqueous, stepwise deposition process to assemble nanoscale thin film coatings with the Bombyx mori silk fibroin, which were stable under physiological conditions and supported human bone marrow stem cell adhesion, growth, and differentiation. Silk fibroin fiber scaffolds

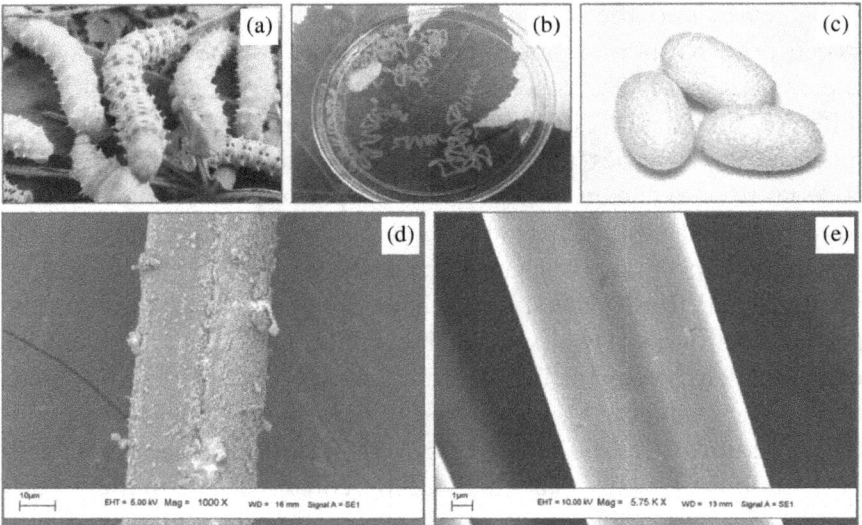

Fig. 2. Representative images of (a) silkworms, (b) dissected silk glands, (c) cocoons, (d) pristine cocoon fiber (sericin encased fibroin) and (e) purified cocoon fiber (sericin-free fibroin fiber). (Kasoju and Bora, 2012)

containing bone morphogenetic protein 2 (BMP-2) prepared via electrospinning were used for *in vitro* bone formation from human bone marrow-derived mesenchymal stem cells (hMSCs), and the electrospun silk-fibroin-based scaffolds showed highest calcium deposition and upregulation of BMP-2 transcript levels when compared with the other systems (Li *et al.*, 2006). In the *in vivo* test for evaluation of bone regenerative efficacy, the silk fibroin (SF) nanofiber membranes were used to impair the calvarial defect of rabbits, then a complete bony union across the defects was observed after eight weeks and the defect had completely healed with a new bone at 12 weeks (Kim *et al.*, 2005). Meinel L. *et al.* (2005) explored the use of novel porous silk fibroin scaffolds as templates for the engineering of bone tissues, whose implantation into calvarial critical size defects in mice demonstrated the capacity of these systems to induce advanced bone formation within five weeks. In critical sized mid-femoral segmental defects in nude rats, it was found that the newly formed bone almost bridged defects after eight weeks implantation of the silk scaffolds with hMSCs, which also resulted in greater maximal load and torque (Meinel *et al.*, 2006). While the bulk of reports have involved the use of silkworm silks, there is some attention directed at spider silks. The recombinant spider silk with RGD encoded into the protein support enhanced the differentiation of hMSCs to osteogenic cells, which illustrated the potential of bioengineer spider silk proteins into new biomaterial matrixes and highlighted the importance of subtle differences in silk sources in terms of tissue-specific outcomes (Bini *et al.*, 2006).

Similar to collagen, silks always use composites with other materials to advance their biological properties for better bone repair. The silk sericin-immobilized titanium surfaces had been proven to promote osteoblast cells' adhesion, proliferation, and alkaline phosphatase activity, showing potential applications combating biomaterial-centered infection and promoting osseointegration (Zhang *et al.*, 2008). Yan L.P. *et al.* (2013) indicated that the silk/nano-CaP scaffolds prepared through a combination of salt-leaching/lyophilization approaches could be suitable candidates for bone tissue engineering applications, and a novel hydroxyapatite/regenerated silk fibroin scaffold also displayed a positive effect on osteoinductivity and osteoconductivity (Jiang *et al.*, 2013). Biomimetic bone substitutes of collagen-silk fibroin/hydroxyapatite were synthesized via a bi-template-induced co-assembly strategy, and *in vitro* stimulation on BMSCs to differentiate into the osteoblast cell lineage and the good biocompatibility and strong ability of the new bone formation in comparison with the control of single-template material *in vivo* were demonstrated (Wang *et al.*, 2012).

However, silk fibroin exhibits low immunogenicity and elicits a foreign body response following implantation *in vivo*, comparable to the most popular synthetic materials in use today as biomaterials (Horan *et al.*, 2005). In rare cases a granuloma may form as a result of an abandoned phagocytic response to silk by macrophages and giant body cells (Altman *et al.*, 2003). To solve these problems would help silk to advance further in the way of application in bone tissue engineering.

2.3. *Zein*

Zein is a class of prolamine protein from corn, which has been considered as one of the best understood plant proteins and has a variety of industrial and food uses. Pure zein is clear, odorless, tasteless, hard, water-insoluble, and edible, making it invaluable in processed foods and pharmaceuticals, and used as adhesives and biodegradable plastics, chewing gum, coating for candy, nuts, fruit, pills, other encapsulated foods, and drug-delivery matrices (Shukla and Cheryan, 2001).

Recently, Wang J.Y.'s group has developed a three-dimensional (3D) zein scaffold and studied its porous structure, mechanical properties, biocompatibility, and biodegradation, and all the results demonstrated that it would be used as a promising biomaterial for the development of tissue (Dong *et al.*, 2004; Liu *et al.*, 2005; Wang *et al.*, 2007). As a 3D porous scaffold, zein was also proven to provide a comparatively good biocompatibility, which allowed rat mesenchymal stem cells (MSCs) to adhere, proliferate, and undergo osteogenic differentiation in the presence of dexamethasone. As the following research for bone repair, our group investigated the effect of zein/inorganic composite on the physical and biological properties of porous zein scaffolds. Using scanning electron microscopy, it was established that the morphology of pores located on the surface and within the porous scaffolds showed equally good pore interconnectivity with zein. From the *in vitro* test with human bone marrow stroma cells, the osteoblastic differentiation on the surface of the HA-coated zein scaffold was increased, as expressed by the alkaline phosphatase activity and reverse transcription-polymerase chain reaction analysis for marker genes, which suggested that the HA-coated zein scaffold might be the optimal biomaterial for bone tissue engineering (Qu *et al.*, 2008). Moreover, we investigated ectopic bone formation in nude mice, and also implanted the composite scaffolds into the radius defects of rabbits and assessed whether they could be helpful in the repair of critical-sized bone defects. The results showed that the complexes of zein scaffolds and rabbit MSCs could undergo ectopic bone formation in the thigh muscle pouches of nude mice, and the complexes could lead to the repair of critical-sized radius defects in rabbits accompanied with blood vessels' formation, which clearly demonstrates promise for the treatment of bone defects (Tu *et al.*, 2009).

3. Polysaccharides

Polysaccharides are long carbohydrate molecules of monosaccharide units joined together by glycosidic bonds, ranging in structure from linear to highly branched. They consist of a large variety of polymers biosynthesized not only in wood, plants, algae and marine crustaceans, but also produced by bacteria and fungi. Polysaccharides may be structural components, provide carbon and energy reserves for cells or may be excreted as plant exudates or as microbial exopolysaccharides. They are often quite heterogeneous, containing slight modifications of the repeating units. Depending on the structure, these macromolecules can have distinct properties from their monosaccharide building blocks. These excellent properties, such as non-toxicity,

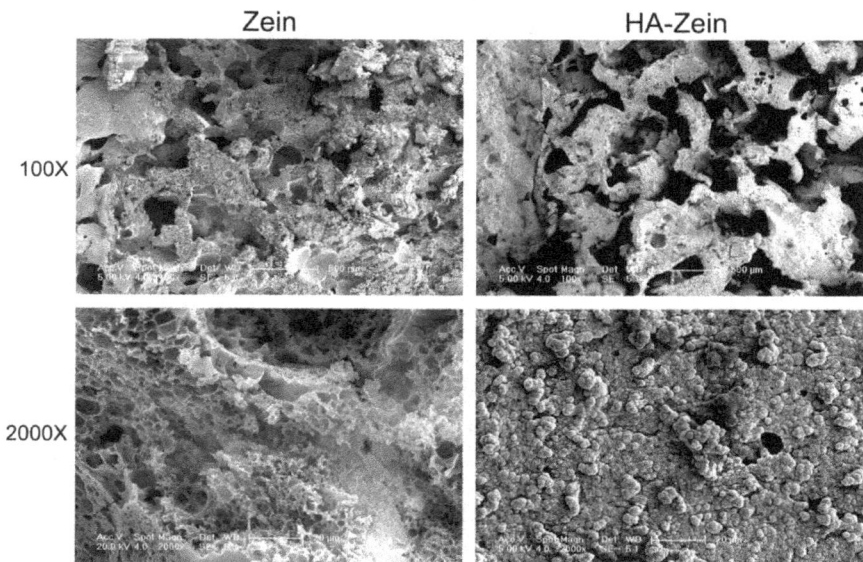

Fig. 3. SEM evaluation of the zein scaffold microstructure. (Qu *et al.*, 2008)

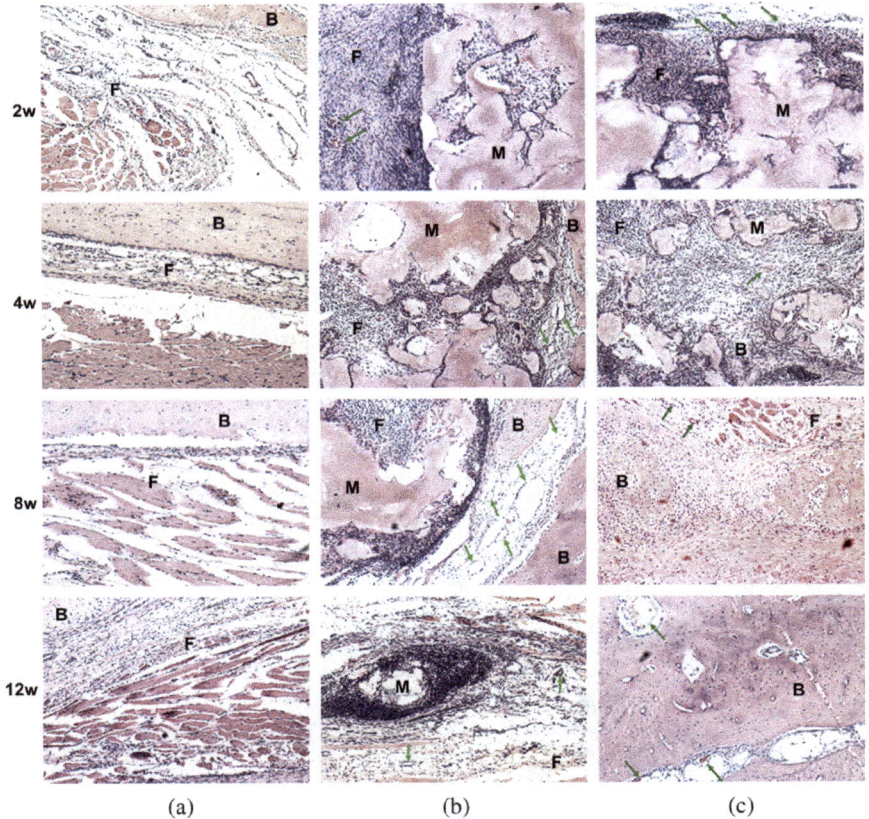

Fig. 4.   H&E staining (100x) of rabbit radius bone defects. (a) Bone defect with no implantation; (b) bone defect with implantation of zein; (c) bone defect with implantation of zein and rabbit MSCs. "M" means material, "B" means bone, "F" means fibrous tissue, and the green arrows indicate blood vessels. (Tu *et al.*, 2009)

water solubility or high swelling ability by simple chemical modification, stability to pH variations, and a broad variety of chemical structures etc., make them applicable in a wide range in the medical field (Miyamoto *et al.*, 1989).

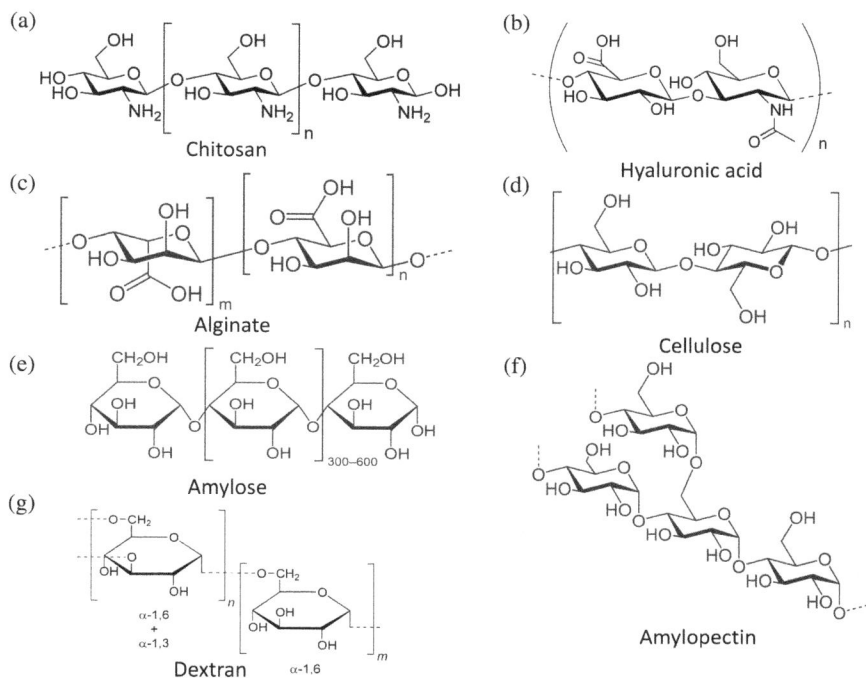

Fig. 5. Representative chemical structure of some polysaccharides: (a) chitosan, (b) hyaluronic acid, (c) alginate, (d) cellulose, (e) amylose, (f) amylopectin, (g) dextran.

3.1. *Chitosan*

Chitosan is a linear biodegradable cationic polysaccharide composed of randomly distributed β-(1–4)-linked D-glucosamine (deacetylated unit) and N-acetyl-D-glucosamine (acetylated unit), which is always made by treating shrimp and other crustacean shells with the alkali sodium hydroxide. Chitosan are homopolymers, containing varying fractions of the two residues, with a hydrophilic surface promoting cell adhesion, proliferation and differentiation. Moreover, it is antibacterial, often used in agriculture as a seed treatment and biopesticide, helping plants to fight off fungal infections. In medicine, it may be useful in bandages to reduce bleeding as an antibacterial agent, and used to help deliver drugs through the skin. In addition, the wellbiocompatibility of chitosan makes the implants based on it evoke minimal foreign body reaction, with little or no fibrous encapsulation. Besides, chitosan can be molded in various forms with a fairly well designed porous structure by means of different techniques, such as freeze-drying, rapid prototyping and internal bubbling process, which help chitosan to be widely used for biomedical applications, such as sutures, wound dressings, bone substitutes, tissue engineering and drug and gene delivery vehicles (Vande Vord *et al.*, 2002; Di Martino *et al.*, 2005).

The enzymatic degradability of chitosan associated to its structural similarity to extracellular matrix glycosaminoglycans makes it an attractive biopolymer for bone tissue repair. It has been proven that chitosan is osteoconductive, enhancing bone formation both *in vitro* and *in vivo* (Muzzarelli *et al.*, 1994), and the adequate chitosan properties depending on its source and preparation procedure shows a well-structured subchondral bone and noticeable cartilaginous tissue regeneration for osteochondral defect regeneration (Abarrategi *et al.*, 2010). However, the mechanical weakness and instability of chitosan, together with its incapacity to maintain a predefined shape, narrows its application field.

To overcome the problems brought by its natural drawbacks, chitosan has recently been developed to combine with a variety of other materials, such as hydroxyapatite, calcium phosphate, collagen, gelatin, alginate, and poly(lactic-co-glycolic acid) (PLGA) for potential application in orthopaedics for bone repair (Hu *et al.*, 2004). Biomimetic composite nanofibrous scaffolds of hydroxyapatite/chitosan supported and enhanced the adhesion, proliferation, and particularly osteogenic differentiation of the mesenchymal stem cells (Peng *et al.*, 2012; Kim *et al.*, 2012). The femoral condyle bone defects were repaired by implanting nanohydroxyapatite/chitosan compositions, and complete healing of the segmental bone defect was observed 12 weeks after surgery (Zhang *et al.*, 2012). Human embryonic stem cells had good viability and osteogenic differentiation on the novel calcium phosphate cement-chitosan-RGD scaffold, which also greatly enhanced their attachment, proliferation, and bone mineral synthesis (Chen *et al.*, 2013). The calcium phosphate cements reinforced with phosphorylated chitosan were implanted into rabbit radial defects, and histological and histomorphological studies proved that P-chitosan containing cements are biocompatible, bioabsorbable and osteoinductive, with the progressive substitution taking place at the interface of implants and host bones (Wang *et al.*, 2002). Pallela R. *et al.* (2011) suggested a novel scaffold, containing chitosan, hydroxyapatit and marine sponge collagen prepared using freeze-drying and lyophilization method, was promising as biomaterials for matrix-based bone repair and bone augmentation. hBMSC was found to be a good attachment and proliferation occurred in three-dimensional matrices composed of chitosan/collagen, which supported osteogenic differentiation in response to stimulation (Wang and Stegeman, 2011). Bone marrow mesenchymal stem cells were seeded on a chitosan-gelatin scaffold and then implanted in the rats' sockets, and the researchers found these composites contributed to bone, epithelial, and vascular repair with the biomaterial resorbed in a typical foreign body reaction (Miranda *et al.*, 2012). Chitosan/-alginate hybrid scaffolds displayed improved mechanical strength and structural stability and were shown to stimulate new bone formation and rapid vascularization during *in vivo* experiments (Li *et al.*, 2005). Modified poly(DL-lactic-co-glycolic acid) (PLGA) substrate by coating chitosan could increase osteoblast adhesion and differentiation on PLGA scaffolds, which seemed to provide a useful strategy for bone tissue regeneration (Wu *et al.*, 2006).

Moreover, due to its excellent characters, currently chitosan has also been used widely for systemic and local delivery of drugs and vaccines, for application in wound healing and bone regeneration. The use of chitosan/TCP sponges as a delivery system for growth factors demonstrated an osteogenic effect on bone regeneration *in vivo* (Lee *et al.*, 2000), and the collagen/chitosan microspheres composite scaffold had been proven to be a promising carrier of BMP-2 for the treatment of segmental bone defects (Hou *et al.*, 2012).

3.2. *Hyaluronic acid*

Hyaluronic acid (HA) is an anionic, non-sulfated glycosaminoglycan found widely throughout connective, epithelial, and neural tissues, which forms in the plasma membrane and can be very large, with its molecular weight often reaching the range of millions. As one of the chief components of the extracellular matrix, hyaluronan not only plays the role of structural element, but also interacts with binding proteins, proteoglycans and other bioactive molecules, contributing to the regulation of water balance, cell proliferation and migration, protecting articular cartilage surface and scavenging molecule for free radicals (Fraser *et al.*, 1997). Additionally, HA is recognized by specific cell receptors that regulate cell behavior, inflammation, angiogenesis and healing processes, and act as a selective and protective coat around the cell membrane (Campoccia *et al.*, 1998).

Due to the good biocompatibility and viscoelastic properties, HA has been extensively studied and applied in the biomedical field for delivery system, cell encapsulation and tissue engineering (Ji *et al.*, 2006; Wieland *et al.*, 2007; Kang *et al.*, 2009). For bone reconstruction, the hyaluronic acid-supplemented bone graft had been implanted into the cavities of rabbits' tibia and the results showed a higher score than the control group during the period of healing (Aslan *et al.*, 2006). A HA gel associated scaffold improved new bone formation in critical-size defects, even though it was without complete closure (de Brito Bezerra *et al.*, 2012). Just like the chitosan, hyaluronic acid is also a suitable carrier for the delivery of growth factors in bone repair, because of their ability to retain the factors, release their low levels to the local environment in a sustained manner, and stimulate differentiation of pluripotent stem cells (Kim and Valentini, 2002). The *in vivo* research demonstrated that the hydrogels with BMP-2 made the highest expression of osteocalcin and mature bone formation with vascular markers for rat calvarial defect regeneration, and also a good effect in repairing defects of distal femur in rabbits (Kim *et al.*, 2007; Peng *et al.*, 2008).

But the physical and biological characteristics of HA with purified form, such as water solubility, rapid resorption and short residence time in the tissue, limit its application as a biomaterial. Covalent crosslinking and photocross-linking have been used to modify its molecular structure and improve its properties for overcoming these limitations and providing long term stability and increased mechanical strength (Allison and Grande-Allen, 2006). It was reported as a new type of covalent synthetic ECM, disulfide-crosslinked HA-gelatin hydrogels, constituted biocompatible and biodegradable substrata for cell culture *in vitro* (Shu *et al.*, 2003).

3.3. *Alginate*

Alginate is an anionic polysaccharide distributed widely in the cell walls of brown algae, where it always forms a viscous gum through binding with water. It belongs to a family of linear copolymers of β-Dmannuronic acid and α-L-guluronic acid residues, which can be arranged in different proportions and sequences along the polymer chain (Wee and Gombotz, 1998). Physical properties of alginates depend on molecular weight, composition and extent of the sequences. The native alginates are mainly present as insoluble Ca^{2+} cross-linked gels through ionotropic gelation, but they can also form relatively stable hydrogels in the presence of other multivalent cations. The cross-linking process can be carried out under very mild conditions, at low temperature and in the absence of organic solvents, and hydrogels of different shapes can be prepared. Due to these advantages, several therapeutic agents, including antibiotics, enzymes, growth factors and DNA, have already been successfully incorporated in alginate gels, retaining a high percentage of biological activity. Together with other special characteristics, such as the presence of a relatively inert aqueous environment within the matrix, the dissolution and biodegradation of the system under normal physiological conditions and the high and tunable porosity, allowing high diffusion rates of macromolecules, alginates have now been widely used for biomedical applications such as drug delivery, cells, encapsulation, anti-adhesion materials and tissue engineering scaffolds (Simmons *et al.*, 2004; Jeon *et al.*, 2009; Yasuda *et al.*, 2006).

In orthopedics, He X. *et al.* (2013) had used the osteoconductive and mechanical properties of nanoscale calcium sulfate (nCS) and the biocompatibility of alginate to develop the injectable nCS/alginate (nCS/A) paste, which promoted the MSCs-mediated bone formation and vascularization for the repair of critical-sized calvarial bone defects in a rat model. A hybrid growth factor delivery system that consists of an electrospun nanofiber mesh tube for guiding bone regeneration combined with peptide-modified alginate hydrogel injected inside the tube for sustained growth factor release had been created, and the results of the testing ability of this system to deliver rhBMP-2 for the repair of critically-sized segmental bone defects in a rat model indicate that it induced a consistent bony bridging of the challenging bone defects (Kolambkar *et al.*, 2011). hESC-derived mesenchymal stem cell encapsulation in hydrogel microbeads in macroporous calcium phosphate cement was investigated for bone tissue engineering, and researchers found that it showed good cell viability, osteogenic differentiation and mineral synthesis, which might be used in a wide range of orthopedic and maxillofacial applications (Tang *et al.*, 2012).

3.4. *Starch-based material*

Starch is a carbohydrate consisting of a large number of glucose units joined by glycosidic bonds. As a most common and major polysaccharide produced by all green plants as an energy store, starch is also a major carbohydrate in the human diet and is contained in large amounts in staple foods such as potatoes, wheat, maize (corn), rice, and cassava. Composed with a mixture of a linear poly(1,4-α-D-glucopyranose) (amylose) and a branched poly(1,4-α-D-glucopyranose) with branches of (1,6-α-D-glucopyranose) (amylopectin) occurring nearly every 25 glucosidic moieties, starch generally contains 75 to 80% amylopectin in the plants and a more branched version of amylopectin in glycogen considered as the glucose store of animals (Revedin *et al.*, 2010). With the biodegradable nature, inexpensive price and the ability to be processed by various methods (Gomes *et al.*, 2001; Sousa *et al.*, 2003; Elvira *et al.*, 2002) into diverse shapes (such as three-dimensional porous scaffolds, microparticles, bone cements and so on) (Espigares *et al.*, 2002), the starch-based materials possess a wide range of properties that support their potential for biomedical applications and are very attractive to be utilized. Starch-based materials for bone repair, which are commonly blended with other ceramics or other polymers to better resist degradation, and make them less brittle and more easily processed, have been studied and shown to be very attractive for several biomedical applications, such as bone scaffolds and drug release system (Elvira *et al.*, 2002; Silva *et al.*, 2005).

3D porous scaffolds based on starch have been shown to be biocompatible and to possess excellent *in vitro* and *in vivo* behavior. Various studies conducted on scaffolds obtained from different starch-based materials (SEVA-C, SCA) showed that the systems possess adequate porous structure, with pores size varying into the believed optimal range for bone cell culturing and bone tissue ingrowth. A biodegradable starch-based polymer was used to produce polymer/hydroxyapatite (HA) composites (with or without the use of coupling agents) with mechanical properties matching those of the human bone, and to obtain 3D structures generated by solid blowing agents, that are suitable for tissue engineering applications. The biocompatibility evaluation of them had shown that all of the different materials based on a blend of corn starch with ethylene vinyl alcohol, as well as all the additives (including the novel coupling agents) and different processing methods required to obtain the different properties/products, could be used without inducing a cytotoxic behavior to the developed biomaterials (Gomes *et al.*, 2001). The porous polymeric system composed of cornstarch blended with poly(epsilon-caprolactone) (SPCL) was compared with the porous hydroxyapatite granules controls to evaluate as scaffolds for bone repair. Rat bone marrow cells seeded on them were proven to be able to proliferate, differentiate, and form extracellular matrix and additionally induced abundant formation of bone and bone marrow after four weeks of ectopic implantation in a nude mouse model. Though the amount of bone marrow and the degree of bone contact were higher on HA scaffolds, the findings still suggested that SPCL systems were excellent candidates to be used as scaffolds for a cell therapy approach in the treatment of bone defects (Mendes *et al.*, 2003). A polymer (starch/ethylene vinyl alcohol blend, SEVA-C) and a composite of SEVA-C reinforced with hydroxyapatite (HA) particles were evaluated *in vitro* cell culture assay and in an intramuscular and intracortical bone implantation on goats *in vivo*. In both models, the SEVA-C-based materials did not induce adverse reactions, which in addition to their bone-matching mechanical properties made them promising materials for bone replacement fixation (Mendes *et al.*, 2001). Salgado AJ *et al.* (2007) compared and evaluated the *in vivo* endosseous response to three starch-based scaffolds in distal femurs proximal to the epiphyseal plate of rats, such as a blend of corn starch and ethylene-vinyl alcohol (SEVA-C), the same composition coated with a biomimetic calcium phosphate (Ca-P) layer (SEVA-C/CaP), and a blend of corn starch and cellulose acetate (SCA), which were all produced by extrusion with blowing agents; and at last they concluded that all materials exhibited a favorable bony response and that the rapidly forming initial "connective tissue" seen around all scaffolds was a very early form of bone formation (Salgado *et al.*, 2007). The starch poly (ε-caprolactone) (SPCL) fiber meshes were tested to assess the effect of these scaffolds alone or combined with osteoblast-like cells in the regeneration of a critical-sized cranial defect in male Fisher rats, and the results proved that SPCL fiber meshes might be an osteoconductive material to use for bone regeneration purpose (Link *et al.*, 2013).

As the same as other polysaccharides, starch-based microparticles were used as the drug delivery system for bone repair. A novel injectable drug delivery system consisting of starch-poly-epsilon-caprolactone microparticles was shown to induce osteogenesis, reduce the amount of BMP-2 needed and allow more sustained osteogenic effects (Balmayor *et al.*, 2009). A polymeric blend of starch with polycaprolactone (SPCL) was used to produce a microparticle carrier for the controlled release of dexamethasone (DEX), and those developed microparticles were shown to be completely released as a consequence of an increasingly permeable matrix and faster diffusion of the drug, which indicated the possibility to be used as biomaterial due to their reduced cytotoxic effects (Balmayor *et al.*, 2008).

3.5. *Cellulose*

Cellulose is an organic compound with the formula $(C_6H_{10}O_5)_n$, a polysaccharide consisting of a linear chain of several hundred to over ten thousand $\beta(1\rightarrow4)$ linked D-glucose units (Updegraff, 1969). It constitutes the most abundant and renewable polymer resource available worldwide, which could be obtained from plants, algae, bacteria biosynthesis and chemosynthesis. As a natural polymer with good biocompatibility, cellulose usually exhibits relatively low protein adsorption and cell adhesion (particularly blood cells), low immune response (low phagocytosis by macrophages and low interleukin-1 release), and induces comparatively higher activation of the complement system. Though with poor biodegradability in the body and indigestible properties, cellulose can still be made hydrolysable by changing its higher order structure (Miyamoto *et al.*, 1989).

Märtson M. *et al.* (1998) investigated the biocompatibility of viscose cellulose sponge (VCS) with bone in the femur implantation of rats, and then suggested VCS as a compatible matrix for osseous tissue ingrowth which might be useful as a scaffold for bone tissue engineering in experiments and possibly also in clinical practice. Furthermore, it had been discovered that it was possible to control the structure of the sponge by certain changes, which could increase the number of invading cells and the production of granulation tissue in the sponge (Pajulo *et al.*, 1996).

To improve the osseointegration, the cellulose phosphate (CP) was then investigated as a biomaterial for orthopedic applications. Due to its high Ca binding capacity, associated with lack of toxicity and indigestibility, CP has been used for decades in the treatment of Ca metabolism-related diseases (Mizusawa and Burke, 1996). Moreover, the CP also specifically binds to biologically active species, such as enzymes, peptides and growth factors. Because of these unique characters, CP can be proposed as a promising alternative biomaterial, capable of promoting an adequate healing response once implanted. *In vitro* studies showed that the phosphorylated cellulose induced the formation of a calcium phosphate layer, and *in vivo* histomorphometry and the measurement of the amount of 45Ca incorporated in the tissue surrounding the rabbits' implantations also indicated a good osseointegration of phosphorylated cellulose. However, CP promoted poor rates of cell attachment and proliferation, which were attributed to the negative charge, associated with the high hydrophilicity of the cellulose derivative (Fricain *et al.*, 2002).

Bacterial cellulose (BC) is a nanomaterial synthesized extracellularly by bacteria, principally of the genera Acetobacter, Sarcina ventriculi and Agro bacterium. Different from the properties of plant cellulose, the bacterial cellulose is more chemically pure, containing no hemicellulose or lignin but significantly smaller and thinner microfibrils, characterized by high purity, strength, crystallinity, moldability and tensile strength, increased water holding ability as well as good biocompatibility (Bae and Shoda, 2004; Fang *et al.*, 2009).

In more modern applications, bacterial cellulose has become relevant in the medical sector, which has been tested and successfully used in blood vessels engineering (Klemm *et al.*, 2001) and wound healing (Czaja *et al.*, 2006), especially in burn cases. Studies have shown that burns treated with microbial cellulose coverings healed faster than traditional treatments and had less scaring (Helenius *et al.*, 2006). It has also been proven that biodegradable cellulose-based polymers have a range of properties suitable for use in a wide array of biomedical applications ranging from bone replacement to engineering of tissue scaffolds. In recent years, in orthopedic, BC has been also proposed and studied as potential scaffold for bone repair. Fang B. *et al.* (2009) designed the hydroxyapatite/bacterial cellulose (HAp/BC) nanocomposite scaffolds utilizing the biomimetic technique, and investigated the proliferation and osteoblastic differentiation of stromal cells derived from human bone marrow (hBMSC) on them. The results showed that seeded cells exhibited good adhesion and activity, and excellent osteogenic activity including the high alkaline phosphatase (ALP) activity of hBMSC as well as expression of other osteogenic marker genes such as osteopontin, osteocalcin and bone sialoprotein, which suggested the scaffold might have potential use for bone tissue engineering (Fang *et al.*, 2009). Microporous bacterial cellulose (BC) scaffolds were prepared by incorporating 300–500 microm paraffin wax microspheres into the fermentation process, and the MC3T3-E1 osteoprogenitor cells seeded on them were found to cluster within the pores of microporous BC and formed denser mineral deposits, showing microporous BC as a promising biomaterial for bone tissue engineering applications (Zaborowska *et al.*, 2010). *In vivo* biocompatibility of BC membranes was analyzed through long-term subcutaneous implants in the mice, and a tendency to calcify the implants over time was observed (Pertile *et al.*, 2011). The bacterial cellulose-hydroxyapatite (BC-HA) nanocomposite membranes were incubated in solutions of $CaCl_2$ followed by Na_2HPO_4, then the formation of HA crystals on BC nanofibres were proven, with no inflammatory reaction and filled defects in by new bone tissue (Saska *et al.*, 2011). Collagen production, as the initial step of tissue mineralization, was detected to be produced by the osteoprogenitor cells seeded on microporous bacterial cellulose during the first days of growth, and then the developed collagen fiber networks inside compact regions of cells were also found to be located in the cellulose micropores (Brackmann *et al.*, 2012). An osteogenic growth peptide (OGP) and its C-terminal pentapeptide OGP functionalized BC membranes had been studied, and the results of cell viability/proliferation, total protein content, alkaline phosphatase activity and mineralization assays indicated that BC-OGP membranes enabled the highest development of the osteoblastic phenotype *in vitro*, with the conclusion that these new membranes could be employed in bone tissue regeneration (Saska *et al.*, 2012). Furthermore, Shi Q. *et al.* (2012) found that BC had good biocompatibility and is able to induce differentiation of mouse fibroblast-like C2C12 cells into osteoblasts in the presence of BMP-2 *in vitro*, as demonstrated by alkaline phosphatase (ALP) activity assays. In *in vivo* subcutaneous implantation studies, BC scaffolds carrying BMP-2 showed more bone formation and higher calcium concentration, both suggesting BC as a good localized delivery system for BMPs and a potential candidate in bone tissue engineering (Shi *et al.*, 2012). In addition, it was proven that bacterial cellulose might retain antibiotics and the cellulose composited to antibiotic bone cement would improve mechanical strength and antibiotic release, which might advance the application of bacterial cellulose in clinic bone defects therapies (Mori *et al.*, 2011).

3.6. *Dextran*

Dextran is a complex and branched glucan (polysaccharide made of many glucose molecules) composed of chains of varying lengths (from 3 to 2000 kilodaltons), usually synthesized from sucrose by certain lactic-acid bacteria, such as the *Leuconostoc mesenteroides* and *Streptococcus mutans*. As the poly(ethylene glycol) (PEG), dextran has been shown to be resistant to both protein adsorption and cell adhesion and allowing one to design a scaffold with specific sites for cell recognition (McLean *et al.*, 2000; Massia and Stark, 2001). The dextran hydrogel has been investigated as the microcarrier for cell culture, drug delivery vehicle and tissue engineering scaffold, which could be obtained by either physical and chemical cross-linking, or by partial oxidation of glucose hydroxyl groups to aldehydes that were then cross-linked with gelatin (Pawlowski *et al.*, 1984; Chiu *et al.*, 1999).

Biodegradable macroporous cryogels were produced from dextran modified with oligo L-lactide bearing hydroxyethylmethacrylate (HEMA) end groups in moderately frozen solutions. With highly open and interconnected pore structures, they were considered as novel tissue engineering scaffolds, which then were evaluated in rats after dorsal subcutaneous implantation, iliac submuscular implantation, auricular implantation, or in calvarial defect model. Histological analyses showed these scaffolds integrated with the surrounding tissue, with no necrosis or foreign body reaction in any cases, and the new tissue formation were observed to accompany significant ingrowth of connective tissue cells and new blood vessels into the cryogel, which in general showed that these materials are highly biocompatible. In combination with other properties, like mechanical strength significant for a gelious material and very rapid swelling, the cryogel scaffolds might be used as attractive novel candidates for further clinical applications in bone regeneration (Bolgen *et al.*, 2009). Moreover, as a microcarrier, the dextran-based beads in spinner flask cultures increased the rabbits' MSCs colonization, and the multipotentiality of cells were preserved, with successfully differentiated osteogenic and chondrogenic lineages (Boo *et al.*, 2011).

4. Microbial Origin Polyesters

Polyester is a category of polymers which contains the ester functional group in their main chain. They are the most commonly researched polymers for bone regeneration applications, likely because there are several FDA-approved polyesters with extensive clinical history. Natural polyesters are biodegradable and synthesized by plants as structural components of the cuticle covering the aerial part of plants, such as cutin and suberin, or by prokaryotic microorganisms as water-insoluble intracellular storage compounds.

Polyhydroxyalkanoates (PHAs) are linear polyesters produced in nature by bacterial fermentation of sugar or lipids, used for carbon and energy store. More than 150 different monomers can be combined within and resulting in extremely different properties, of which, the mechanical and biocompatibility can also be changed by blending, modifying the surface or combining with other polymers, enzymes and inorganic materials, making PHAs possible for a wider range of applications (Puppi *et al.*, 2010). Particularly poly(3-hydroxybutyrate)

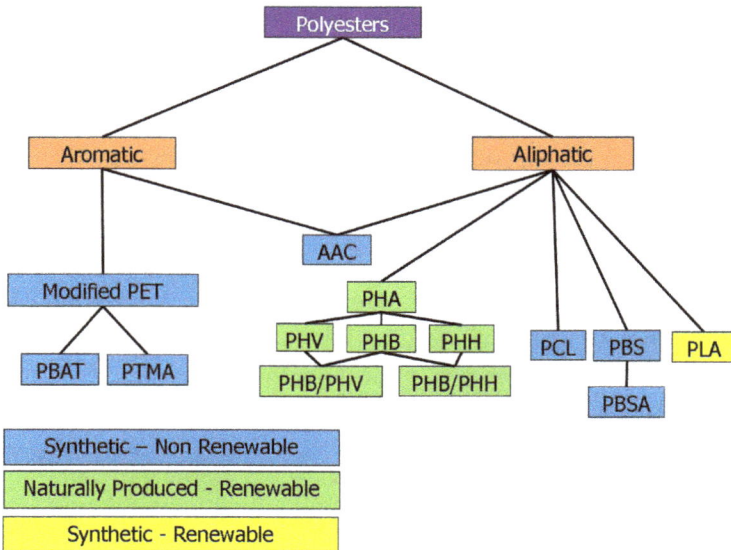

Fig. 6. Polyesters family.

(PHB), poly(4-hydroxybutyrate), copolymers of 3-hydroxybutyrate and 3-hydroxyvalerate (PHBV), copolymers of 3-hydroxybutyrate, 3-hydroxyhexanoate (PHBHHx) and poly-3-hydroxyoctanoate, have attracted medical interest of many researchers (Freier, 2006).

Doyle C and his coworkers presented some findings regarding the degradation and biological properties of polyhydroxybutyrate and composites reinforced with particulate hydroxyapatite. The strength and stiffness of these materials reduce in *in-vitro* environment exposure in phosphate-buffered saline, and the degradation rate is a function of composition and processing conditions. They also demonstrated that those materials based on PHB produced a consistent favourable bone tissue adaptation response with no evidence of an undesirable chronic inflammatory response after implantation periods up to 12 months, and the formed bone showed close to the material and subsequently highly organized, with up to 80% of the implant surface lying in direct apposition to new bone (Doyle *et al.*, 1991). The mechanical push-out test was performed on implants in the femur of mature Japanese White rabbits for the evaluation of tissue bonding to polyhydroxybutyrate (PHB) based composites, and the results indicated enhanced endosteal bone growth for the HA/PHB (Knowles *et al.*, 1992). Boeree NR *et al.* (1993) had reported that the injection-moulded composite material comprising polyhydroxybutyrate with hydroxyapatite had considerable potential for use in orthopaedic surgery, both as a material to construct certain orthopaedic implants and as an alternative to corticocancellous bone graft. Kostopoulos L.'s group explored the possibility of obtaining bone regeneration in jaw bone defects in rats after coverage of the defects with a polyhydroxybutyrate resorbable membrane, with results that concluded selective repopulation of bone defects with bone-forming cells could be ensured by excluding surrounding soft tissues from the wound area with this special membrane (Kostopoulos and Karring, 1994). Two particulate bioactive ceramics, hydroxyapatite and tricalcium phosphate (TCP), were incorporated into polyhydroxybutyrate-polyhydroxyvalerate (PHB-PHV), to produce new biomaterials for potential medical applications. The degradation temperature of PHB-PHV was significantly reduced by the incorporation of bioceramics, while the storage modulus and loss modulus of the composites increased with the increase in HA or TCP content. Additionally *in vitro* study indicated enhanced ability of the composites to induce the formation of bone-like apatite on their surfaces (Chen and Wang, 2002). *In vivo*, the morphological and compositional structure of bone-implant interfaces were examined, after the materials composed of biodegradable polyhydroxybutyrate/polyhydroxyvalerate copolymer reinforced with synthetic hydroxyapatite particles implanted into the tibias of rabbits. A strong tendency to rebuild the bone structure was found at the interface after implantation, but direct bone bonding with the implant depended on the bioactive nature of the interface. Lamellar bone formed at the interface and replaced degrading polymer matrix, and the bone region displayed an osteon organization, with osteoblasts and osteocytes identified throughout the interface region, which all suggested these composite materials might be useful in some bone replacement therapies (Luklinska and Schluckwerder, 2003). Furthermore, Carlo E.C. *et al.* (2009) compared three composites made of hydroxyapatite and polyhydroxybutyrate to determine their biocompatibility, osteointegration, and osteoconduction in rabbits bone defects, and finally found that the 25:75 (Hydroxyapatite-polyhydroxybutyrate in vol/vol ratio) and 50:50 composites showed osteoconductive, displaying best characteristics for bone substitution. However, a recent study investigated the influence PHB and PHBV exerted on cell cycle progression of mesenchymal stem cells commonly used in the engineering of bone tissues, and analysis revealed that the biomaterials induced significant cell cycle progression, showing significantly higher percentages of cells cycled at synthesis (S) phase of the cycle on PHBV films compared to PHB. Therefore, the researchers suggested that application of these biomaterials in tissue engineering are specific to cell type and requires a detailed investigation at the cell-material interface (Ahmed *et al.*, 2010).

In summary, bone defect results in severe pain and disability for millions of people worldwide, with profound economic impact on the healthcare system. Here, we reviewed some common natural polymers, which possess several inherent advantages such as low toxicity, low manufacture and disposal costs, renewability, good bioactivity, the ability to present receptor-binding ligands to cells, susceptibility to cell-triggered proteolytic degradation and natural remodeling. These have attracted increasing interest from most scientists and surgeons. However, the inherent bioactivity of these natural polymers shows its own downsides, including low mechanical, thermal and chemical stability, possible loss of biological properties during formulation, immunogenic response associated with most polymers, the complexities associated with their purification and the possibility of disease transmission. To overcome the drawbacks and to make the natural polymers become the perfect right substitutes for wide applications in the clinic therapy, all researchers need to work hard and be more detailed.

Acknowledgments

This work was supported by grants from The Ministry of Science and Technology of China (No. 2011DFA30790), National Natural Science Foundation of China (No. 81190133, 31101056), Chinese Academy of Sciences (No. XDA01030404, KSCX2-EW-Q-1-07), Science and Technology Commission of Shanghai Municipality (No. 12411951100).

References

Abarrategi A, *et al.* Chitosan scaffolds for osteochondral tissue regeneration. *J Biomed Mater Res A* 2010;95(4):1132–41.

Ahmed T, *et al.* Polyhydroxybutyrate and its copolymer with polyhydroxyvalerate as biomaterials: Influence on progression of stem cell cycle. *Biomacromolecules* 2010;11(10):2707–15.

Allison DD, Grande-Allen KJ. Review. Hyaluronan: A powerful tissue engineering tool. *Tissue Eng* 2006;12(8):2131–40.

Altman GH, *et al.* Silk-based biomaterials. *Biomaterials* 2003;24(3):401–16.

Andrade AL, *et al.* Influence of recovering collagen with bioactive glass on osteoblast behavior. *J Biomed Mater Res B Appl Biomater* 2007;83(2):481–9.

Angele P, *et al.* Influence of different collagen species on physico-chemical properties of crosslinked collagen matrices. *Biomaterials* 2004;25(14):2831–41.

Aslan M, Simsek G, Dayi E. The effect of hyaluronic acid-supplemented bone graft in bone healing: Experimental study in rabbits. *J Biomater Appl* 2006;20(3):209–20.

Bae S, Shoda M. Bacterial cellulose production by fed-batch fermentation in molasses medium. *Biotechnol Prog* 2004;20(5):1366–71.

Balmayor ER, *et al*. A novel enzymatically-mediated drug delivery carrier for bone tissue engineering applications: Combining biodegradable starch-based microparticles and differentiation agents. *J Mater Sci Mater Med* 2008;19(4): 1617–23.

Balmayor ER, *et al*. Starch-poly-epsilon-caprolactone microparticles reduce the needed amount of BMP-2. *Clin Orthop Relat Res* 2009;467(12):3138–48.

Bini E, *et al*. RGD-functionalized bioengineered spider dragline silk biomaterial. *Biomacromolecules* 2006;7(11):3139–45.

Boeree NR, *et al*. Development of a degradable composite for orthopaedic use: Mechanical evaluation of an hydroxyapatite-polyhydroxybutyrate composite material. *Biomaterials* 1993;14(10):793–6.

Bolgen N, *et al*. Tissue responses to novel tissue engineering biodegradable cryogel scaffolds: an animal model. *J Biomed Mater Res A* 2009;91(1):60–8.

Boo L, *et al*. Expansion and preservation of multipotentiality of rabbit bone-marrow derived mesenchymal stem cells in dextran-based microcarrier spin culture. *J Mater Sci Mater Med* 2011;22(5):1343–56.

Brackmann C, *et al*. *In situ* imaging of collagen synthesis by osteoprogenitor cells in microporous bacterial cellulose scaffolds. *Tissue Eng Part C Methods* 2012;18(3):227–34.

Burchardt H. The biology of bone graft repair. *Clin Orthop Relat Res* 1983;(174): 28–42.

Campoccia D, *et al*. Semisynthetic resorbable materials from hyaluronan esterification. *Biomaterials* 1998;19(23):2101–27.

Carlo EC, *et al*. Comparison of *in vivo* properties of hydroxyapatite-polyhydroxybutyrate composites assessed for bone substitution. *J Craniofac Surg*, 2009;20(3):853–9.

Chen LJ, Wang M. Production and evaluation of biodegradable composites based on PHB-PHV copolymer. *Biomaterials* 2002;23(13):2631–9.

Chen W, *et al*. Human embryonic stem cell-derived mesenchymal stem cell seeding on calcium phosphate cement-chitosan-RGD scaffold for bone repair. *Tissue Eng Part A* 2013;19(7–8):915–27.

Cheung HY, *et al*. A critical review on polymer-based bio-engineered materials for scaffold development. *Composites Part B-Engineering* 2007;38(3):291–300.

Chiu HC, *et al*. Synthesis and characterization of pH-sensitive dextran hydrogels as a potential colon-specific drug delivery system. *J Biomater Sci Polym Ed* 1999;10(5):591–608.

Ciardelli G, *et al*. Enzymatically crosslinked porous composite matrices for bone tissue regeneration. *J Biomed Mater Res A* 2010;92(1):137–51.

Czaja W, *et al*. Microbial cellulose--the natural power to heal wounds. *Biomaterials* 2006;27(2):145–51.

Daamen WF, *et al.* Preparation and evaluation of molecularly-defined collagen-elastin-glycosaminoglycan scaffolds for tissue engineering. *Biomaterials* 2003;24(22):4001–9.

de Brito Bezerra B, *et al.* Association of hyaluronic acid with a collagen scaffold may improve bone healing in critical-size bone defects. *Clin Oral Implants Res,* 2012;23(8):938–42.

Di Lullo GA, *et al.* Mapping the ligand-binding sites and disease-associated mutations on the most abundant protein in the human, type I collagen. *J Biol Chem* 2002;277(6):4223–31.

Di Martino A, Sittinger M, Risbud MV. Chitosan: A versatile biopolymer for orthopaedic tissue-engineering. *Biomaterials* 2005;26(30):5983–90.

Dong J, Sun Q, Wang JY. Basic study of corn protein, zein, as a biomaterial in tissue engineering, surface morphology and biocompatibility. *Biomaterials* 2004; 25(19):4691–7.

Doyle C, Tanner ET, Bonfield W. *In vitro* and *in vivo* evaluation of polyhydroxybutyrate and of polyhydroxybutyrate reinforced with hydroxyapatite. *Biomaterials* 1991;12(9):841–7.

Elvira C, *et al.* Starch-based biodegradable hydrogels with potential biomedical applications as drug delivery systems. *Biomaterials* 2002;23(9):1955–66.

Espigares I, *et al.* New partially degradable and bioactive acrylic bone cements based on starch blends and ceramic fillers. *Biomaterials* 2002;23(8):1883–95.

Fang B, *et al.* Proliferation and osteoblastic differentiation of human bone marrow stromal cells on hydroxyapatite/bacterial cellulose nanocomposite scaffolds. *Tissue Eng Part A* 2009;15(5):1091–8.

Finkemeier CG. Bone-grafting and bone-graft substitutes. *J Bone Joint Surg Am* 2002;84-A(3):454–64.

Fraser JR, Laurent TC, Laurent UB. Hyaluronan: Its nature, distribution, functions and turnover. *J Intern Med* 1997;242(1):27–33.

Freier T. Biopolyesters in tissue engineering applications. *Polymers for Regenerative Medicine* 2006:1–61.

Fricain JC, *et al.* Cellulose phosphates as biomaterials. *In vivo* biocompatibility studies. *Biomaterials* 2002;23(4):971–80.

Giannoudis PV, Dinopoulos H, Tsiridis E. Bone substitutes: An update. *Injury* 2005;36(Suppl 3):S20–7.

Glowacki J, Mizuno S. Collagen scaffolds for tissue engineering. *Biopolymers* 2008;89(5):338–44.

Gomes ME, *et al.* A new approach based on injection moulding to produce biodegradable starch-based polymeric scaffolds: Morphology, mechanical and degradation behaviour. *Biomaterials* 2001;22(9):883–9.

Gomes ME, *et al.* Cytocompatibility and response of osteoblastic-like cells to starch-based polymers: Effect of several additives and processing conditions. *Biomaterials* 2001;22(13):1911–7.

Green J, *et al.* Cell-matrix interaction in bone: Type I collagen modulates signal transduction in osteoblast-like cells. *Am J Physiol* 1995;268(5 Pt 1):C1090–103.

He X, *et al.* Integration of a novel injectable nano calcium sulfate/alginate scaffold and BMP2 gene-modified mesenchymal stem cells for bone regeneration. *Tissue Eng Part A* 2013;19(3–4):508–18.

Helenius G, *et al.* *In vivo* biocompatibility of bacterial cellulose. *J Biomed Mater Res A* 2006;76(2):431–8.

Horan RL, *et al.* *In vitro* degradation of silk fibroin. *Biomaterials* 2005;26(17): 3385–93.

Hou J, *et al.* Segmental bone regeneration using rhBMP-2-loaded collagen/chitosan microspheres composite scaffold in a rabbit model. *Biomed Mater* 2012;7(3):035002.

Hu Q, *et al.* Preparation and characterization of biodegradable chitosan/hydroxyapatite nanocomposite rods via *in situ* hybridization: A potential material as internal fixation of bone fracture. *Biomaterials* 2004;25(5):779–85.

Jeon O, *et al.* Photocrosslinked alginate hydrogels with tunable biodegradation rates and mechanical properties. *Biomaterials* 2009;30(14):2724–34.

Ji Y, *et al.* Electrospun three-dimensional hyaluronic acid nanofibrous scaffolds. *Biomaterials* 2006;27(20):3782–92.

Jiang J, *et al.* Hydroxyapatite/regenerated silk fibroin scaffold-enhanced osteoinductivity and osteoconductivity of bone marrow-derived mesenchymal stromal cells. *Biotechnol Lett* 2013;35(4):657–61.

Kang JY, *et al.* Novel porous matrix of hyaluronic acid for the three-dimensional culture of chondrocytes. *Int J Pharm* 2009;369(1–2):114–20.

Karageorgiou V, Kaplan D. Porosity of 3D biomaterial scaffolds and osteogenesis. *Biomaterials* 2005;26(27):5474–91.

Karp JM, Shoichet MS, Davies JE. Bone formation on two-dimensional poly(DL-lactide-co-glycolide) (PLGA) films and three-dimensional PLGA tissue engineering scaffolds *in vitro*. *J Biomed Mater Res A* 2003;64(2):388–96.

Kasoju N, Bora U. Silk fibroin in tissue engineering. *Adv Healthc Mater*, 2012; 1(4):393–412.

Kim BS, Mooney DJ. Development of biocompatible synthetic extracellular matrices for tissue engineering. *Trends Biotechnol* 1998;16(5):224–30.

Kim BS, Mooney DJ. Engineering smooth muscle tissue with a predefined structure. *J Biomed Mater Res* 1998;41(2):322–32.

Kim BS, *et al.* Growth and osteogenic differentiation of alveolar human bone marrow-derived mesenchymal stem cells on chitosan/hydroxyapatite composite fabric. *J Biomed Mater Res A* 2012.

Kim HD, Valentini RF. Retention and activity of BMP-2 in hyaluronic acid-based scaffolds *in vitro*. *J Biomed Mater Res* 2002;59(3):573–84.

Kim HW, Song JH, Kim HE. Bioactive glass nanofiber-collagen nanocomposite as a novel bone regeneration matrix. *J Biomed Mater Res A* 2006;79(3):698–705.

Kim J, *et al.* Bone regeneration using hyaluronic acid-based hydrogel with bone morphogenic protein-2 and human mesenchymal stem cells. *Biomaterials* 2007;28(10):1830–7.

Kim KH, *et al.* Biological efficacy of silk fibroin nanofiber membranes for guided bone regeneration. *J Biotechnol* 2005;120(3):327–39.

Klemm D, *et al.* Bacterial synthesized cellulose — artificial blood vessels for microsurgery. *Prog Polym Sci* 2001;26(9):1561–1603.

Knowles JC, *et al.* Development of a degradable composite for orthopaedic use: *In vivo* biomechanical and histological evaluation of two bioactive degradable composites based on the polyhydroxybutyrate polymer. *Biomaterials* 1992; 13(8):491–6.

Kolambkar YM, *et al.* An alginate-based hybrid system for growth factor delivery in the functional repair of large bone defects. *Biomaterials* 2011;32(1):65–74.

Kostopoulos L, Karring T. Guided bone regeneration in mandibular defects in rats using a bioresorbable polymer. *Clin Oral Implants Res* 1994;5(2):66–74.

Lee CH, Singla A, Lee Y. Biomedical applications of collagen. *Int J Pharm* 2001; 221(1–2):1–22.

Lee YM, *et al.*, The bone regenerative effect of platelet-derived growth factor-BB delivered with a chitosan/tricalcium phosphate sponge carrier. *J Periodontol* 2000;71(3):418–24.

LeGeros RZ. Properties of osteoconductive biomaterials: Calcium phosphates. *Clin Orthop Relat Res* 2002;(395):81–98.

Leong KF, Cheah CM, Chua CK. Solid freeform fabrication of three-dimensional scaffolds for engineering replacement tissues and organs. *Biomaterials* 2003;24(13):2363–78.

Li C, *et al.* Electrospun silk-BMP-2 scaffolds for bone tissue engineering. *Biomaterials* 2006;27(16):3115–24.

Li Z, *et al.* Chitosan-alginate hybrid scaffolds for bone tissue engineering. *Biomaterials* 2005;26(18):3919–28.

Liao S, *et al.* Processing nanoengineered scaffolds through electrospinning and mineralization suitable for biomimetic bone tissue engineering. *J Mech Behav Biomed Mater* 2008;1(3):252–60.

Link DP, *et al.* Osteogenic properties of starch poly(epsilon-caprolactone) (SPCL) fiber meshes loaded with osteoblast-like cells in a rat critical-sized cranial defect. *J Biomed Mater Res A* 2013.

Liu LS, *et al.* An osteoconductive collagen/hyaluronate matrix for bone regeneration. *Biomaterials* 1999;20(12):1097–108.

Liu X, *et al.* Microspheres of corn protein, zein, for an ivermectin drug delivery system. *Biomaterials* 2005;26(1):109–15.

Liu Y, Chan-Park MB. A biomimetic hydrogel based on methacrylated dextran-graft-lysine and gelatin for 3D smooth muscle cell culture. *Biomaterials* 2010; 31(6):1158–70.

Luklinska ZB, Schluckwerder H. *In vivo* response to HA-polyhydroxybutyrate/polyhydroxyvalerate composite. *J Microsc* 2003;211(Pt 2):121–9.

Malafaya PB, *et al.* Polymer based scaffolds and carriers for bioactive agents from different natural origin materials. *Adv Exp Med Biol* 2003;534: 201–33.

Mano JF, *et al.* Natural origin biodegradable systems in tissue engineering and regenerative medicine: Present status and some moving trends. *J R Soc Interface* 2007;4(17):999–1030.

Mao JS, *et al.* Structure and properties of bilayer chitosan-gelatin scaffolds. *Biomaterials* 2003;24(6):1067–74.

Martson M, *et al.* Biocompatibility of cellulose sponge with bone. *Eur Surg Res* 1998;30(6):426–32.

Massia SP, Stark J. Immobilized RGD peptides on surface-grafted dextran promote biospecific cell attachment. *J Biomed Mater Res* 2001;56(3):390–9.

McLean KM, *et al.* Method of immobilization of carboxymethyl-dextran affects resistance to tissue and cell colonization. *Colloids Surf B Biointerfaces* 2000;18(3–4):221–234.

Meinel L, *et al.* Engineering bone-like tissue *in vitro* using human bone marrow stem cells and silk scaffolds. *J Biomed Mater Res A* 2004;71(1):25–34.

Meinel L, *et al.* Silk implants for the healing of critical size bone defects. *Bone* 2005;37(5):688–98.

Meinel L, *et al.* Silk based biomaterials to heal critical sized femur defects. *Bone* 2006;39(4):922–31.

Mendes SC, *et al.* Biocompatibility testing of novel starch-based materials with potential application in orthopaedic surgery: A preliminary study. *Biomaterials* 2001;22(14):2057–64.

Mendes SC, *et al.* Evaluation of two biodegradable polymeric systems as substrates for bone tissue engineering. *Tissue Eng* 2003;9(Suppl 1):S91–101.

Miranda SC, *et al.* Mesenchymal stem cells associated with porous chitosan-gelatin scaffold: A potential strategy for alveolar bone regeneration. *J Biomed Mater Res A* 2012;100(10):2775–86.

Miyamoto T, *et al.* Tissue biocompatibility of cellulose and its derivatives. *J Biomed Mater Res* 1989;23(1):125–33.

Mizuno M, *et al.* Osteogenesis by bone marrow stromal cells maintained on type I collagen matrix gels *in vivo*. *Bone* 1997;20(2):101–7.

Mizusawa Y, Burke JR. Prednisolone and cellulose phosphate treatment in idiopathic infantile hypercalcaemia with nephrocalcinosis. *J Paediatr Child Health*, 1996;32(4):350–2.

Mori R, *et al.* Increased antibiotic release from a bone cement containing bacterial cellulose. *Clin Orthop Relat Res* 2011;469(2):600–6.

Muller WE. The origin of metazoan complexity: porifera as integrated animals. *Integr Comp Biol* 2003;43(1):3–10.

Murphy WL, *et al.* Salt fusion: An approach to improve pore interconnectivity within tissue engineering scaffolds. *Tissue Eng* 2002;8(1):43–52.

Muzzarelli RA, *et al.* Stimulatory effect on bone formation exerted by a modified chitosan. *Biomaterials* 1994;15(13):1075–81.

O'Brien FJ, *et al.* The effect of pore size on cell adhesion in collagen-GAG scaffolds. *Biomaterials* 2005;26(4):433–41.

Pajulo Q, *et al.* Viscose cellulose sponge as an implantable matrix: Changes in the structure increase the production of granulation tissue. *J Biomed Mater Res* 1996;32(3):439–46.

Pallela R, *et al.* Biophysicochemical evaluation of chitosan-hydroxyapatite-marine sponge collagen composite for bone tissue engineering. *J Biomed Mater Res A* 2011.

Pawlowski R, *et al.* Primary culture of chick embryo skeletal muscle on dextran microcarrier. *Eur J Cell Biol* 1984;35(2):296–303.

Peng H, *et al.* Electrospun biomimetic scaffold of hydroxyapatite/chitosan supports enhanced osteogenic differentiation of mMSCs. *Nanotechnology* 2012;23(48): 485102.

Peng L, *et al.* Implanting hydroxyapatite-coated porous titanium with bone morphogenetic protein-2 and hyaluronic acid into distal femoral metaphysis of rabbits. *Chin J Traumatol* 2008;11(3):179–85.

Pertile RA, *et al.* Bacterial Cellulose: Long-term biocompatibility studies. *J Biomater Sci Polym Ed* 2011.

Pohunkova H, Adam M. Reactivity and the fate of some composite bioimplants based on collagen in connective tissue. *Biomaterials* 1995;16(1):67–71.

Porter JR, Ruckh TT, Popat KC. Bone tissue engineering: A review in bone biomimetics and drug delivery strategies. *Biotechnol Prog* 2009;25(6): 1539–60.

Puppi D, *et al.* Polymeric materials for bone and cartilage repair. *Prog Polym Sci* 2010;35(4):403–440.

Qu ZH, *et al.* Evaluation of the zein/inorganics composite on biocompatibility and osteoblastic differentiation. *Acta Biomater* 2008;4(5):1360–8.

Rao JK, Ramesh DV, Rao KP. Implantable controlled delivery systems for proteins based on collagen–pHEMA hydrogels. *Biomaterials* 1994;15(5):383–9.

Revedin A, *et al.* Thirty thousand-year-old evidence of plant food processing. *Proc Natl Acad Sci USA* 2010;107(44):18815–9.

Rezwan K. *et al.* Biodegradable and bioactive porous polymer/inorganic composite scaffolds for bone tissue engineering. *Biomaterials* 2006;27(18):3413–31.

Rocha LB, Goissis G. Rossi MA. Biocompatibility of anionic collagen matrix as scaffold for bone healing. *Biomaterials* 2002;23(2):449–56.

Salgado AJ, *et al. In vivo* response to starch-based scaffolds designed for bone tissue engineering applications. *J Biomed Mater Res A* 2007;80(4):983–9.

Saska S, *et al.* Bacterial cellulose-hydroxyapatite nanocomposites for bone regeneration. *Int J Biomater* 2011;(2011):175362.

Saska S, *et al.* Characterization and *in vitro* evaluation of bacterial cellulose membranes functionalized with osteogenic growth peptide for bone tissue engineering. *J Mater Sci Mater Med* 2012;23(9):2253–66.

Shea LD, *et al.* Engineered bone development from a pre-osteoblast cell line on three-dimensional scaffolds. *Tissue Eng* 2000;6(6):605–17.

Shi Q, *et al.* The osteogenesis of bacterial cellulose scaffold loaded with bone morphogenetic protein-2. *Biomaterials* 2012;33(28):6644–9.

Shi S, Kirk M, Kahn AJ. The role of type I collagen in the regulation of the osteoblast phenotype. *J Bone Miner Res* 1996;11(8):1139–45.

Shibata Y, Yamamoto H, Miyazaki T. Colloidal beta-tricalcium phosphate prepared by discharge in a modified body fluid facilitates synthesis of collagen composites. *J Dent Res* 2005;84(9):827–31.

Shoulders MD, Raines RT. Collagen structure and stability. *Annu Rev Biochem* 2009;78:929–58.

Shu XZ, *et al.* Disulfide-crosslinked hyaluronan-gelatin hydrogel films: A covalent mimic of the extracellular matrix for *in vitro* cell growth. *Biomaterials* 2003; 24(21):3825–34.

Shukla R, Cheryan M. Zein: The industrial protein from corn. *Industrial Crops and Products* 2001;13(3):171–192.

Silva GA, *et al.* Entrapment ability and release profile of corticosteroids from starch-based microparticles. *J Biomed Mater Res A* 2005;73(2):234–43.

Simmons CA, *et al.* Dual growth factor delivery and controlled scaffold degradation enhance *in vivo* bone formation by transplanted bone marrow stromal cells. *Bone* 2004;35(2):562–9.

Sofia S, *et al.* Functionalized silk-based biomaterials for bone formation. *J Biomed Mater Res* 2001;54(1):139–48.

Sousa RA, *et al.* Bi-composite sandwich moldings: Processing, mechanical performance and bioactive behavior. *J Mater Sci Mater Med* 2003;14(5):385–97.

Sutherland TD, *et al.* Insect silk: One name, many materials. *Annu Rev Entomol* 2010;55:171–88.

Taboas JM, *et al.*, Indirect solid free form fabrication of local and global porous, biomimetic and composite 3D polymer-ceramic scaffolds. *Biomaterials* 2003; 24(1):181–94.

Takahashi Y, Yamamoto M, Tabata Y. Enhanced osteoinduction by controlled release of bone morphogenetic protein-2 from biodegradable sponge composed of gelatin and beta-tricalcium phosphate. *Biomaterials* 2005;26(23):4856–65.

Tampieri A, *et al.* Design of graded biomimetic osteochondral composite scaffolds. *Biomaterials* 2008;29(26):3539–46.

Tang M, *et al.* Human embryonic stem cell encapsulation in alginate microbeads in macroporous calcium phosphate cement for bone tissue engineering. *Acta Biomater* 2012;8(9):3436–45.

Teo WE, Ramakrishna S. A review on electrospinning design and nanofibre assemblies. *Nanotechnology* 2006;17(14):R89-R106.

Thomas V, *et al.* Nanostructured biocomposite scaffolds based on collagen coelectro-spun with nanohydroxyapatite. *Biomacromolecules* 2007;8(2):631–7.

Tu J, *et al.* The *in vivo* bone formation by mesenchymal stem cells in zein scaffolds. *Biomaterials* 2009;30(26):4369–76.

Updegraff DM. Semimicro determination of cellulose in biological materials. *Anal Biochem* 1969;32(3):420–4.

VandeVord PJ, *et al.* Evaluation of the biocompatibility of a chitosan scaffold in mice. *J Biomed Mater Res* 2002;59(3):585–90.

Vepari C, Kaplan DL. Silk as a Biomaterial. *Prog Polym Sci* 2007;32(8–9): 991–1007.

Wakitani S, *et al.* Mesenchymal cell-based repair of large, full-thickness defects of articular cartilage. *J Bone Joint Surg Am* 1994;76(4):579–92.

Wang HJ, *et al.* In *vivo* biocompatibility and mechanical properties of porous zein scaffolds. *Biomaterials* 2007;28(27):3952–64.

Wang J, *et al.* Osteogenic differentiation of bone marrow mesenchymal stem cells on the collagen/silk fibroin bi-template-induced biomimetic bone substitutes. *J Biomed Mater Res A* 2012;100(11):2929–38.

Wang L, Stegemann JP. Glyoxal crosslinking of cell-seeded chitosan/collagen hydro-gels for bone regeneration. *Acta Biomater* 2011;7(6):2410–7.

Wang X, *et al.* Biomaterial coatings by stepwise deposition of silk fibroin. *Langmuir* 2005;21(24):11335–41.

Wang X, *et al.* Bone repair in radii and tibias of rabbits with phosphorylated chitosan reinforced calcium phosphate cements. *Biomaterials* 2002;23(21): 4167–76.

Wee S, Gombotz WR. Protein release from alginate matrices. *Adv Drug Deliv Rev* 1998;31(3):267–285.

Weinberg CB, Bell E. A blood vessel model constructed from collagen and cultured vascular cells. *Science* 1986;231(4736):397–400.

Wieland JA, Houchin-Ray TL, Shea LD. Non-viral vector delivery from PEG-hyaluronic acid hydrogels. *J Control Release* 2007;120(3):233–41.

Wu YC, *et al.* Bone tissue engineering evaluation based on rat calvaria stromal cells cultured on modified PLGA scaffolds. *Biomaterials* 2006;27(6):896–904.

Yan LP, *et al.* Bioactive macro/micro porous silk fibroin/nano-sized calcium phosphate scaffolds with potential for bone-tissue-engineering applications. *Nanomedicine (Lond)* 2013;8(3):359–78.

Yasuda A, *et al.* In *vitro* culture of chondrocytes in a novel thermoreversible gelation polymer scaffold containing growth factors. *Tissue Eng* 2006;12(5): 1237–45.

Yoon JJ, Park TG. Degradation behaviors of biodegradable macroporous scaffolds prepared by gas foaming of effervescent salts. *J Biomed Mater Res* 2001;55(3):401–8.

Zaborowska M, *et al.* Microporous bacterial cellulose as a potential scaffold for bone regeneration. *Acta Biomater* 2010;6(7):2540–7.

Zhang F, *et al.* Silk-functionalized titanium surfaces for enhancing osteoblast functions and reducing bacterial adhesion. *Biomaterials* 2008;29(36):4751–9.

Zhang X, *et al.* Repair of rabbit femoral condyle bone defects with injectable nanohydroxyapatite/chitosan composites. *J Mater Sci Mater Med* 2012; 23(8):1941–9.

Chapter 5

Native Polymer-based 3D Substitutes in Plastic Surgery

Jing Wang M.D.[1], *Xiaoling Zhang PH.D.*[2],
Qingfeng Li M.D., PH.D.[1,*]

[1]*Department of Plastic and Reconstructive Surgery,
Shanghai Ninth People's Hospital, Shanghai
Jiao Tong University School of Medicine, 639 Zhizaoju
Road, Shanghai 200011, P.R. China*
[2]*The Key Laboratory of Stem Cell Biology,
Institute of Health Sciences, Shanghai Jiao Tong University
School of Medicine (SJTUSM) & Shanghai Institutes for
Biological Sciences (SIBS), Chinese Academy of
Sciences (CAS), P.R. China*

Rather than congenital, oncologic, and traumatic tissue damage in surgery or orthopedics, volume defect refers more to the defect of soft and connective tissues in the realm of plastic surgery. Volume defect is the most common conundrum, especially for aesthetic surgeons. Soft tissues are subjected to gravity, UV and undergo progressive atrophy as the body ages, which produce an aged appearance (Rohrich and Pessa, 2007; Truswell, 2013). Soft tissues defect also occur in patients with extensive burn, facial hemiatrophy, scleroderma, atrophic scar and other cutaneous diseases. Suspensory surgical procedures

*Corresponding author: dr.liqingfeng@shsmu.edu.cn

may return the tissues to a more youthful position, but the soft tissues defect and atrophic changes are left uncorrected. Soft tissue augmentation is one of the most popular and effective means to treat wrinkles, folds and other changes caused by aging skin (Rohrich and Pessa, 2008). At the same time, it can increase the volume of lips, nasal dorsum and other normal body parts for aesthetic purposes.

Moreover, volume defect also refers to the loss of bone or cartilage in plastic or aesthetic surgery (Shaw *et al.*, 2010). Bone defect could appear in developmental dysplasia patients such as facial hemiatrophy, facial cleft, temporal depression (especially in people of the southeast china) and other cutaneous diseases which affect underlying bones. Bone loss also contributes to aged appearances in normal people. Recent researches show that the bony elements of the mandible change significantly with age for both genders and that these changes, coupled with soft-tissue changes, lead to the appearance of the aged lower third of the face. Cartilage loss is seen in microtia, trauma, and nasal septum dysplasia in mongoloid race.

Soft-tissue augmentation requires biological or synthesized materials fillers for the loss volume. It dates back more than 100 years ago, when autologous fat grafts were filled to restore facial volume defects (Klein and Elson, 2000). Then, paraffin and liquid silicone were used for tissue augmentation and then banned by the US Food and Drug Administration (FDA) over concern of its possible toxicity and dangerous complications. Considering collagen and hyaluronic acid, the major structural component of the dermis found in humans, the introduction of hyaluronic acid (HA) gel fillers was a turning point. Using these fillers in the deeper layer of the face, subcutaneous and deeper, brought subtle, yet definitive, rejuvenation to the patients' faces. The initial movement into three-dimensional correction with injectable fillers began with the lip volume augmentation with the collagen fillers carried out by Arnold W. Klein, MD, in the 1980s. This volume correction achieved in the lip far exceeded the impact of wrinkle and scar correction, and gendered an enormous impact on clinical practice.

Cosmetic injection has become the most popular aesthetic surgery as it is a minimally invasive procedure and makes itself the most suitable for facial rejuvenation. Fillers in aesthetic surgery always refer to injectable fillers such as hyaluronic acid (HA), collagen and liquid silicone, calcium hydroxylapatite (CaHA). The latter two non-biodegradable materials were bioengineered with gel carrier into injectable filler nowadays. These agents, most of which were synthesized in the last half decade, are divided into semi-permanent and permanent fillers across several categories. The semi-permanent fillers includes hyaluronic acid, calcium hydroxylapatite, poly-L-lactic acid and the longer-lasting, so-called "permanent fillers," consist of polymethyl methacrylate microspheres (PMMA), which are highly purified forms of liquid silicone and hydrogel polymers. The permanent fillers are meant to seek extended duration of effect, exempting patients from continuous injection. However companied with their long duration, more potential complications are much more troublesome to solve than the semi-permanent fillers mentioned above. In this chapter, we will focus on bioengineered native polymer-based 3D materials such as hyaluronic acid, collagen and their derivatives which are approved by FDA and widely used in plastic surgery.

1. Bioengineered Hyaluronic Acid and its Derivatives

1.1. *Introduction*

Hyaluronic acid (HA, also called hyaluronate) is a non-sulfated, linear polysaccharide with the repeating disaccharide, β-1,4-D-glucuronic acid–β-1,3-N-acetyl-D-glucosamine. Its structural and biological properties mediate its activity in cellular signaling, wound repair, morphogenesis, and matrix organization. It is a non-sulfated glycosaminoglycan, and is found throughout the body, from the eye to the extracellular matrix. It is well suitable for biomedical applications. A significant property of HA is its capacity to bind huge amounts of water (1000 fold of its own weight) (Toole, 2004). Therefore, HA functions as a biological lubricant in joints, reducing friction during movement and providing resiliency under static conditions. Hyaluronic acid is also involved in the transport of essential nutrients to the skin's viable cells. It also helps to maintain water within the extracellular space, increasing its volume and density. It provides volume for the contribution to the skin's overall appearance (Jiang *et al.*, 2007; Prestwich and Kuo, 2008).

1.2. *Category and characteristic*

Since HA is rapidly biodegraded in the body by hyaluronidase, with tissue half-lives ranging from hours to days, it is modified in many ways to alter its properties for long-lasting effects, which include modifications leading to hydrophobicity and biological activity. Chemical modifications of HA target three functional groups: the glucuronic acid carboxylic acid, the primary and secondary hydroxyl groups, and the N-acetyl group. Hylaform®, Juvederm®, Perlane® and Restylane® are the most common injectable HA derivatives in plastic surgery. Juvederm, Restylane and Perlane are chemically cross-linked with BDDE (1,4-butanediol diglycidyl ether), stabilized and sus-pended in phosphate buffered saline. Hylaform is cross-linked with divinylsulphone (DVS). Cross-linking quality has to be in the right balance between duration and the biocompatibility of the HA filler. Each cross-linking agent has its characteristics that affect the perfor-mance of filler. Juvederm is produced by a proprietary manufacturing process referred to as "Hylacross technology", which means that Juvederm is not "sized" in contrast to the other HA fillers (Restylane, Perlane) which use sizing technology. "Sizing" is the process in which cross-linked HA is pushed through a sized screen and broken into pieces (Bogdan Allemann and Baumann, 2008). The medium size pieces of HA are made into Restylane while the larger ones are made into Perlane. It is not known what effect the sizing technology or the Hylacross technology have on a filler's performance (Skardal *et al.*, 2010).

Hyaluronic acid can be refined from bacterial or avian sources, and each product has its own specific characteristics. Cross-linked hyaluronic acid of avian origin became the first noncollagen filler to be widely used (Burdick and Prestwich, 2011). The first hyaluronan biomedical product was developed in the 1970s and 1980s, and is approved for use in eye surgery. In 2003, the FDA approved hyaluro-nan injections for filling soft tissue defects under the trade name

Restylane®. The Hylaform product family is based on hyaluronic acid derived from rooster combs. Typical examples for bacterial hyaluronic acid products are the Restylane and Juvederm families. Perlane is a more highly cross-linked product than Restylane. Cross-linking is the procedure that endows these products the ability to persist in the tissue and provide long term benefits. In the case of Restylane, this product may last for about a year or more. Perlane persists a bit longer and that duration time seems to be four to six months longer than the Restylane. These hyaluronic acid fillers are proven to have a longer-lasting effect than traditional bovine collagen. The nonanimal-based fillers can be administered without pretesting so that hyaluronic acid products surpassed collagen to become the new "gold standard" for soft tissue fillers. HA fillers have the major share of the US market place for injectable fillers, with Juvederm and Restylane dominating the market (Table 1).

The pivotal FDA trials and duration trials for Restylane, Juvederm have gained similar data about its clinical effect. All three products seem to produce a similar result in the nasolabial fold (NLF) despite differences in elastic modulus, cohesivity, concentration, cross-linking agents, and amounts of free HA. Specifically, randomized, double-blinded trials comparing each product with Zyplast collagen show that the effect of each HA fillers product is superior to Zyplast at six months (Baumann *et al.*, 2007; Narins *et al.*, 2003; Narins *et al.*, 2010).

Table 1. HA dermal fillers of the most common injected.

Brand	Corporation	Crosslinking agent	Resource	Lasting time
Restylane, Perlane	Medicis, Scottsdale, Arizona	BDDE	generated by Streptococcus in a labratory	up to 6 months for Restylane; up to 12 months longer for Perlane
Juvederm	Allergan, Irvine, California	BDDE	Same as above	Up to one year
Hylaform	INAMED Aesthetics	DVS	purified from rooster comber	up to 6 months

2. Bioengineered Collagen and Its Derivatives

2.1. *Introduction*

Collagen is the major structural component of the dermis and is responsible for providing strength and support to human skin. In addition to glycosaminoglycans and elastin fibers, dermal matrix in adult skin is mainly composed of type I (80–85%) and type III collagen (10–15%) (Baumann *et al.*, 2006). Collagen types I and III are initially synthesized by dermal fibroblasts as α procollagens and then hydroxylation of proline and lysyl residues is performed by prolyl and lysyl hydroxylase in the presence of copper and ascorbic acid (vitamin C). The alpha chains form the triple helix and then get secreted into the extracellular space, where the protocollagen N-proteinase and protocollagen C-proteinase cleave the amino and carboxy terminal domains of the propeptides. Collagen type I triple helices comprise of two α1 and one α2 chains while collagen III triple helices contain three identical α1 chains. Collagen molecules then assemble to form fibrils with other noncollageneous molecules (Fenske and Lober, 1986; Lavker, 1979).

The molecular subunit of collagen is tropocollagen. It is a long, narrow rod with a molecular weight of ~300,000 (Homicz and Watson, 2004). The non-helical telopeptide end is believed to be the most antigenic portion of the collagen molecule. There are naturally occurring covalent cross-links within and between molecules. Release of collagenase subsequent to phagocytosis is responsible for its degradation (Bailey, 2000). Cross-linking and increasing the concentration of the product have been the main strategies for decreasing the effectiveness of phagocytosis of injectable collagen. Further increasing this natural cross-linking in the molecule prolongs the duration time of this product. In animal studies, there was a typical polymorphonuclear response between 12 and 72 hours that released in five to seven days and was followed by an influx of vascular channels and fibroblasts. During a six-month period, the implant retained its original random phenotype. There was also no evidence of migration or encapsulaton. Early animal studies showed persistence at 22 weeks to six months of products. In early studies, biopsies of injected areas in rodents showed no persistent inflammation and collagen indistin-guishable from the surrounding host collagen (Sclafani and Romo, 2001; Webster *et al.*, 1984).

2.2. *Category and characteristic*

There are currently three kinds of FDA-approved injectable collagen: bovine calf skin (Zyderm and Zyplast), human genetically engineered dermal fibroblasts (CosmoDerm and CosmoPlast), and porcine gastrocnemius tendons (Evolence). Zyderm I was approved in 1981, Zyderm II and Zyplast in 1983, CosmoDerm and CosmoPlast in 2003, and Evolence in 2008. Zyderm I, Zyplast, CosmoDerm I, CosmoPlast, and Evolence are all have a concentration of 3.5% collagen (35 mg/mL). Zyderm II and CosmoPlast II are 6.5% collagen (65 mg/mL). Zyderm I and CosmoPlast I are designed for injection into the superficial dermis. Zyderm II, Zyplast, CosmoDerm II, CosmoPlast, and Evolence are intended to be injected into the mid to deep dermis. These products are all buffered to be approximately pH neutral in saline and contain lidocaine to decrease the pain of injection. Zyplast and CosmoPlast are cross-linked with glutaraldehyde. Evolence is cross-linked with D-ribose. (Table 2).

Porcine collagen more closely resembles human collagen. Because it so closely resembles human collagen and because the N-terminals are removed in the production process, it should have significantly decreased immunogenicity. In a study of 519 patients, there were no hypersensitivity reactions (Shoshani *et al.*, 2007). In one small study comparing Evolence with Zyplast (12 patients), better persistence of Evolence was achieved at 18 months and no elevation of immunoglobulins was found during the study (Monstrey *et al.*, 2007). In a larger study comparing melolabial fold augmenta-tion of porcine collagen and Restylane (Medicis Aesthetics, Scottsdale, AZ), there was no difference in the effect at six months. One percent of the subjects developed asymptomatic IgG antibodies to the porcine collagen. It has also been reported persistence of aug-mentation at 12 months. Within Zyderm II and CosmoDerm II, it was initially believed that increasing the concentration helps to

Table 2. Collagen dermal fillers of the most common injected.

Brand	Corporation	Crosslinking agent	Resource	Concentration	Lasting time
Zyplast	(Allergan, Irvine, CA)	glutaraldehyde	Bovine	35 mg/mL(3.5% collagen)	3 months
Zyderm	(Allergan, Irvine, CA)	glutaraldehyde	Bovine	35 mg/mL,65 mg/mL for Zyderm II	3 months
Cosmoderm	(Allergan, Irvine, CA)	glutaraldehyde	Human	35 mg/mL	4–6 months
Cosmoplast	(Allergan, Irvine, CA)	glutaraldehyde	Human	35 mg/mL, 65 mg/mL for Cosmoplast II	4–6 months
Evolence	(Colbar, Herzliya Israel)	D -ribose	Porcine	65 mg/mL	6 months
Artefill	Artefill (Artes, San Diego, CA)	Polymethylmethacrylate (PMMA)	Bovine	3.5% bovine collagen; 6 million PMMA nanoparticle/1 mL	permanent

decrease the rate of phagocytosis. The higher concentration required deeper augmentation, mid to deep dermis injection instead for the superficial dermis.

Immunogenicity is an issue with all collagen implants (DeLustro *et al.*, 1987; Frank *et al.*, 1991; Moscona *et al.*, 1993). Late local and systemic reactions to bovine collagen and elevations of immunoglobulins have been reported. Bovine collagen is significantly immunogenic. The incidence of allergic reaction to skin tests is ~3 to 5%. During its purification and manufacturing process, cleavage of many of the telopeptides decreases but does not eliminate its immunogenicity. Although cross-linking also decreases the number of telopeptides, it may also decrease the immunogenicity, but it has not been proven definitively. Patients should be skin tested twice within one month between the first and second tests and receive treatment no sooner than two weeks after the second skin test. A small percentage of patients can still react to the product even though negative skin tests are achieved. Hypertrophic scar may occasionally form at the site of an initial skin test which required injection with triamcinolone to help it resolve. Patients with documented autoimmune disease should not be injected, and atopic individuals are expected to have a higher incidence of allergic reactions. The FDA has not required skin testing for human and porcine collagen, but there is also evidence of immunogenicity with these products. The Inamed Human Collagen Immunogenicity Clinical Study showed, with a 95% upper confidence level, that the chance of experiencing a hypersensitivity reaction to CosmoDerm and CosmoPlast was less than 1.3% (Takasaki *et al.*, 1995). Stolman (2005) reported on two cases of reactions to CosmoDerm that persisted three weeks and three months, respectively. One of the patients had a history of a previous skin hypersensitivity to Zyplast. The author has seen localized swelling and persistent erythema with human collagen in one patient previously treated successfully with Zyderm. The areas improved some with four weeks of topical steroid but essentially took four months to resolve. The areas were raised and red but not itchy. The literature on porcine collagen is still less extensive, but elevations of IgG have been documented (Cockerham and Hsu, 2009).

ArteFill: although both patients and physicians accepted the inconvenience and repeated expense of short-term fillers, they sought a dermal filler substance that promised long-lasting results with extremely low incidences of complications and adverse events. With the development of ArteFill (Suneva Medical, San Diego, CA), a viscous liquid of polymethylmethacrylate (PMMA) microspheres suspended in solubilized bovine collagen, and its approval by the FDA in 2006, a long-lasting subdermal filler is now available (Cohen and Holmes, 2004b). After seeking the optimal collagen/PMMA ratios in laboratory animals, trials on humans were initiated. The first-generation product, called Arteplast, as well as its successor, Artecoll, proved efficacious, although adverse events did emerge (Lemperle *et al.*, 2006a; Lemperle and Gauthier-Hazan, 2009b; Lemperle *et al.*, 2009). Most of these adverse events were firm nodularities at injection sites, occasionally with an associated inflammatory response. In several instances, surgical removal of the implant was required. Further investigation led to the conclusion that there is a specific threshold of PMMA microsphere size that is critical to avoid phagocytosis by macrophages and giant cell formation with resulting granulomatous inflammation. Associated observations suggested that small PMMA microspheres, less than 20 μm in diameter, engendered a foreign body response. The combination of two widely used and proven biocompatible materials, bovine collagen (sutures, hemostatic agents, implants) and PMMA (orthopedic bone cement, craniotomy plates), satisfies biocompatibility issues. The microscopically small particles of PMMA in the bovine collagen carrier are enveloped by autologous collagen as the byproduct of natural connective tissue turnover, leaving a pliable and permanent tissue residual. The key to ArteFill's biocompatibility and safety, as documented in animal experiments, is the extremely uniform, round and smooth PMMA microspheres, and especially the absence of particles less than 20 μm in diameter (Lemperle *et al.*, 2004). The smooth surface morphology of ArteFill's PMMA microspheres also appears to mitigate an inflammatory response. Microscopically, macrophages and foreign body giant cells can be detected around particles with an irregular surface. This may explain the rather high rate of granulomas after injection of Dermalive, whose particles are characterized by an irregular, rugged surface. It has also been observed that small, irregular particulate materials such as polyurethane foam and the silicone particles on the surface of textured breast implants often elicit a chronic granulomatous tissue reaction.

The reaction to foreign material is the formation of "granulation tissue," composed of macrophages, fibroblasts, capillaries, and collagen that fill the interstitial spaces between the microspheres. The bovine collagen appears to maintain the separation between the microspheres and facilitates tissue ingrowth. Approximately four weeks after implantation, the ArteFill implant consists of 20% inert microspheres and 80% granulation tissue. This ratio may vary at four weeks depending on the volume of the material implanted. Subsequently, over time the connective tissue matures through a natural process similar to scar formation and the interstices are filled with fibroblasts and autologous collagen fibers. Histology at three months demonstrates that all of the PMMA microspheres are completely encapsulated and are surrounded by fibroblasts and collagen fibers. Macrophages are rare and capillary in-growth is evident. Human histology after 10 years revealed strong bands of mature collagen fibers with fully intact capillary vasculature surrounding intact PMMA microspheres. In essence, the ArteFill injection serves as a scaffold to promote a "living implant." The PMMA components of ArteFill become fully integrated into the connective tissue, whether dermis or subdermal spaces. As in normal tissue with sufficient blood supply, there appears to be constant turnover of cells, including fibroblasts and macrophages (Nicolau, 2007).

A sustained tissue augmentation effect appears achievable only with a non-absorbable synthetic component to the filler (Cohen *et al.*, 2006; Cohen and Holmes, 2004a). ArteFill's 30–50 µm diameter PMMA microspheres seem to be the ideal size for dermal injections — large enough to escape phagocytosis yet small enough to be smoothly delivered through a fine 26G or 30G needle. There are other specific advantages to microspheres in this size range. The smaller the microspheres (to the threshold of phagocytosis), the larger their combined surface area in a given volume and the greater the total amount of new collagen deposition. Microspheres with a diameter of 100 lm, for example, promote the ingrowth of only about 56% connective tissue, whereas microspheres with a mean diameter of 40 µm promote the ingrowth of about 80% connective tissue. The PMMA/collagen ratio and the 30–50 µm PMMA microsphere sizing for ArteFill are direct applications of these observations. Two key features appear to prevent dissipation of ArteFill into the surrounding fine network of collagen fibers of the deep dermal and subdermal layers following injection.

3. Bioengineered Poly-L-Lactic Acid (PLLA)

PLLA is a synthetic polymer that is biodegradable and resorbable. Polylactic acids do not occur naturally, but were synthesized by French chemists in 1954. Polylactic acid and polyglycolic acid have been used safely in suture materials; in resorbable plates and screws; in guided bone regeneration; in orthopedic, neurologic, and craniofacial surgery; and as drug delivery devices. Unlike HA fillers, the effects of PLLA are gradually achieved as Sculptra induces an expansion of dermal thickness. The substance is degraded by conversion to lactic acid monomers that are subsequently metabolized to glucose and CO_2 (Dayan and Bassichis, 2008; Moyle *et al.*, 2004).

Polylactic acid is metabolized to carbon dioxide and water. Long-term tissue filling effects are caused by ingrowth of type I collagen into the areas of accumulated particles as the polylactic acid microspheres undergo dissolution, which takes place weeks to months after injection. Nine months after implantation, no polymer or remnant cicatricial fibrosis could be detected histologically, demonstrating good biocompatibility of the polylactic acid microspheres. Sculptra contains no animal proteins, so allergies are not expected; however, tumor necrosis factor-alpha (TNF-α) is produced by phagocytes (Morhenn *et al.*, 2002; Vleggaar and Bauer, 2004).

Sculptra obtained U.S. Food and Drug Administration approval in 2004 for use as a soft-tissue filler into lipoatrophy of cheeks and temples of human immunodeficiency virus patients who are under highly active antiretroviral therapy (Alam *et al.*, 2008). Sculptra has demonstrated efficacy by increasing dermal thickness of up to three times baseline, which correlates with a clinically visible volume deficit correction. Quality of life and anxiety and depression were also improved in human immunodeficiency virus–positive patients. Sculptra has demonstrated safety in human immunodeficiency virus patients, with short-term adverse events consisting of localized ecchymosis and edema at the site of injection and long-term events that included nonvisible subcutaneous papules occurring within up to two years after injection and granulomas at nine to 14 months after injection (Moyle *et al.*, 2004; Valantin *et al.*, 2003).

Adverse events include palpable but nonvisible nodules that can be effectively dissipated with daily massage. Concerns over delayed-type hypersensitivity reactions occurring months following injections may be hindering its widespread acceptance as a cosmetic agent. Overall, the delayed results, pain on injection, and high price contribute to a product that is not as "user-friendly" as some of the other materials used for HIV lipoatrophy and aesthetic correction.

4. Clinical Indications

Fillers must be injected in the subcutaneous or supraperiosteal plane when volumizing the mid-face. Intradermal or too-superficial injection may create persistent dermal contour irregularities. Hyaluronic acid dermal fillers can help to temporarily replace the lost hyaluronic acid and restore the skin's volume and smooth natural appearance. They are indicated for injection into the mid to deep dermis for correction of moderate to severe facial wrinkles and folds (such as nasolabial folds). They are very useful for deeper folds, lips, and irregularities such as soft acne scars, nasal deformities, and areas that require more sculpting. One of the main indications for hyaluronic acid is treatment of the perioral area and augmentation of the lips. The application in the bags (tear trough deformity) under the eyes is reserved for specialists specifically trained in this technique and possessing a sound knowledge of the physiology in that particular area. Hyaluronic acid should not be injected into the blood vessels (intravascular) and not be used in patients who tend to develop hypertrophic scarring. In association with Botox, hyaluronic acid can produce an excellent rejuvenation effect on the face. It should not be used in association with laser therapy, chemical peeling, or dermal abrasion. For surface peeling, it is recommended not to inject the product if the inflammatory reaction generated is significant (Cohen *et al.*, 2013; Dayan and Bassichis, 2008; Glogau, 2012; Prestwich and Kuo, 2008).

All currently FDA-approved HAs have only been studied in NLFs, and carry the specific indication on package inserts that they are approved for dermal injection for correction of moderate to severe facial wrinkles and folds, such as NLFs. Although not specifically FDA-approved, available HAs have been studied off-label for correction of glabellar rhytides, oral commissures (meilolabial folds), lips, mid-face volumizing, infraorbital or naso-jugal grooves (tear troughs), and augmentation of the dorsal nose (Carruthers *et al.*, 2008).

4.1. *Rhinoplasty*

4.1.1. *Anatomy*

Rhinoplasty and blepharoplasty are probably the most popular cosmetic surgeries performed in most of the Far East. Compared with Caucasians, Asians generally have a shorter, wider, and less projecting nose. While reduction rhinoplasty with dorsal hump reduction and some form of lower lateral cartilage reduction is more popular in Caucasians, augmentation rhinoplasty is most frequently performed on Asian patients (Liao *et al.*, 2007; Shirakabe *et al.*, 2003; Tham *et al.*, 2005). There are many types of Oriental nose morphologies. There is, however, a common set of characteristics that differentiates them from Caucasian noses (Aung *et al.*, 2000). To better analyze these differences, the nose can be divided into thirds. In its upper third, the nasal bridge is lower. In the middle third, the dorsum is less projecting. Japanese patients also frequently present a convex dorsum. In the lower nasal third, there is less tip projection, a poorly defined or absent supratip break, and a rounder nasal tip with infratip fullness. The labionasal angle is sharper, and the nasal base is wider compared with nasal height. Alar flaring among some Koreans has been attributed to increased skin thickness, hypertrophy of the dilator naris anterior, and a more anterior insertion of the dilator naris posterior (Chun *et al.*, 2008). Most patients requesting rhinoplasty ask for dorsum augmentation and tip projection improvement.

4.1.2. *Clinical usage*

Silicone implant augmentation is the most widely performed type of augmentation rhinoplasty in Asia. Filler rhinoplasty, also known as augmentation rhinoplasty, has been proposed to patients interested in less invasive techniques than traditional rhinoplasty (Fernandez and Mackley, 2006). The volume of material used for filler-based rhinoplasty usually ranges from 0.1 to 0.4 mL. This feature makes filler-based rhinoplasty an excellent indication for patients who will be treated with fillers for improvement of the lips and nasal labial folds and are unsatisfied with their nose (de Lacerda and Zancanaro, 2007; Han *et al.*, 2006).

Multiplane hyaluronic acid injection technique was recommended: simulating silicone nasal augmentation techniques, the authors sequentially inject the filler under the nasodorsal fascia, on the perichondrium, and between the alar cartilages. Injection into the correct plane of the nose is the key to a successful treatment. Traditionally, hyaluronic acid was injected into the deep dermis, with material shifting often occurring because of low surrounding tissue density. Therefore, we inject filler material into specific planes according to different anatomical features of different parts of the nasal dorsum: under nasodorsal fascia at the bony dorsum area, on the perichondrium at the nasal cartilage area, and between the alar cartilages at the nasal tip area. The rigidity of these structures can help pressurize the filler material, thereby restraining it at designated places and providing a vividly natural appearing result. At the same time, material shifting and local area swelling are avoided (Xue *et al.*, 2012) (Fig. 1).

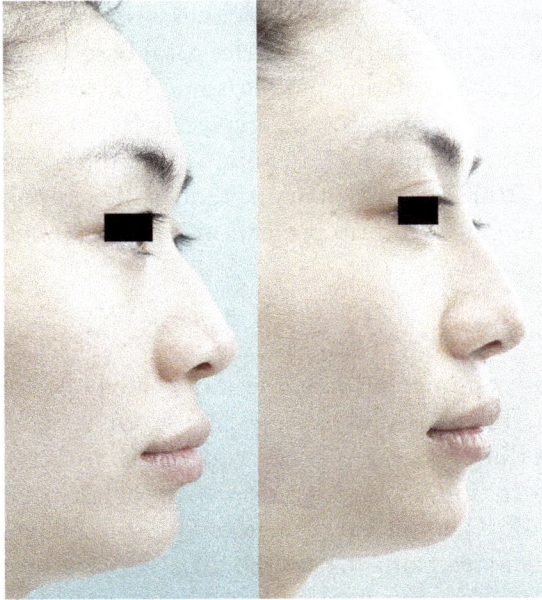

Fig. 1. pretreatment (left) and posttreatment (right) HA injection for rhinoplasty.

4.2. *Nasolabial fold*

4.2.1. *Anatomy*

The nasolabial folds, commonly known as "smile lines" or "laugh lines," are facial features. They are the two skin folds that run from each side of the nose to the corners of the mouth. They separate the cheeks from the upper lip. The nasolabial fold is made up of the following: dense fibrous tissue, muscle fibers branching from the elevators of the upper lip muscle, elevators of the upper lip muscles passing through the fold on the way to the upper lip vermilion, muscle fibers originating in the labialfold fascia (fold muscle). The nasolabial fold is the boundary between the cheek and the lip, which seems to be a true anatomical border, a transition zone between the fatty cheek above and the lean upper lip below. In the cheek, a generous layer of fat covers the mimetic muscles. In contrast, medial to the fold over the upper lip, the dermis rests against the perioral sphincter, with little or no intervening fat (Barton and Gyimesi, 1997; Pogrel *et al.*, 1998; Yousif *et al.*, 1994).

The fold is delineated by the presence of the buccal fat pad superior to the fold, which tends to bulge over the fold laterally, giving it further definition. The buccal fat pad appears to be held in place by a combination of muscle bundles running parallel and at right angles to the nasolabial fold and also by a combination of fibrous septae running through the buccal fat pad itself. Where the muscles are not in direct contact or where the fibrous septae are incomplete or running in an abnormal direction, herniation inferiorly of the buccal fat pad occurs, causing loss of definition of the nasolabial fold. The amount and relative content of the buccal fat pad may be a major determinant in the depth of the nasolabial fold (Rubin, 1999).

The sum of the midface change and related ptosis produces a deepening of the nasolabial folds. Maxillary skeletal regression and loss of dental vertical dimension are contributors. The dermal muscle insertions of the levator labii superioris, zygomatic major/minor, the risorius cross the superficial musculoaponeurotic systems (SMAS) to form the nasolabial fold. The nasolabial fold is also formed by the muscular SMAS (modiolus) and the nonmuscular SMAS (subcutaneous tissue and fat). A muscular and tonic modiodus and SMAS support the nasolabial fold in youths. The SMAS supports the buccal fact pad and holds it in place, preventing forward protrusion and contributes to a deep (concave) nasolabial fold. The loss of muscle tone of the modiolus and SMAS, with the resultant ptosis, contributes to the deepening of the nasolabial fold with aging. This, along with overlying fat and dermal atrophy and osseous contraction, add to the gradual deepening of the nasolabial fold over time. With the weakening of the cheek-retaining ligament systems, the malar tissue come ptotic, but fat cannot "cross" the nasolabial fold because of the dense facia-to-dermis adherence within the fold (Barton, 1992).

The standard rhytidectomy does not flatten the fold because elevating the skin over the fold does not alter nasolabial fold position, as it is tethered to the lip levator muscles. Weight increase is accompanied by the deposition of fat in the cheek, which raises the superior wall of the fold, meanwhile subacute atrophy of the cheek fat and subcutaneous tissue of the cheek leaves loose and wrinkled skin drooping over the fold. The nasolabial fold has been assumed to be merely a valley that can be filled with collagen, dermis, or fat grafts. These efforts introduce padding between the dermis of the lip and the superolateral border of the orbicularis oris, and succeed only in forcing the fold medially, albeit softening a hitherto sharp crease.

4.2.2. *Clinical usage*

Both hyaluronic acid and collagen can be used to correct deepen nasolabial fold (Arlette and Trotter, 2008). An assessment of the depth and character of the nasolabial fold is critical to a successful outcome. Lemperle *et al.* provide a useful classification system for grading nasolabial fold depth. A concomitant face lift will also affect the degree to which the fold will require soft-tissue filler augmentation (blunting). The fold will never be fully correct, and this would be unnatural even if it were possible. Nevertheless, a 50 percent or more correction is attainable with proper technique and patient selection. Soft-tissue augmentation of a deep fold can be a powerful tool when combined with a midface or face lift technique that also addresses the nasolabial region. A combination of serial puncture and linear threading in the mid to deeper dermis is used in this region, while the nasolabial fold is held taut. Serial puncture injections should be aimed medially away from the large cheek fold, beginning inferiorly and moving superiorly. The dermis is often thicker as one moves superiorly. Beveling the needle up assists with etched-in lines, as do injection of smaller-particle hyaluronic acid in a layered fashion. Molding of the hyaluronic acid by immediately massaging it will help soften and smooth out the blunted fold. Overcorrection can result in an awkward, paradoxically aged appearance when smiling, as the softness in the upper one-third of the nasolabial fold is natural and youthful. Approximately 0.5 to 2.0 cc is used per patient for the nasolabial region. As stated above, complete fold correction is not desirable, is difficult to attain, and should not be the goal in this region. Subsequent touch-up layering or further blunting of fold depth can be accomplished at follow-up visits (Narins *et al.*, 2008) (Fig. 2).

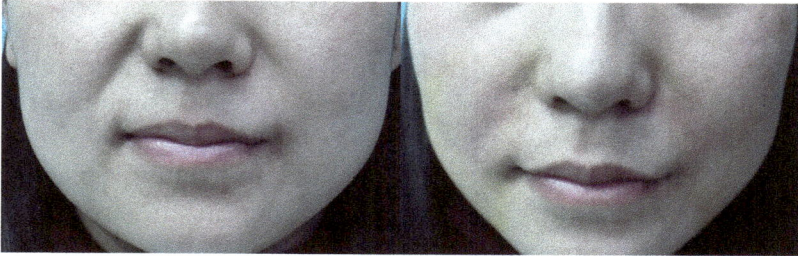

Fig. 2. pretreatment (left) and posttreatment (right) HA injection for correction of the deepen nasolabial folds.

4.3. *Glabellar rhytides*

4.3.1. *Anatomy*

The glabellar occupies a relatively central position in the face and therefore easily attracts the attention of patients and their observers. Rhytids in this region, which range from fine lines to deep furrows, may result in the patient being incorrectly seen as angry, anxious, fatigued, fearful, or of advanced age (Lemperle *et al.*, 2001; Macdonald *et al.*, 1998). Most commonly, glabellar rhytids are dynamic in nature. These hyperfunctional lines are the result of a pull on the skin by the underlying facial musculature. This is in contrast to facial wrinkles in other areas, which result from age-induced changes in the collagen of the dermis. While the latter are frequently sequelae of sun damage and aging in an older population, hyperfunctional glabellar lines may be seen in younger patients, aged 20 to 50 years. Individuals present different patterns of rhytid formation according to their habits of facial expression and resting facial posture. The vertically oriented procerus muscle is thought to make the greatest contribution to the formation of horizontal glabellar furrows, while the corrugator supercilii muscles produce the vertical rhytids in this region. There may be some contribution from a muscle identified as the depressor supercilii, but descriptions of this muscle are few, bringing its existence as a distinct entity into question. The elastic properties of the skin and factors associated with photoaging contribute to rhytid formation but play a much less important role.

4.3.2. *Clinical usage*

Optimal treatment of the glabellar region often requires combined treatment with both hyaluronic acid and Botox. This combined treatment modality can increase the longevity of the treatment to as long as nine months. A serial puncture technique is used in this region for the deep and/or wider folds, while staying in the mid-dermis. Injection along the rhytide(s) is performed while the needle is being pulled out. The finer etched-in lines can be treated with a dermal-epidermal level injection of small-particle hyaluronic acid (or collagen), with a precise serial puncture technique and subsequent linear injections to disrupt the fold and provide complete geometric filling. It is important to compress the supratrochlear vessels with the nondominant hand while injecting, to prevent inadvertent intravascular injection and minimize bruising. Approximately 0.5 cc is used per patient in this region. In wrinkle treatment, the resulting structural support of Arte Fill's six million microspheres per milliliter prevents further wrinkling and folding. This allows the diminished thickness of the corium to recover. This wrinkle recovery process appears similar to the well-known phenomenon after facial palsy in older patients and in stroke victims (or repeat Botox patients), whose facial wrinkles and furrows on the paralyzed side completely disappear over time due to the lack of movement (Fig. 3).

4.4. *Lip enhancement*

Optimal lip rejuvenation involves two main components: volume enhancement and vermiliocutaneous border enhancement. Volume filling is often required in older patients and those who have thin lips. Vermiliocutaneous enhancement is usually required in younger individuals who have enough volume, but is also indicated in older patients, along with volume augmentation.

Linear threading or serial puncture techniques are implemented starting at the oral commissures and proceeding in a lateral to medial direction. Marionette lines are a key element in overall lip enhancement; otherwise, results are destined to be disappointing to both

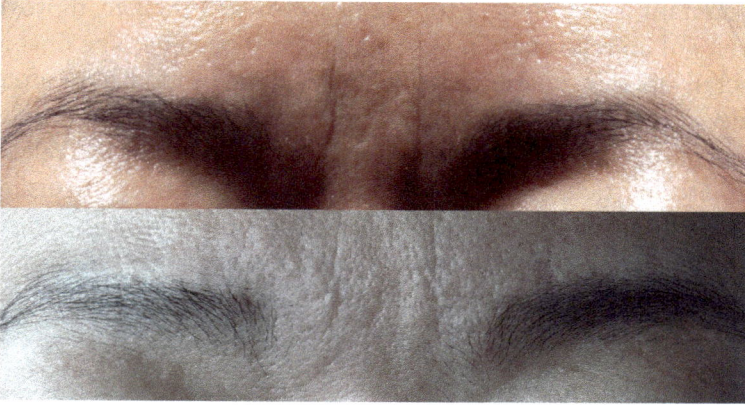

Fig. 3. pretreatment (upper) and posttreatment (down) HA injection for correction of glabellar rhytides.

the patient and the physician. A cross-radial technique is used around the oral commissure and marionette line to enhance and "lift," or fill in, the corners of the mouth. Botox injection into the depressor angulioris can further enhance this lifting effect. The dermal level is, once again, the mid-dermis. A range of 0.5 to 1.5 cc is often needed for each lip. Care should be taken to avoid superficial injection in this region, as a light blue hue may become visible. Intrainjection and postinjection palpation for surface irregularities are important. If material tracks away from the intended injection plane and created tunnel, then immediate massage is necessary to recontour the area. Massaging should be instituted immediately by the physician, as this is the best time to achieve molding and shaping. This avoids later discomfort which can be present if the patient is given that task. Injection of the lip itself can be accomplished at the submucosal level, within the superficial or bicularis oris muscle mass. Placing the hyaluronic acid in this deeper level decreases its visibility and augments lip volume. Minimal augmentation of the philtral columns can further enrich the periorbital and lip augmentation. Restylane can also be used to refine the white rolls, which will enhance the overall aesthetic result. More superficial, finer vertical rhytides are augmented with smaller-particle hyaluronic acid (Restylane Fine Line) or collagen.

Concomitant injection of 2 to 4 U of Botox will further improve the longevity of lip rejuvenation by as much as 50 percent. The final result of overall lip rejuvenation should be evident immediately after the injections, unless excess bruising and edema are present. Immediate swelling is uncommon and may be a result of histamine release or immediate particle expansion by water absorption. Bruising, if present, should be controlled with compression during the injection so that there is no compromise of the final result from blood staining or volume due to extravasated blood (Rohrich *et al.*, 2007).

4.5. *Nasojugal grooves (tear troughs)*

4.5.1. *Anatomy*

The nasojugal groove, commonly known as the tear trough, is a distinct cutaneous groove that extends inferolaterally from the medial canthus to approximately the medial pupillary line and becomes progressively more prominent with aging (Flowers, 1993). Extending laterally from this point is the palpebromalar groove. With more advanced aging, these two grooves connect to become a continuous groove that sharply demarcates the bulging orbital fat above, from the retruded midcheek below (Mendelson and Jacobson, 2008).

However, its exact anatomical origin remains essentially unknown, with various authors offering conflicting descriptions of its anatomy. These include the following: it is the prominence of the orbital rim resulting from descent of the malar fat pad; it is the attachment of the orbital septum to the arcus marginalis; it is the result of a loss of fat in the tear trough or herniation of the orbital fat superior to the trough; it is the result of a triangular confluence of the origins of the orbicularis oculi, the levator labii superioris alaeque nasi, and the levator labii superioris; and it is a "cleft" between the palpebral and orbital parts of the orbicularis oculi (Haddock *et al.*, 2009; Loeb, 1993).

The main components of the tear trough are believed to be the hollow itself; the fat bulge just superior to it; and the very distinct change of skin quality, color, and thickness between the lid and the cheek. The intrinsic loss at the tear trough is magnified by the overlying fat pad which as it enlarges, increases the shadowing and apparent depth of the tear trough. For reasons probably related to septal containment and ligamentous support, the fat pads do not transgress the borders of the tear trough. Firstly, a true osteocutaneous ligament, called the tear trough ligament, exists in the medial suborbital region of the maxilla, extending from the level of the insertion of the medial canthal tendon, just inferior to the anterior lacrimal crest, to approximately the medial pupil line. Secondly, the tear trough ligament is the main etiologic factor responsible for the tear trough deformity. (Muzaffar *et al.*, 2002). Thirdly, the tear trough ligament is continuous laterally as the bilayered orbicularis retaining ligament. This is the anatomical basis for the clinical manifestation of a deep lid-cheek junction. Nonsurgical correction of the tear trough deformity with hyaluronic acid is effective and safe and is associated with high patient satisfaction. Hyaluronic acid can be injected preperiosteally on each side, with placement of the filler strictly below the tear trough ligament. This effectively softens the tear trough deformity. The cause of the tear trough deformity is most likely multifactorial. Volume loss seems to predominate. However, orbital fat herniation, skin laxity, and possible ptosis of tissues below secondary to volumetric changes or other reasons could all play a role. (Lucarelli *et al.*, 2000). The tear trough is in fact a dynamic area. It would stand to reason, given the fixed attachment of the muscle to the rim and contraction of muscles, notably the orbicularis oculi, that the tear trough would show the consequences of volume loss earlier and more dramatically than the rest of the face. It is unclear which component of the tear trough is the most responsible in general, but fortunately, as with most facial cosmetic surgery, understanding the cause of a deformity is not necessary to undertake its correction. (Wong *et al.*, 2012).

4.5.2. *Clinical usage*

Although fat injections can be used easily in this area, along with surgical modalities, hyaluronic acid augmentation is an alternative that may be more predictable, with less risk of postinjection irregularities

(Hirmand, 2010). This is particularly true of the tear trough region. Serial and linear threading is used in this region starting from a lateral to medial direction, with molding of the filler as one proceeds. The injection plane is supraperiosteal. Often, approximately 0.5 to 1.0 cc is all that is required in the tear trough region, but up to 2.0 cc of hyaluronic acid may be used for malar augmentation. The malar injection plane is also just superficial to the periosteum. Larger-particle hyaluronic acids, such as Perlane, once approved for use in the United States, will be especially useful for deeper augmentation planes. Light massage of the area after injection allows the implanted material to conform to the contours of adjacent tissues. The tear trough region is prone to bruising because of the thin skin and increased vascularity of the periorbital area. Placing cold compresses over the area of injection for the first 24 hours can reduce ecchymosis and swelling (Lambros, 2007).

5. Injection Techniques

Several injection techniques are commonly described, although the effectiveness of various techniques has not been shown in clinical trials. With serial puncture/droplet, small aliquots are deposited using multiple injections along the wrinkle or fold. With linear threading, the injectable is deposited as the needle is advanced, anterograde, or withdrawn, retrograde/retrotracking. Anterograde injection may be less painful and may cause less bruising. This may be useful for enhancing the nasolabial folds and the vermiliocutaneous border by finding the potential spaces with limited needle movement. Retrograde injection does not create additional tracks and may avoid intravascular injection, as in the glabellar region. Fanning involves multiple passes in different directions or layers without withdrawing the needle, which is believed to achieve superior results for the nasolabial fold. Crosshatching involves linear injection in an evenly spaced grid pattern. This is effective for large areas, three-dimensional filling, and the oral commissures. Injecting large volumes in the subepidermal plane increases local adverse events as opposed to multiple or deep injections (Baumann *et al.*, 2006; Carruthers and Carruthers, 2007).

5.1. *Tunneling technique*

Locating the correct plane for the injection of ArteFill permanent filler is of the utmost importance. The thickness of the facial dermis varies between 0.4 mm in lids and 1.2 mm in the forehead and cheeks. In a deep crease, the thickness of the dermis may be reduced to only one-third of its normal thickness. The outer diameter of a 26G needle is 0.45 mm and can be used as a depth gauge to estimate the thickness of the dermis. The facial dermis is only about twice as thick as a 26G needle. With fine blunt microneedles of 26G, one cannot achieve longitudinal penetration intradermally. Therefore, with this type of needle it is easy to establish the correct plane for injection by manipulating the cannula tip along the dermis from beneath. This technique minimizes trauma and prevents bruising and intradermal ridges, especially in nasolabial folds (Coleman, 2002). It requires only a single cannula penetration through a tiny stab incision. For deep dermal injections in patients with facial lipodystrophy syndrome, blunt needles of 23G or larger will facilitate nudging venules and arterioles aside and prevent bruising, intravascular deposition of ArteFill, and potential embolism. Prior to inserting the needle in the dermis, one may stretch the skin tightly to create a firm surface. The needle is inserted into the dermis at an approximately 10° angle, parallel to the length of the wrinkle or fold. While maintaining constant thumb pressure on the plunger, the needle (with bevel either up or down) is inserted into the skin at the dermal/subdermal interface along the line of the wrinkle. ArteFill is implanted as the needle is withdrawn along the course of the wrinkle by placing a continuous strand of material under consistent pressure (resistance will be noted) into the junction of the reticular dermis and subcutaneous fat. If the needle placement is too superficial, the gray of the needle will be visible through the skin and tissue blanching will be observed upon injection. This indicates improper needle placement. The needle should be withdrawn and reinserted one needle diameter deeper. Blanching is a sign that the papillary plexus is compressed. If blanching occurs, distribute the implant into the surrounding dermis using fingernail massage. If needle placement is too deep, the needle will be felt to "pop" the subcutaneous fat and no resistance will be felt when injecting the ArteFill. If the needle is in the correct plane, i.e., the junction of the dermis and subdermal fat, it will be possible to pull the dermis superficially causing a ridge or push the dermis deeper causing a depression or groove. An even, continuous strand should be delivered while withdrawing the needle. Using pressure on the plunger during the forward movement of the needle, a protruding droplet of ArteFill may create a "blunt needle tip" and preserve some of the capillaries in its way, e.g., it may prevent bruising. Immediately after implant placement, the injector should palpate the implant by gently applying pressure and eventually massage with the fingernail evenly to facilitate uniform distribution. Overly vigorous massage may spread ArteFill deeper into the fatty tissue where it does not achieve the desired effect, and massage may cause unwanted swelling and bruising. Using the smallest needle possible that allows accurate injection may reduce pain, trauma, adjacent tissue trauma, and risk of infection: 30 gauge for less viscous, 27 gauge for more viscous to avoid clumping or clogging. Alternatively, a 26-gauge needle will provide visual clues to needle depth: intradermal, the gray of the needle can be seen; superficial subdermal, the shape of the needle can be seen; and deep subdermal, fat is pressed down with the tip of the needle. Despite classic teachings of dermal injection, the majority of product is found in the subcutis (Lemperle *et al.*, 2010).

5.2. *Serial puncture*

Serial puncture is optimal for the glabella, philtral column enhancement, and fine rhytides. It can also be used for the nasolabial folds. Multiple injections are made serially along the fine wrinkle or fold. The injection sites should be close together, so that the injected material merges into a smooth, continuous line that ultimately lifts the wrinkle or fold. It is helpful to pull the skin slightly away and out from the injection area while injecting. No spaces should remain between the serially injected material. If some minimal gaps are present, postinjection molding and massage can be used to blend the material into a smooth layer (Rohrich *et al.*, 2007).

6. Complications

Early complications are commonly inflammatory in nature; late complications are usually secondary to granuloma formation; and both of these may be infectious in cause and may be complicated by a phenomenon that is collectively referred to as biofilm (Rohrich *et al.*, 2010), a structured community of microorganisms encapsulated within a self-developed polymeric matrix and irreversibly adherent to a living or inert surface now being found on every indwelling foreign body (Christensen *et al.*, 2005). More serious early adverse events include allergic reactions, infections, vascular compromise, and placement of product that is too superficial. Antihistamines, topical immunomodulators and steroid injections are effective in managing immunologic reactions (Lemperle *et al.*, 2006b). Injections around the lips may trigger herpetic outbreaks. If the patient has a history of herpes, prophylactic antiviral treatment should be considered, especially when injecting around the lips (Glaich *et al.*, 2006). To avoid vascular compromise, aspirate before injection, use lower volumes in high-risk areas, remembering vascular anatomy and depth planes, treat one side at a time, and occlude the vessels at their origin (Cohen and Brown, 2009). Late complications include migration, discoloration, scarring, atrophy, and foreign body granulomas. Causes of granuloma formation include the volume of injected material, size of the filler particles used, impurities, and biofilms. Steroid injections are proven treatments of granulomas. Delayed complications of injectables potentially occur from biofilms (Cohen, 2008; Lemperle and Gauthier-Hazan, 2009a; Lemperle *et al.*, 2009).

The most common procedure or device-related events are injection-site erythema, swelling, pain, and bruising, which all usually resolve within a few days. More serious complications can sometimes occur, but most can be avoided with appropriate injection techniques. Inappropriate and superficial placements are among the most frequent reasons for patient dissatisfaction. Too-superficial placement of HA in the dermis can result in a Tyndall effect, which is a blue discoloration caused by the refraction of light from the clear gel visible superficially in the skin. To avoid superficial injection, the metal barrel of the needle should not be visible through the skin in the plane of injection when injecting in a linear fashion.

Because of the reversibility of HA, complications from these fillers can be easily corrected. The use of ovine testicular hyaluronidase (Vitrase) can dissolve injected HA, which is highly useful if the product is misplaced, a complication occurs postinjection (eg, vascular occlusion, delayed granulomatous reactions) or if there is an impending vascular necrosis (Brody, 2005; Hirsch *et al.*, 2007). Numerous reports have described a prolonged hypersensitivity with a granulomatous-like, foreign-body reaction. Biopsy specimens usually demonstrate a granulomatous foreign-body reaction with multinucleated cells surrounding a blue, amorphous material. Brody reported a case involving a granulomatous-like reaction with persistent nodularity that was recalcitrant to a steroid regimen (injection and oral therapy) and antibiotic therapy, even at five months after injection. The nodularity finally responded to an injection of 15 U of hyaluronidase. Complete resolution without recurrence was noted within 24 hours. The hyaluronidase was prepared by diluting 0.5 cc of 150 U/cc (75 U) with 1.5 cc of 1% lidocaine with epinephrine. Lambros and Vartanian *et al.* have also reported on the benefits of hyaluronidase injection for both a chronic granulomatous-like reaction and misplacement of material. Hyaluronidase is a soluble protein enzyme that hydrolyzes hyaluronic acid by breaking the glucosamine bond between C1 of the glucosamine moiety and C4 of glucuronic acid. It is often used to augment the affected area during injection of local anesthesia as well as to increase the hypotonic effect of local anesthesia in ophthalmologic procedures. Although the bovine-derived hyaluronidase is no longer available, recent forms of the drug have been approved by the U.S. Food and Drug Administration.

Two cases of injection site necrosis were reported and attributed to compression of the vascular supply from excessive use of the product at the time of injection. Manna *et al.* demonstrated transient adverse events in 12 to 13 percent of patients treated with Restylane and showed a higher protein load per milliliter of gel in Restylane versus Hylaform. However, no long-term sequelae resulted. The majority of these reports predates improved purification by manufacturers and therefore may not be relevant to the product line presently available.

7. Summary

The use of injectable dermal fillers has advanced significantly from its beginnings with fat grafts and liquid silicone in the early 20th century to the full array of bioengineered biodegradable materials now available. These native macromolecule-based biomaterials with great bio-compatiblility and biodegradability have been widely used in plastic surgery for facial rejuvenation and soft augmentation. Many novel dermal types of filler that have longer duration time are now undergoing FDA testing, and some of these will likely be approved for use in the United States within the next few years. These fillers will expand the choices available to patients and physicians and promise to increase longevity and minimize adverse events.

Reference

Alam M, Gladstone H, Kramer EM, Murphy JP, Jr., Nouri K, Neuhaus IM, *et al.* ASDS guidelines of care: Injectable fillers. *Dermatologic surgery: Official publication for American Society for Dermatologic Surgery [et al]*, 2008;34(Suppl 1):S115–S148.

Arlette JP, Trotter MJ. Anatomic location of hyaluronic acid filler material injected into nasolabial fold: A histologic study. *Dermatologic surgery: Official publication for American Society for Dermatologic Surgery [et al]*, 2008;34(Suppl 1): S56–S62; discussion S62–S53.

Aung SC, Foo CL, Lee ST. Three dimensional laser scan assessment of the Oriental nose with a new classification of Oriental nasal types. *Brit J Plast Surg* 2000;53:109–116.

Bailey AJ. Perspective article: The fate of collagen implants in tissue defects. *Wound repair and regeneration: Official publication of the Wound Healing Society [and] the European Tissue Repair Society* 2000;8:5–12.

Barton FE, Jr. The SMAS and the nasolabial fold. *Plast Recon Surg* 1992;89:1054–1057; discussion 1058–1059.

Barton FE, Jr., Gyimesi IM. Anatomy of the nasolabial fold. *Plast Reconstr Surg* 1997;100:1276–1280.

Baumann L, Kaufman J, Saghari S. Collagen fillers. *Dermatol Ther* 2006;19: 134–140.

Baumann LS, Shamban AT, Lupo MP, Monheit GD, Thomas JA, Murphy DK, *et al.* Comparison of smooth-gel hyaluronic acid dermal fillers with cross-linked bovine collagen: A multicenter, double-masked, randomized, within-subject study. *Dermatologic surgery: Official publication for American Society for Dermatologic Surgery [et al]*, 2007;33(Suppl 2):S128–S135.

Bogdan Allemann I, Baumann L. Hyaluronic acid gel (Juvederm) preparations in the treatment of facial wrinkles and folds. *Clinical Interv Aging* 2008;3:629–634.

Brody HJ. Use of hyaluronidase in the treatment of granulomatous hyaluronic acid reactions or unwanted hyaluronic acid misplacement. *Dermatol Surg* 2005;31:893–897.

Burdick JA, Prestwich GD. Hyaluronic acid hydrogels for biomedical applications. *Adv Mater* 2011;23:H41–H56.

Carruthers A, Carruthers J. Non-animal-based hyaluronic acid fillers: Scientific and technical considerations. *Plast Reconstr Surg* 2007;120:33S–S40S.

Carruthers JD, Glogau RG, Blitzer A. Advances in facial rejuvenation: Botulinum toxin type a, hyaluronic acid dermal fillers, and combination therapies–consensus recommendations. *Plast Reconstr Surg* 2008;121:5S–30S; quiz 31S–36S.

Christensen L, Breiting V, Janssen M, Vuust J, Hogdall E. Adverse reactions to injectable soft tissue permanent fillers. *Aesthet Plast Surg* 2005;29:34–48.

Chun KW, Kang HJ, Han SK, Lee ES, Chang H, Kim SB, *et al.* Anatomy of the alar lobule in the Asian nose. *J Plastic Reconstr Aesthet Surg: JPRAS* 2008;61:400–407.

Cockerham K, Hsu VJ. Collagen-based dermal fillers: Past, present, future. *Facial Plast Surg: FPS* 2009;25:106–113.

Cohen JL. Understanding, avoiding, and managing dermal filler complications. *Dermatologic surgery: Official publication for American Society for Dermatologic Surgery [et al]*, 2008;34(Suppl 1):S92–S99.

Cohen JL, Brown MR. Anatomic considerations for soft tissue augmentation of the face. *J Drugs Dermatol: JDD* 2009;8:13–16.

Cohen JL, Dayan SH, Brandt FS, Nelson DB, Axford-Gatley RA, Theisen MJ, *et al.* Systematic review of clinical trials of small- and large-gel-particle hyaluronic acid injectable fillers for aesthetic soft tissue augmentation. *Dermatologic surgery: Official publication for American Society for Dermatologic Surgery [et al]*, 2013;39:205–231.

Cohen SR, Berner CF, Busso M, Gleason MC, Hamilton D, Holmes RE, *et al.* ArteFill: A long-lasting injectable wrinkle filler material — Summary of the US Food and Drug Administration trials and a progress report on 4- to 5-year outcomes. *Plast Reconstr Surg* 2006;118:64s–76s.

Cohen SR, Holmes RE Artecoll: A long-lasting injectable wrinkle filler material: Report of a controlled, randomized, Multicenter clinical trial of 251 subjects. *Plast Reconstr Surg* 2004a;114:964–976.

Cohen SR, Holmes RE. Artecoll: A long-lasting injectable wrinkle filler material: Report of a controlled, randomized, multicenter clinical trial of 251 subjects. *Plast Reconstr Surg*, 2004b;114:964–976; discussion 977–969.

Coleman SR. Avoidance of arterial occlusion from injection of soft tissue fillers. *Aesthet Surg J/the American Society for Aesthetic Plastic surgery*, 2002;22:555–557.

Dayan SH, Bassichis BA. Facial dermal fillers: Selection of appropriate products and techniques. *Aesthet Surg J/the American Society for Aesthetic Plastic surgery* 2008;28:335–347.

de Lacerda DA, Zancanaro P. Filler rhinoplasty. *Dermatologic surgery: Official publication for American Society for Dermatologic Surgery [et al]*, 2007;33 (Suppl 2):S207–S212; discussion S212.

DeLustro F, Smith ST, Sundsmo J, Salem G, Kincaid S, Ellingsworth L. Reaction to injectable collagen: Results in animal models and clinical use. *Plast Reconstr Surg* 1987;79:581–594.

Fenske NA, Lober CW. Structural and functional changes of normal aging skin. *J Am Acad Dermatol* 1986;15:571–585.

Fernandez EM, Mackley CL. Soft tissue augmentation: A review. *J Drugs Dermatol: JDD* 2006;5:630–641.

Flowers RS. Tear trough implants for correction of tear trough deformity. *Clin Plast Surg* 1993;20:403–415.

Frank DH, Vakassian L, Fisher JC, Ozkan N. Human antibody response following multiple injections of bovine collagen. *Plast Reconstr Surg* 1991;87:1080–1088.

Glaich AS, Cohen JL, Goldberg LH. Injection necrosis of the glabella: Protocol for prevention and treatment after use of dermal fillers. *Dermatologic surgery: Official publication for American Society for Dermatologic Surgery [et al]*, 2006;32:276–281.

Glogau RG. Fillers: From the past to the future. *Seminars in Cutaneous Medicine and Surgery* 2012;31:78–87.

Haddock NT, Saadeh PB, Boutros S, Thorne CH. The tear trough and lid/cheek junction: Anatomy and implications for surgical correction. *Plast Reconstr Surg* 2009;123:1332–1340; discussion 1341–1332.

Han SK, Shin SH, Kang HL, Kim WK. Augmentation rhinoplasty using injectable tissue-engineered soft tissue — A pilot study. *Ann Plast Surg* 2006; 56:251–255.

Hirmand H. Anatomy and nonsurgical correction of the tear trough deformity. *Plast Reconstr Surg* 2010;125:699–708.

Hirsch RJ, Cohen JL, Carruthers JDA. Successful management of an unusual presentation of impending necrosis following a hyaluronic acid injection embolus and a proposed algorithm for management with hyaluronidase. *Dermatol Surg* 2007;33:357–360.

Homicz MR, Watson D. Review of injectable materials for soft tissue augmentation. *Facial Plast Surg: FPS* 2004;20:21–29.

Jiang D, Liang J, Noble PW. Hyaluronan in tissue injury and repair. *Annu Rev Cell Dev Biol* 2007:23:435–461.

Klein AW, Elson ML. The history of substances for soft tissue augmentation. *Dermatologic surgery: Official publication for American Society for Dermatologic Surgery [et al]*, 2000;26:1096–1105.

Lambros VS. Hyaluronic acid injections for correction of the tear trough deformity. *Plast Reconstr Surg* 2007;120:74S-80S.

Lavker RM. Structural alterations in exposed and unexposed aged skin. *J Invest Dermatol* 1979;73:59–66.

Lemperle G, de Fazio S, Nicolau P. ArteFill: A third-generation permanent dermal filler and tissue stimulator. *Clin Plast Surg* 2006a;33:551–565.

Lemperle G, Gauthier-Hazan N. Foreign body granulomas after all injectable dermal fillers: Part 2. Treatment options. *Plast Reconstr Surg* 2009a;123: 1864–1873.

Lemperle G, Gauthier-Hazan N. Foreign body granulomas after all injectable dermal fillers: Part 2. Treatment options. *Plast Reconstr Surg* 2009b;123: 1864–1873.

Lemperle G, Gauthier-Hazan N, Wolters M, Eisemann-Klein M, Zimmermann U, Duffy DM. Foreign Body Granulomas after All Injectable Dermal Fillers: Part 1. Possible Causes. *Plast Reconstr Surg* 2009;123:1842–1863.

Lemperle G, Holmes RE, Cohen SR, Lemperle SM. A classification of facial wrinkles. *Plast Reconstr Surg* 2001;108:1735–1750; discussion 1751–1732.

Lemperle G, Knapp TR, Sadick NS, Lemperle SM. ArteFill permanent injectable for soft tissue augmentation: I. Mechanism of action and injection techniques. *Aesthet Plast Surg* 2010;34:264–272.

Lemperle G, Morhenn VB, Pestonjamasp V, Gallo RL. Migration studies and histology of injectable microspheres of different sizes in mice. *Plast Reconstr Surg* 2004;113:1380–1390.

Lemperle G, Rullan PP, Gauthier-Hazan N. Avoiding and treating dermal filler complications. *Plast Reconstr Surg* 2006b;118:92S-107S.

Liao WC, Ma H, Lin CH. Balanced rhinoplasty in an oriental population. *Aesthetic Plast Surg* 2007;31:636–642; discussion 643–634.

Loeb R. Naso-jugal groove leveling with fat tissue. *Clin Plast Surg* 1993;20:393–400; discussion 401.

Lucarelli MJ, Khwarg SI, Lemke BN, Kozel JS, Dortzbach RK. The anatomy of midfacial ptosis. *Ophthal Plast Reconstr Surg* 2000;16:7–22.

Macdonald MR, Spiegel JH, Raven RB, Kabaker SS, Maas CS. An anatomical approach to glabellar rhytids. *Arch Otolaryngol Head Neck Surg* 1998;124:1315–1320.

Mendelson BC, Jacobson SR. Surgical anatomy of the midcheek: Facial layers, spaces, and the midcheek segments. *Clin Plast Surg* 2008;35:395–404; discussion 393.

Monstrey SJ, Pitaru S, Hamdi M, Van Landuyt K, Blondeel P, Shiri J, *et al.* A two-stage phase I trial of Evolence30 collagen for soft-tissue contour correction. *Plast Reconstr Surg* 2007;120:303–311.

Morhenn VB, Lemperle G, Gallo RL Phagocytosis of different particulate dermal filler substances by human macrophages and skin cells. *Dermatologic surgery: Official publication for American Society for Dermatologic Surgery [et al]*, 2002;28:484–490.

Moscona RR, Bergman R, Friedman-Birnbaum R. An unusual late reaction to Zyderm I injections: A challenge for treatment. *Plast Reconstr Surg* 1993;92:331–334.

Moyle GJ, Lysakova L, Brown S, Sibtain N, Healy J, Priest C, *et al.* A randomized open-label study of immediate versus delayed polylactic acid injections for the cosmetic management of facial lipoatrophy in persons with HIV infection. *HIV medicine* 2004;5:82–87.

Muzaffar AR, Mendelson BC, Adams WP. Jr. Surgical anatomy of the ligamentous attachments of the lower lid and lateral canthus. *Plast Reconstr Surg* 2002;110:873–884; discussion 897–911.

Narins RS, Brandt F, Leyden J, Lorenc ZP, Rubin M, Smith S. A randomized, double-blind, multicenter comparison of the efficacy and tolerability of Restylane versus Zyplast for the correction of nasolabial folds. *Dermatologic surgery: Official publication for American Society for Dermatologic Surgery [et al]*, 2003;29:588–595.

Narins RS, Coleman W, Donofrio L, Jones DH, Maas C, Monheit G, *et al.* Nonanimal Sourced Hyaluronic Acid-Based Dermal Filler Using a Cohesive Polydensified Matrix Technology is Superior to Bovine Collagen in the Correction of Moderate to Severe Nasolabial Folds: Results from a 6-Month, Randomized, Blinded, Controlled, Multicenter Study. *Dermatol Surg* 2010;36:730–740.

Narins RS, Dayan SH, Brandt FS, Baldwin EK. Persistence and improvement of nasolabial fold correction with nonanimal-stabilized hyaluronic acid 100,000 gel particles/mL filler on two retreatment schedules: Results up to 18 months on two retreatment schedules. *Dermatologic surgery: Official publication for American Society for Dermatologic Surgery [et al]*, 2008;34(Suppl 1):S2–S8; discussion S8.

Nicolau PJ. Long-lasting and permanent fillers: Biomaterial influence over host tissue response. *Plast Reconstr Surg* 2007;119:2271–2286.

Pogrel MA, Shariati S, Schmidt B, Faal ZH, Regezi J. The surgical anatomy of the nasolabial fold. *Oral Surg, Oral Med Oral Pathol, Oral Radiol Endod* 1998;86:410–415.

Prestwich GD, Kuo JW. Chemically-modified HA for therapy and regenerative medicine. *Curr Pharm Biotechnol* 2008;9:242–245.

Rohrich RJ, Ghavami A, Crosby MA. The role of hyaluronic acid fillers (Restylane) in facial cosmetic surgery: Review and technical considerations. *Plast Reconstr Surg* 2007;120:41S-54S.

Rohrich RJ, Monheit G, Nguyen AT, Brown SA, Fagien S. Soft-tissue filler complications: The important role of biofilms. *Plast Reconstr Surg* 2010;125:1250–1256.

Rohrich RJ, Pessa JE. The fat compartments of the face: Anatomy and clinical implications for cosmetic surgery. *Plast Reconstr Surg* 2007;119:2219–2227; discussion 2228–2231.

Rohrich RJ, Pessa JE. The retaining system of the face: Histologic evaluation of the septal boundaries of the subcutaneous fat compartments. *Plast Reconstr Surg* 2008;121:1804–1809.

Rubin LR. The anatomy of the nasolabial fold: The keystone of the smiling mechanism. *Plast Reconstr Surg* 1999;103:687–691; discussion 692–684.

Sclafani AP, Romo T, 3rd Collagen, human collagen, and fat: The search for a three-dimensional soft tissue filler. *Facial Plast Surg: FPS* 2001;17:79–85.

Shaw RB, Jr., Katzel EB, Koltz PF, Kahn DM, Girotto JA, Langstein HN. Aging of the mandible and its aesthetic implications. *Plast Reconstr Surg* 2010;125:332–342.

Shirakabe Y, Suzuki Y, Lam SM. A systematic approach to rhinoplasty of the Japanese nose: A thirty-year experience. *Aesthet Plast Surg* 2003;27:221–231.

Shoshani D, Markovitz E, Cohen Y, Heremans A, Goldlust A. Skin test hypersensitivity study of a cross-linked, porcine collagen implant for aesthetic surgery. *Dermatologic surgery: Official publication for American Society for Dermatologic Surgery [et al]*, 2007;33(Suppl 2):S152–S158.

Skardal A, Zhang J, McCoard L, Oottamasathien S, Prestwich GD. Dynamically crosslinked gold nanoparticle — hyaluronan hydrogels. *Adv Mater* 2010; 22:4736–4740.

Stolman LP. Human collagen reactions. *Dermatologic surgery: Official publication for American Society for Dermatologic Surgery [et al]*, 2005;31:1634.

Takasaki S, Fujiwara S, Shinkai H, Ooshima A. Human type VI collagen: Purification from human subcutaneous fat tissue and an immunohistochemical study of morphea and systemic sclerosis. *J Dermatol* 1995;22:480–485.

Tham C, Lai YL, Weng CJ, Chen YR. Silicone augmentation rhinoplasty in an oriental population. *Ann Plast Surg* 2005;54:1–5; discussion 6–7.

Toole BP. Hyaluronan: From extracellular glue to pericellular cue. *Nat Rev Cancer* 2004;4:528–539.

Truswell WH, 4th. Aging changes of the periorbita, cheeks, and midface. *Facial Plast Surg: FPS* 2013;29:3–12.

Valantin MA, Aubron-Olivier C, Ghosn J, Laglenne E, Pauchard M, Schoen H, *et al.* Polylactic acid implants (New-Fill) to correct facial lipoatrophy in HIV-infected patients: Results of the open-label study VEGA. *AIDS* 2003;17:2471–2477.

Vleggaar D, Bauer U. Facial enhancement and the European experience with Sculptra (poly-l-lactic acid). *J Drugs Dermatol: JDD* 2004;3:542–547.

Webster RC, Kattner MD, Smith RC. Injectable collagen for augmentation of facial areas. *Arch Otolaryngol* 1984;110:652–656.

Wong CH, Hsieh MK, Mendelson B. The tear trough ligament: Anatomical basis for the tear trough deformity. *Plast Reconstr Surg* 2012;129:1392–1402.

Xue K, Chiang CA, Liu K, Gu B, Li Q. Multiplane hyaluronic acid rhinoplasty. *Plast Reconstr Surg* 2012;129:371e-372e.

Yousif NJ, Gosain A, Matloub HS, Sanger JR, Madiedo G, Larson DL. The nasolabial fold: An anatomic and histologic reappraisal. *Plast Reconstr Surg* 1994;93:60–69.

Chapter 6

Nanofabrication Techniques in Native Polymer-based 3D Substitutes

Yangchao Luo, Qin Wang

Department of Nutrition and Food Science,
University of Maryland, College Park, MD 20742, USA

1. Introduction

Tissue engineering is one of the fastest-growing technologies with the goal to facilitate the repair and replacement of injured or compromised tissues or organs. The tissue substitute functions as a scaffold that not only supports cell growth but also facilitates cell adhesion, proliferation, and extracellular matrix (ECM) production. Based on these requirements, the substitute, therefore, must mimic *in vivo* microenvironments and offer chemical and physical cues to regulate cellular functions (Ng *et al.*, 2012). Three-dimensional (3D) substitutes have been proven to provide a better engineered construct than conventional two-dimensional structures, because they offer spatial cell organization for interactions between individual cells and organs. The challenges of 3D substitutes fabrication remain in the scaffold structural design, including the mechanical properties, ability to direct cell-matrix and cell-cell interactions, and porous structures for efficient mass transport (Coutinho *et al.*, 2011). Because the native tissues and the associated ECM are in fact made of nanoscale structures, which are cooperatively organized and controlled, there has been increasing attention to develop novel techniques to fabricate 3D substitute at micro- and nanoscale for tissue engineering applications (Ng *et al.*, 2012; Coutinho *et al.*, 2011; Kelleher and Vacanti, 2010). The 3D substitutes with nanostructural features have shown promising advantages, such as increased cell-surface interactions, cell attachment, and cell alignment on the surface of nanotopography.

In recent decades, the different fabrication techniques have been developed to design 3D nanostructured scaffolds using various materials, including both native and synthetic polymers. The basic requirements for the ideal polymeric materials are the biocompatibility, appropriate mechanical and biological properties, as well as low cost and availability. Generally, the native polymers are not necessarily superior to the synthetic polymers, because they both have their own limitations. However, the native biopolymers have potentially better clinical applications due to their low toxicity, low chronic inflammatory response, and biodegradability, and more importantly the abundant biological cell-recognition signals, which are significantly lacking in synthetic polymers (Ng *et al.*, 2012). Therefore, this chapter will mainly focus on the recent development of nanofabrication techniques that have been studied to fabricate nanostructured 3D scaffolds from native polymers, including collagen, gelatin, silk fibroin, and chitosan, etc.

2. Electrospinning

2.1. *Introduction of electrospinning technique*

Electrospinning is a well-known and ubiquitous technique to produce untrathin fiber with diameters ranging from submicrometer level to less than 100 nm. The history of application of electrospinning can be traced back to the year of 1902, when it was first described as electrostatic spraying (Morton, 1902). The electrospinning process was accomplished by a strong electric field generated by high voltage to continuously draw a polymer liquid to a screen collector through a conductive capillary. The typical electrospinning setup is composed of a high voltage power supply, a nozzle (spinneret), a reservoir for polymer solutions, and a grounded collector plate. When the high voltage is applied, the body of the droplet becomes highly charged. The electrostatic repulsion on the surface counteracts the surface tension and the droplet becomes stretched; at a critical point, the polymer droplet changes its shape from spherical to conical and a stream is then formed which erupts from the surface. This critical point is the so-called Taylor cone. As the electrostatically charged polymer jet erupts, the solvent evaporates quickly and the dried ultrathin fiber is formed which is then collected at the grounded collector plate. The schematic illustration of a typical electrospinning setup is shown in Fig. 1. This process has been considered as a noninvasive process to produce fibers at nanoscale without any severe chemical reactions or high temperatures involved (Wang and Zhang, 2012; Torres-Giner *et al.*, 2008).

2.2. *Modifications in electrospinning*

Besides the aforementioned electrospinning setup on a laboratory level, some innovative electrospinning setups have been recently

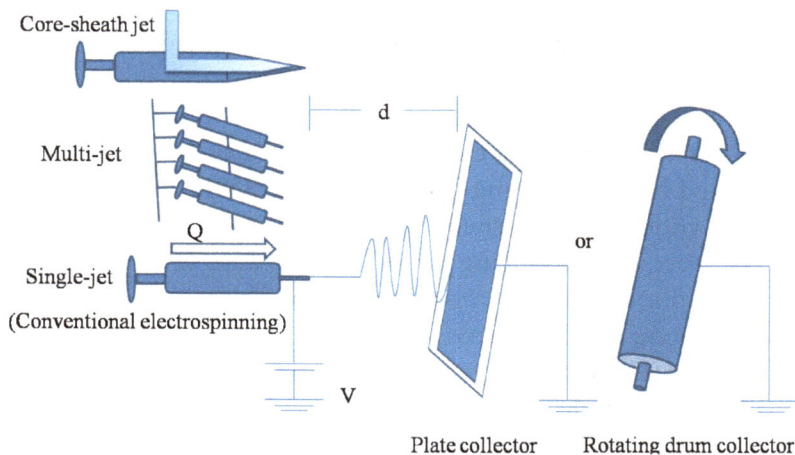

Fig. 1. The schematic illustration of electrospinning and its modifications. Q, flow rate of polymer solution; d, distance between plate (drum) collector and needle; V, applied voltage.

developed with many advantageous applications. As shown in Fig. 1, the conventional electrospinning is quite simple and has been widely adapted in many research laboratories around the world. However, it has many limitations, such as random alignment, low productivity, and poor mechanical properties of nanofibers, which will be discussed in the following paragraphs. With the aim to further facilitate the electrospinning process and produce the nanofiber in more controlled and tailored ways, several modified setups have been invented (Li and Xia, 2004).

The alignment of nanofibers produced by electrospinning is hard to be controlled, while with the help of an aluminum rotating drum collector (Fig. 1), the partially aligned electrospun fibers can be successfully prepared. With this method, by modulating the flow rate and polymer concentration, it was shown that human β-nerve growth factor, stabilized in bovine serum albumin, was encapsulated into the aligned nanofibers, although the encapsulated protein existed in aggregated form that was randomly dispersed in the fibrous mesh (Chew *et al.*, 2005). The protein released from nanofibers was shown to survive during this electrospinning process with great bioactivity to stimulate cell growth in the *in vitro* test. It clearly demonstrated the feasibility to encapsulate proteins via modified electrospinning process to produce biofunctional tissue scaffolds from aligned nanofibers. Several other modifications have also been invented to produce aligned nanofibers, including two needles with opposite voltages spray (Pan *et al.*, 2006) and a collector composed of two conductive strips separated by an insulating gap (Li *et al.*, 2004).

The conventional electrospinning process is considered as a low production process in producing nanofibers, because the flow rate of polymer solution is kept low in order to obtain ultra-thin fibers. To solve this problem, a new electrospinning setup by setting an array of multiple needles as the spinneret was developed. In this method, it is crucial to ensure the electric field strength of arrayed needles to be identical in order to produce uniform nanofibers. Several research groups investigated electrospinning technology using a multiple-jet setup (Fang *et al.*, 2006; Ding *et al.*, 2004). This technology is accomplished by the incorporation of secondary electrodes to isolate the electric field distribution of the primary electrode spinnerets, making it possible not only to increase production yield, but also to fabricate membranes with composite nanofibers, tailor-designed composition variations, and 3D pattern formation with nonwoven fibrous structure from multiple polymer solutions. A new study also confirmed that using a secondary electrode reduced the divergence angle between multiple jets (Varesano *et al.*, 2009). Different configurations may be needed to produce nanofibers with best mechanical quality, depending on the charging of the solution and collector. Park and coworkers developed a new prototype electrospinning setup based on the multi-jet techniques, but they used a robotic-controlled movable dual-nozzle with two separate power supplies to each nozzle (Park *et al.*, 2013). The research group further applied this tone-step angled two-nozzle electrospinning technique to produce curly and randomly-oriented nanofibers with interfiber bonding to obtain improved mechanical properties comparing to that fabricated from the single-nozzle electrospinning process (Tijing *et al.*, 2013).

Coaxial electrospinning or co-electrospinning method has gained increased interests in recent years. In this technique, a plastic syringe with two compartments containing different polymer solutions or a polymer solution (shell) and a non-polymeric Newtonian fluid or even a powder (core) is used to initiate a core-shell structure (Yarin, 2011). The illustration of this setup is also shown in Fig. 1. Through the combination of different materials in the axial and radical direction, novel properties and functionalities can be produced. For instance, the core material provides certain required mechanical properties or is used as encapsulation compartment while the sheath material provides the desirable chemical resistance. It has been shown that this technique is highly dependent on the polymer solution used. This is because during this electrospinning process, the rapid solvent evaporation requires systems with high molecular mobility for the formation of core-sheath structures, the polymer blends with lower molecular weight were considered as better materials than those with high molecular weight (Wei *et al.*, 2006). This process has also been reported to provide a better control on the structure and morphology of resultant fibers and prevent the blockade of the nozzle by using highly volatile solvents in outer capillary (Larsen *et al.*, 2004). This modified electrospinning technique is able to produce nanofibers with core-shell structure, which may possess new applications in tissue engineering, especially the encapsulation in the nanofiber core or wrapping as a shell of non-spinnable polymers. For instance, a novel drug delivery platform has been developed by Viry and coworkers recently using coaxial spinneret electrospinning (Viry *et al.*, 2012). In this study, the highly water soluble small molecular drug was first emulsified with polymers to obtain reverse emulsion; and then, this reverse emulsion was used as a core feed in the coaxial electrospinning process. The internal fiber architecture can be tailored by adjusting the emulsion dispersed-to-continuous phase ratio. The encapsulated drug showed a nearly linear sustained release for over 18 days by the emulsion/coaxial electrospinning, compared with only four days for nanofibers fabricated with classical coaxial electrospinning (not using emulsion as core feed). This emulsion/coaxial electrospinning combined technique may be promising for the development of sustained drug-release nanofibrous scaffolds for regenerative medicine.

Another well-recognized shortcoming of conventional electrospinning is that it is hard to fabricate nanofibers from viscous polymers even at relatively low concentrations, due to the limited electric force. Therefore, a new and unique technique has been developed by implementing air blowing feature to the conventional electrospinning systems to expand the capability of the process. In this electro-blowing technique, the electrospinning apparatus is modified by attaching an air blowing system, so that the setup has two simultaneously applied forces (an electrical force and an air-blowing shear force) to produce nanofiber from polymer solution (Um and Fang, 2004). The detailed setup of air-blowing assisted electrospinning device is shown in Fig. 2. The air flow system consists of two components, i.e. a heater and a blower, so that the air blow can be heated by controlling the heater and air flow rate. In this setup, two pulling forces are applied to the polymer droplet, i.e. electric force and blowing force, to form nanofibers. With this additional force, the technique is used to prepare nanofibers with uniform diameters in tens of nanometer from

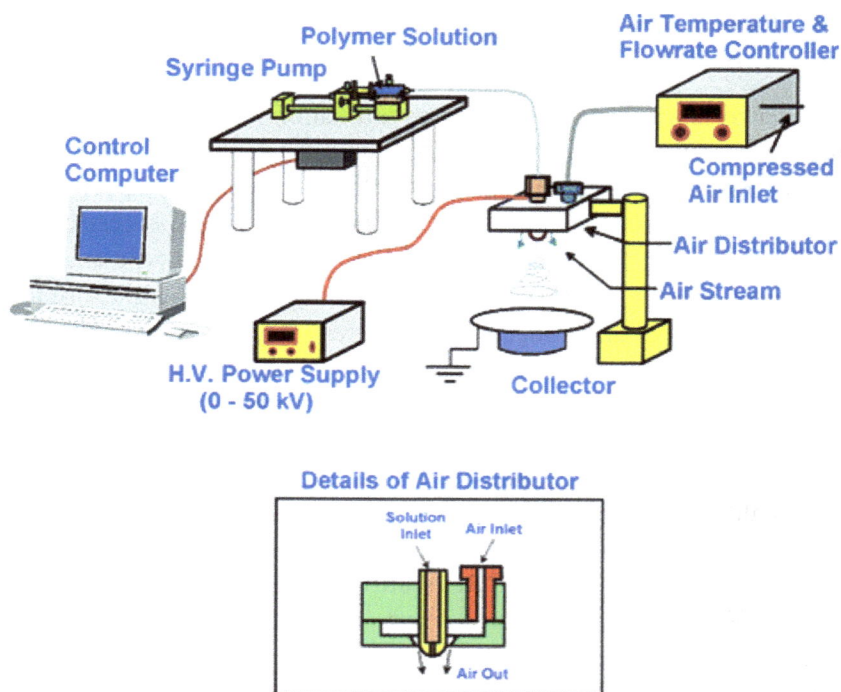

Fig. 2. Scheme diagram for air-blowing assisted electrospinning setup. Reprinted with permission from (Um and Fang, 2004). Copyright (2004) American Chemical Society.

high viscous native polymer fluid, since the fast evaporation rate of the solvent can be accomplished with help of hot air flow. Various experimental parameters in this novel technique, such as air-blowing rate, polymer concentration, feeding rate, applied electric field, and type of collector, have been extensively studied later on, in order to obtain native polymeric nanofibers with optimal quality (Wang *et al.*, 2005; Hsiao *et al.*, 2012). The nanofibers fabricated from air-blowing assisted electrospinning technique have already been demonstrated to possess wound-dressing applications with bacterial-shielding effects (Kim and Yoon, 2008).

2.3. *Parameters affecting production of electrospun nanofibers*

In order to fabricate tailored electrospun nanofiber 3D matrix for specific tissue engineering application, it is critical to fully understand how the electrospinning parameters would affect the quality and property of nanofibers. As introduced before, the electrospinning process can be affected by a number of factors, including the polymer type and concentration, solvent, feeding rate, applied voltage, drop height, distance between spinneret and collector, etc. Torres-Giner and coworkers conducted a comprehensive study on the effects of electrospinning variables and ranges on the properties of electrospun nanofibers, using zein prolamine as a native macromolecule model (Torres-Giner *et al.*, 2008). Among various properties, fiber diameter plays a key role on the final quality of electrospun networks or matrix determining its applications. The smaller diameter a fiber has, the larger surface area it possesses. Therefore, to fabricate 3D scaffold, a small diameter, especially below 100 nm, is preferred to exert better cell-matrix interactions. Both intrinsic and extrinsic parameters affect the fiber size, among which the polymer concentration has been suggested as the most significant factor, taking zein as an example of native polymers (Torres-Giner *et al.*, 2008; Yao *et al.*, 2007). The higher concentration (up to 50%) of the polymer resulted in the production of thicker nanofiber with greater diameter (up to 1250 nm). Although decreasing polymer concentration decreased nanofiber diameter, more nanobeads or nanoparticles were generated, due to the low viscosity of the polymer solution. By increasing the polymer concentration, the nanofiber diameter was dramatically increased and the higher viscosity facilitated the formation of smooth fibers without particulate morphology. The power voltage has been reported to have a linear relationship with feeding flow rate, and thus positively affects the nanofiber diameter. Because increasing applied voltage increases the electrostatic stresses, this results in more polymeric material being drawn from the spinneret. The tip-to-collector distance was shown to have the least influence on the diameter of zein electrospun nanofibers. Besides the diameter, the maximum fiber length also plays an important role in expanding the applications that require longer continuous nanofibers to construct 3D scaffolds. A recent study demonstrated that by using parallel plate method with larger plate size and greater polymer concentration, it was able to fabricate continuous nanofibers with longer length of over 50 cm (Beachley and Wen, 2009).

Another important factor to be considered is the selection of native polymer. Unlike synthetic polymers which can be tailored with appropriate characteristics, electrospinning of native polymers is less facile and often limited by the nature of polymers. Therefore, the selection of native polymer with proper characteristics is very important for the electrospinning process. For instance, chitosan, as a deacetylated product of natural polymer chitin, has been widely studied to prepare nanofibers by electrospinning. However, due to extremely viscous nature at relatively high concentration and poor solubility resulting in limited spinability, chitosan is usually electrospun with the combination of other synthetic polymers (such as polyethylene oxide and polylactic acid) or using toxic organic solvents to decrease viscosity and to achieve solubility as well as to increase solvent volatility (Jayakumar *et al.*, 2010), which pose potential toxicity and limit its applications. Homayoni and coworkers reported a novel method of alkali treatment to prepare electrospun nanofibers from pure chitosan polymer (Homayoni *et al.*, 2009). With proper alkali hydrolysis, chitosan with lower molecular weight and reduced intrinsic viscosity was obtained and therefore was suitable to produce high-quality nanofiber with steady electrospinning processing condition. The authors suggested that success of electrospinning process was due to the fact that alkali treatment yielded low molecular weight chitosan with lower chain length, which was below the threshold required for entanglement coupling formation.

2.4. *Applications of native polymer-based electrospinning technique*

Recent explosion of interest in electrospun nanofibers has focused on the fabrication of 3D nanofibrous scaffolds for a variety of biomedical applications, especially in the field of tissue engineering. Compared with other nanofiber fabrication techniques (self-assembly and phase separation, which will be discussed in following sections), electrospinning can be manipulated in a simpler and more cost-effective manner with less limitations. The 3D nanofibrous scaffolds possess promising physicochemical properties in resembling the native ECM conditions, including the hydrophobicity, surface area, mechanical modulus and strength, specific cell interactions, etc. The native polymer-based 3D nanofibrous scaffolds are of particular interests because they normally exhibit better biocompatibility, lower immunogenicity and limited inflammatory response than synthetic polymers. The native polymers explored for electrospun nanofibrous 3D scaffolds include collagen, gelatin, hyaluronic acid, silk, fibrinogen, chitin and chitosan (Zhang *et al.*, 2008).

In the native ECM *in vivo*, collagen is the major structural element that forms nanoscale multifibrils, therefore it has been widely considered as an excellent native polymer for fabrication of 3D scaffolds *in vitro*. Liao and coworkers have successfully fabricated collagen-based biomimetic bone 3D scaffolds by combining the electrospinning and mineralization methods (Liao *et al.*, 2008). The functional groups of collagen, i.e. carboxyl and carbonyl groups, are important for the mineralization process. The study demonstrated that carbonated hydroxyapatite on nanosize scale was more abundant in the fabricated collagen scaffold than synthetic polymer-based scaffold, indicating that it is a more suitable substitute for bone repair. Cardiovascular and articular tissue engineering, such as creating off-the-shelf bioresorbable vascular graft and cartilage regeneration, also have a high demand in electrospun biopolymer-based 3D scaffolds (Stella *et al.*, 2010; Sell *et al.*, 2009). For instance, Telemeco and coworkers prepared a 3D collagen-based scaffold fabricated from electropsun collagen nanofibers by a modified electrospinning process using a rotating cylindrical mandrel as a ground target (Telemeco *et al.*, 2005). The fiber diameter was controlled by starting concentration of collagen electrospinning solution (80 mg/ml), and the diameter of 700 nm was obtained. The formed matrix consisted of randomly arrayed fibers which were further shaped to form a 3D cylindrical construct with wall thickness of 200–250 µm and length of

20–25 mm. Several biodegradable synthetic polymers were also prepared and evaluated for cellular infiltration, however, only collagen nanofiber scaffolds exhibited excellent cellular infiltration by interstitial and endothelial cells when implanted into the interstitial space of the rat vastus lateralis muscle.

Core-sheath electrospun composite nanofibers prepared from all natural polymers have received increasing attention in recent years. The core-sheath structured composite nanofiber provides better stability in physiological fluids than single native polymer-based nanofibers, and consequently expands their applications in tissue engineering. Alginate electrospun nanofibers are usually prepared by coagulating with calcium ions. However, the calcium ions would be easily extracted from the nanofibers when tested in high concentration of phosphate or citrate buffers, resulting in network structure damage. Therefore, a novel structured nanofiber mat has been developed recently using chitosan as a coagulant and alginate as a core material by electrospinning equipped with a needle and a coagulating sheath bath (Chang *et al.*, 2012). The prepared nanofiber had the average diameter ranging from 600–900 nm depending on the polymer concentration. In this method, ethanol was required to act as coagulation solution, which was used to reduce both the surface tension and specific gravity of the coagulant (chitosan). The optimal concentration of ethanol was required to form quality nanofiber with core-sheath structure, otherwise a flat sheet of coagulant would form and alginate would pile up with rugged surface. When the ethanol concentration reached 50%, the core-sheath structure was clearly revealed by confocal laser scanning microscope (Fig. 3). By adding chitosan as a coagulant, the stability of alginate in core-sheath nanofiber was greatly improved compared with pure alginate nanofiber, showing 40% reduction in degradation rate for three days in physiological environment. Furthermore, similar core-sheath structured electrospun nanofibers have been developed using alginate, chitosan, and collagen as well as hydroxyapatite for bone tissue engineering by the same research group (Yu *et al.*, 2013). The electrospinning setup was the same but the formed alginate-chitosan core-sheath nanofiber was subsequently immersed in collagen-hydroxyapatite solution and then

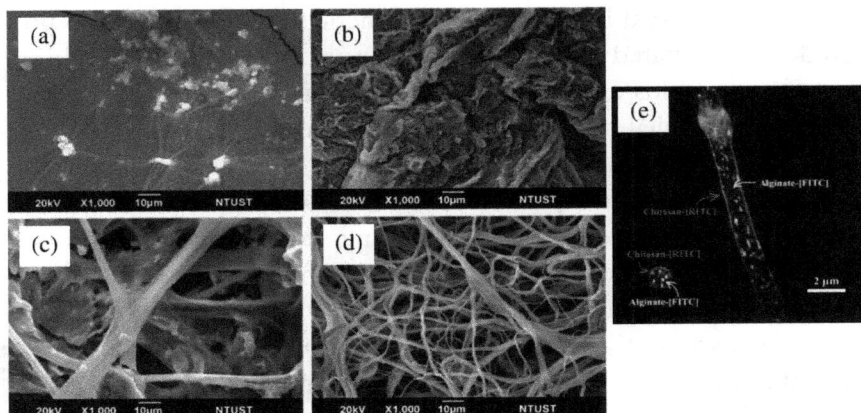

Fig. 3. Effect of adding ethanol in chitosan solution on core-sheath fibrous mats morphology. a, chitosan = 1.0%, ethanol = 0%; b, chitosan = 0.9%, ethanol = 10%; c, chitosan = 0.7%, ethanol = 30%; d, chitosan = 1.0%, ethanol = 50%. Other parameters were kept constant: alginate = 1.5%; glycerol = 49.250%. e, confocal laser scanning microscopic images of chitosan/alginate core-sheath fiber. Reprinted with permission from (Chang *et al.*, 2012). Copyright (2012) Elsevier.

dried in vacuum oven. The composite nanofibers significantly delayed the enzymatic disintegration of collagen by colleganase solution, and the constructed 3D nanoporous matrix showed beneficial effects for cell infiltration and growth. Alternatively, chitosan-alginate composite nanofiber was prepared by using poly(ethylene oxide) as a coagulant during the electrospinning process followed by subsequent removal of coagulant via incubating in water for a few days (Jeong *et al.*, 2010). It was shown that no differences in nanofiber morphology were observed before and after coagulant removal. By increasing chitosan concentration in composite electrospun fiber, the adsorption of serum protein was promoted with great potential for guiding cell behavior in tissue regeneration applications.

2.5. *Challenges in electrospinning technique*

Compared with synthetic polymers, the electrospinning of native polymers is less versatile due to several challenges, such as poor mechanical strength, difficult manipulation, easy denaturation, solvent selection, and so on. Therefore, changes of the experimental design and/or conditions in electrospinning are sometimes necessary in order to produce applicable 3D nanofibrous scaffolds. For instance, the 3D scaffolds fabricated from electrospun pure protein presented insufficient resistance in water and collagenase environment for tissue engineering applications (Friess, 1998; Panzavolta *et al.*, 2011). As a result, scientists recently developed several different techniques to improve its physical and biological performances. Among these techniques is the post-cross-linking technique, which uses different chemical agents (e.g. glutaradehyde, transglutamase, genipin, etc.) or physical methods (e.g. electrostatic cross-linking, ultraviolet exposure, etc.) to bridge and link collagen molecules to construct an interpenetrating and fully water-resistant network (Panzavolta *et al.*, 2011; Torres-Giner *et al.*, 2008). Another widely adopted technique is the electrospinning of biopolymer blends to produce complex nanofibers, which combines the advantages of several individual biopolymers. Various biopolymer composites have been investigated, which consisted of two or more native polymers, such as alginate, chitosan, hyaluronic acid, collagen, silk protein, etc. For instance, collagen has been blended with chitosan (Chen *et al.*, 2010) and hyaluronic acid (Davidenko *et al.*, 2010) to fabricate 3D nanofibrous composite scaffolds with improved mechanical properties and enhanced biological functions. Chitosan has been studied to form composite electrospun nanofiberous 3D structure with alginate to promote protein adsorption and cell guiding behavior (Jeong and Krebs, 2010). Some other studies also used the combination of native polymers and synthetic polymers to improve the modulus of electrospun nanofibers which fabricate biomimetic tissues and organs (McClure *et al.*, 2010; McClure *et al.*, 2012; McCullen *et al.*, 2012).

3. Self-assembly

3.1. *Introduction of self-assembly technique*

Self-assembly is another nanofabrication technique used to develop nanofibrous 3D scaffolds for tissue engineering. It is a bottom-up fabrication strategy to form 3D macroscopic structures from essentially 1D molecules without external intervention. The self-assembly is a thermodynamic minimal process resulting in the generation of stable and robust structures, which are largely driven by multiple, weak, and noncovalent interactions. Although it is a natural process responsible for many essential native and biological components, such as protein synthesis, nucleic acid synthesis, the *in vitro* techniques of self-assembly manipulation are quite complex and require certain specific polymer configurations (Barnes *et al.*, 2007). In this method, molecules are interacting at an atomic level via physical or chemical affinity, which greatly increases the sensitivity and specificity of the entire process. Meanwhile, the nonspecific interactions are energetically unfavorable and less likely to occur during the process, the system therefore is robust and devoid of errors (Madurantakam *et al.*, 2009). The self-assembly was initially developed by Berndt *et al.* (1995) to synthesize the amphiphiles containing peptides from extracellular matrix collagen ligand sequences and investigate the interactions between extracellular ligands and cell surface receptors. Since then, the self-assembly of native macromolecules has gained increasing attention in tissue engineering applications. Compared with electrospinning technique, an important advantage of self-assembly technique is its capability to include functional motif sequences that can promote adherence, differentiation and maturation of cells or other functional ingredients (Dvir *et al.*, 2011). Therefore, the 3D scaffolds fabricated through this technique not only provide good mechanical properties but also give an instructive guidance for tissue development.

3.2. *Parameters affecting production of self-assembly nanostructures*

Preparation of 3D scaffolds by self-assembly methodology does not involve any specific instrument, compared with electrospinning process discussed aforementioned. Therefore, this process can be easily controlled by a limited number of variables, including polymer concentration, ionic strength, pH, as well as addition of other macromolecules. Taking collagen (type I) as an example, its molecules possess the ability to polymerize and form complex 3D supramolecular assemblies *in vitro*, called collagen fibrils, by simply neutralizing the collagen solution with phosphate buffer saline (PBS) followed by incubating at 37°C in a humidified chamber for a specific time period. During this process, collagen concentration, ionic strength and pH of PBS, as well as incubation time have the ability to affect the self-assembly process, leading to varied mechanical strength, morphology, as well as fibril density and diameter of formed nanofibrils (Kokini *et al.*, 2002). As the concentration of collagen increased, the formed fibril density was greatly increased, while the fibril diameter maintained a relatively consistent value of (400–446 nm) with tested range of concentration, i.e. 0.3–3.0 mg/ml. However, both the fibril length and diameter were greatly affected by the pH of the polymerization reaction, showing that higher pH resulted in smaller diameter of collagen fibrils. The mechanical properties were found to be affected by all factors, suggesting that the higher concentration of collagen solution, the more basic pH, and the longer polymerization reaction time could improve the mechanical properties and integrities. Therefore, by controlling the fabrication parameters, the self-assembly 3D fibrous scaffold can be prepared with different mechanical behavior for different tissue regeneration applications. Besides, the addition of polysaccharides plays an important role in self-assembly kinetics and nanostructure (Tsai *et al.*, 2006). The addition of anionic polysaccharides at low concentrations, such as alginate and hyaluronic acid, significantly shortened the lag period and accelerated collagen fibril formation rate, as well as reduced the fibril diameter, whereas the presence of high concentration of polysaccharides exhibited the opposite results. It was suggested that at low concentration of polysaccharides, the collagen molecules were more concentrated by elevated viscosity, promoting the collision-induced interaction, nucleation, and subsequent fibril growth. However, if the concentration of polysaccharides further increased, the viscosity levels became too high to permit free movement of collagen molecules, and thereby decelerate fibril assembly. The addition of cationic polysaccharide, such as chitosan, had no concentration-dependent effect on the collagen fibril formation kinetics, whereas the average diameter of the formed collagen fibril was significantly increased at all concentrations of chitosan, which is possibly due to the electrostatic interactions between collagen and chitosan molecules.

3.3. *Applications of native polymer-based self-assembly technique*

Molecular self-assembly refers to a biomimetic approach to fabricate insoluble nanofibrous 3D scaffolds stabilized by noncovalent bonds. The detailed physics and process mechanisms have been systematically reviewed by Tu and Tirrell (2004). Several types of native marcomolecules have been explored, including peptide, proteins, hybrid biomaterials, as well as DNA (Stepahnopoulos *et al.*, 2013).

The self-assembly structure of proteins are the major support and component of ECM *in vivo*, therefore native proteins are considered as promising biomaterials that are advantageous to fabricate 3D scaffolds mimicking *in vivo* environment. It is widely accepted that the sequence of amino acids in a protein molecule plays a critical role in determining the self-assembly 3D nanostructured conformations, such as α-helix and β-sheet (Cui *et al.*, 2007). Under certain environmental conditions, protein molecules can undergo self-assembly to form 3D scaffolds with controlled structures. Compared with other methods that are widely studied for protein-based 3D scaffolds, such as electrospinning, salt-leaching, freeze-drying, and gas foaming, self-assembly is being reported as a superior technique for fabrication of nano-fibrillar structure similar to *in vivo* ECM. Silk protein and collagen are the two most popular native proteins in the fabrication of self-assembly 3D scaffolds for biomedical applications. A mild self-assembly process has been reported recently to form 3D nanofibrous silk-based scaffolds (Lu *et al.*, 2011). In this novel technique, collagen was blended with silk solution to control the self-assembly of silk. The mixture was lyophilized to achieve nanofibrous silk scaffolds and then water annealing was used to generate insolubility in the obtained silk-collagen scaffolds. By controlling the volume ratio of silk and collagen solutions, different pore sizes were obtained and an optimal ratio was developed to achieve fiber diameter of 20–100 nm, mimicking native collagen in ECM. The organic solvent was avoided in this process by adopting water annealing procedure, and the scaffolds exhibited a very slow dissolution rate in PBS solution, showing high potentials for tissue engineering *in vivo*. The addition of collagen also significantly improved cell compatibility, compared with conventional scaffolds prepared from single component of silk by salt-leached technique. This new technique provides a new way to fabricate native polymer-based 3D scaffolds with nanostructures in mild conditions, i.e. room temperature and pressure, and all-aqueous solution without any toxic solvents or chemicals.

Another newly developed native polymer-based self-assembly technique is called cofibrillogenesis. During this method, two native polymers are mixed together under controlled conditions to form hybrid nanofibrils. Collagen and chitosan are the biopolymers carrying opposite charges, and thus are often studied to form hybrid scaffolds by directly mixing under acidic conditions followed by chemical cross-linking with glutaraldehyde and lyophilization. However, glutaraldehyde may have some toxicity problems in cell compatibility. Wang and coworkers recently reported cofibrillogenesis self-assembly technique to prepare collagen-chitosan 3D nanofibrous scaffolds (Wang *et al.*, 2011). In this technique, collagen and chitosan mixture with different ratios was neutralized to pH 7 at 4°C and then incubated at 37°C for 20 hours, which conditioned thermally triggered cofibrillogenesis. The 3D scaffolds were then obtained by subsequent lyophilization procedure. The thermally triggered cofibrillogenesis process was influenced by the electrostatic interaction of chitosan and collagen. The diameter of hybrid nanofibrils ranged from 50–150 nm and D-periodicity was reported as a consistent value of 64 nm, regardless of polymer concentrations and ratios, suggesting that the collagen/chitosan ratio did not alter the nature of collagen self-assembly. However, the self-assembly process was affected by the mass ratio of chitosan/collagen. With increasing chitosan ratio, more granules at the nanoscale level were formed and part of the fibrillar network became disordered, as shown in Fig. 4. This phenomenon was attributed to a competing

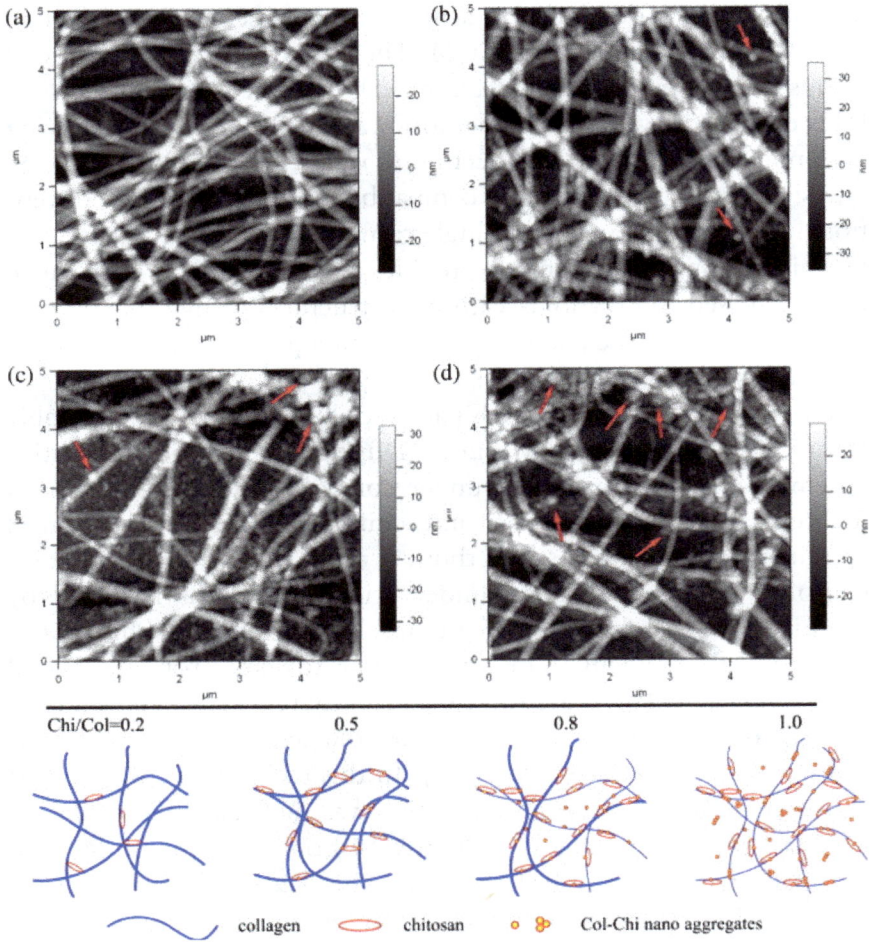

Fig. 4. AFM images of air-dried Col–Chi assemblies with different Chi/Col ratios at (a) 0.2, (b) 0.5, (c) 0.8 and (d) 1.0. Red arrows indicate nanoaggregates of collagen–chitosan complex. Reprinted with permission from (Wang *et al.*, 2011). Copyright (2011) Elsevier.

result of collagen fibrillogenesis and its complexation with chitosan. The scaffolds prepared from cofibrillogenesis exhibited similar tensile strength to the conventional scaffolds cross-linked by glutaraldehyde, indicating the applicability of this green technique in replacing the conventional methods to reduce the toxicity and improve cell biocompatibility. The similar cofibrillogenesis technique has also been widely studied on collagen-heparin self-assembly scaffolds, and the intercalation of heparin is considered as the main trigger for hierarchical formation of polymorphic structures (Stamov *et al.*, 2008; Stamov *et al.*, 2009).

Besides polysaccharides and proteins, another main category of native polymers studied for self-assembly techniques is the peptide, the self-assembly behaviors of which are based on two common natural motifs, i.e. α-helix and β-sheet, and the latter one has been explored more for tissue engineering and regenerative medicine applications (Stephanopoulos *et al.*, 2013; Hauser and Zhang, 2010). Additionally, some specific chemical functionality can be appended to the peptide epitope of β-sheet driven assemblies to obtain the formation of specific 3D nanostructured scaffolds with desired biological functions. In the early 1990s, the self-assembled peptides was achieved with alternating hydrophobic and charged amino acids sequence derived from zoutin, a yeast protein (Zhang *et al.*, 1993). The formed macroscopic membrane showed a network of interwoven filaments with 10–20 nm in diameter, which has been found to have applications in certain neurological disorders. With a similar self-assembly theory, a novel self-assembly peptide nanofiber scaffold has been synthesized by Zhang and coworkers through the assembly of ionic self-complementary peptides with alternating positive and negative charges (L-amino acids) under physiological conditions (Ellis-Behnke *et al.*, 2006). The scaffolds were built as tissue-bridging structures providing a framework with capability not only to regenerate axons through the site of an acute injury but also to knit the brain tissue together, as evident by animal study. It is further reported that several functional motifs, such as cell adhesion, differentiation and bone marrow homing motifs, were attached to the well-controlled 3D nanofiber structured scaffolds formed by self-assembly peptides (Gelain *et al.*, 2006). The functionalized 3D scaffolds were found to not only significantly enhance neural stem cell survival but also promote differentiation towards cells expressing neuronal and glial markers, indicating promising applications in treatment of neuro-trauma and neuro-degeneration diseases.

In addition to self-assembly of native peptides via amino acid sequence, another alternative to guide self-assembly process has been developed, i.e. hybrid peptide materials. Most hybrid materials are amphiphilic in nature with appending of hydrophobic or aromatic moiety onto a hydrophilic peptide, so that additional chemical functionalities can be introduced with known self-assembly properties (Stephanopoulos *et al.*, 2013). Peptide amphiphile (PA) is one of the most common hybrid biomaterials studied in this process. Self-assembly PA nanofibers have diameters as small as 10 nm and pore sizes of 5–200 nm, which is well within the dimensional range of native ECM and significantly smaller than nanofibers fabricated by other methods, such as electrospinning (Ayres *et al.*, 2009). Zhou and coworkers fabricated a biomimetic nanofibrous 3D hydrogel scaffold by self-assembly of a mixture of two aromatic short peptide derivatives (Zhou *et al.*, 2009). The prepared 3D hydrogel provided highly hydrated, stiff and nanofibrous network that uniquely presents bioactive ligands at the fiber surface, mimicking essential features of native ECM. Some other types of native macromolecules, including proteins and polypeptides, have been studied using self-assembly to fabricate nano-scale 3D scaffolds for tissue engineering applications. Recently, DNA has been proven to be a versatile building block for programmable construction of many objects, including 3D scaffolds, through self-assembly. Douglas and coworkers reported a new strategy of molecular self-assembly of DNA to form different 3D customized shapes, including monolith, square nut, railed bridges, genie bottle, stacked cross, and slotted cross, with precisely controlled dimensions ranging from 10–100 nm (Douglas *et al.*, 2009). The self-assembled peptide nanofiber 3D scaffolds have been prepared from natural pure ι-amino acids, RADA16, and systematically tested with the most commonly used scaffolds for tissue engineering, including both synthetic and native polymers (Gelain *et al.*, 2007). The study demonstrated that the 3D scaffolds with native peptides showed higher numbers of living and differentiated cells. It was also found in the study that the laminin surface further improved the overall scaffold performance. The well-defined molecular structure with considerable potential for further functionalization makes the natural peptide-based scaffolds a very promising biological material for various applications. In summary, the self-assembly technique represents an attractive and fertile research area which have been reviewed in more detail elsewhere (Stephanopoulos *et al.*, 2013; Kopecek and Yang, 2009).

3.4. *Challenges of self-assembly technique*

The self-assembly technique for fabrication of 3D scaffolds fabricated from native polymers possesses high potentials in tissue engineering and regenerative medicine. However, it is a relatively new research area, compared with other nanofabrication techniques. For the future of this technique, several challenges need to be addressed to get this technique applied clinically. The first challenge in the future is to not only develop hierarchical structures of self-assembly 3D scaffolds but also make them capable of responding to external stimuli and reconfiguring their physical or chemical properties accordingly (Stephanopoulos *et al.*, 2013). The ideal self-assembly scaffolds should have the ability to mimic the biological systems which can change signals or modulate stiffness during the regenerative stages. Self-assembly itself is the best way to achieve this dynamic response, because the scaffolds are formed by multiple weak bonds, instead of covalently or chemically cross-linking interactions. Although a few examples of the peptide-based scaffolds are shown to have the ability to change shape and structure when triggered by external stimuli, the majority of this class of biomaterials still needs a lot of improvement to make them perform smartly. The second challenge of native polymer-based self-assembly scaffolds will be the scalability of their production. In order to make them applicable clinically, the laboratory based technique needs to achieve large-scale production. However, the self-assembly of natural peptide or native proteins are less scalable than synthetic polymers. Unlike the synthetic polymers which can be produced and tailored upon demands, the native polymers should not be chemically modified in order to not compromise their biocompatibility and biodegradability. The self-assembly scaffolds from naturally-occurring macromolecules, such as peptides and proteins, hold great potential for their clinical trials. As such, a lot of efforts must be put to test their biocompatibility on different animal models to ensure their safety properties before clinical applications.

4. Phase Separation

4.1. *Introduction of phase separation technique*

Phase separation is one of the most popular fabrication methods used to produce highly porous polymeric membranes, but only recently has it been applied to produce nanofibrous scaffold with well-defined 3D porous structure (Barnes *et al.*, 2007). Phase separation is a thermodynamic separation of a polymer solution into two different phases in order to lower the systematic free energy, i.e. a polymer-rich component and a polymer-poor/solvent-rich component. The separation is either induced thermally or by the addition of nonsolvent to the polymer solution to form a gel, and subsequently water is used to extract the solvent from the gel matrix. Then, the formed gel is cooled down to a certain temperature below the glass transition temperature of the polymer and lyophilized under vacuum to obtain the nanofibrous scaffolds. Different from electrospinning technique, phase separation requires no special or advanced equipments and can be easily manipulated with better batch-to-batch consistency. However, this technique is currently limited to only a handful of polymer varieties on a laboratory scale.

4.2. *Parameters affecting production of phase separation nanostructures*

Although this technique has been widely applied to fabricate tissue scaffolds and substitutes using various synthetic polymers, the studies on native polymers are quite limited until recent years. The phase separation technique generally consists of five basic steps: polymer dissolution, phase separation, gelation, solvent extraction, and freeze drying. During the whole process, many variables are there to control the structure of the scaffolds produced, such as solvent, type of polymer and its concentration, thermal treatment, solvent exchange, the order of the procedures, gelation temperature, and so on. For instance, it is well-known that lowering the gelation temperature will produce a better nano-scale fiber scaffold. Additionally, depending on the solvent and polymer used, the thermally induced phase separation can be divided into two different methods, i.e. solid-liquid phase separation and liquid-liquid phase separation. For the solid-liquid phase separation process, the separation is induced by lowering the temperature, resulting in solvent crystallization from a polymer solution and consequent formation of pores after removal of solvent crystals. For the liquid-liquid phase separation process, the polymer solutions with an upper critical temperature form a biocontinuous structure (both polymer-lean phase and polymer-rich phases) (Lu *et al.*, 2013). Furthermore, the addition of certain type of porogen, including sugar, inorganic salt, paraffin spheres, into the polymer solution during the polymer dissolution step can realize better control on the porous architecture, such as pore size, interconnectivity, and geometry (Smith and Ma, 2004).

The 3D nanofibrous chitosan scaffolds have recently been prepared by a thermal-induced solid-liquid phase separation method (Zhao *et al.*, 2011). Chitosan in various concentrations were first dissolved in acetic acid to form the polymer solution, and then the formed solution was soaked in liquid nitrogen for two hours to induce phase separation followed by freeze drying to remove acetic acid. The study pointed out that the fabrication conditions, including chitosan and acetic acid concentrations and phase separation temperature, determined chitosan crystallinity and consequently affected the morphology and dimension of the formed scaffolds. At lower chitosan concentration, some small floccules were observed on the surface of chitosan nanofibrous scaffolds, while increasing chitosan concentration resulted in more uniform nanofibrous structures. The quenching temperature was a critical factor in controlling the formation of nanofibrous structure. The crystallization process of polymer consists of two stages: nucleation and crystal growth. It is well-recognized that lower cooling rate induces a higher nucleation rate and a lower crystal growth rate, and vice versa (Boyer and Haudin, 2010). Therefore, when chitosan was frozen at −18°C or −80°C, the nucleation rate was relatively low but crystal growth rate was relatively high, conditions which were not appropriate to produce nanofibrous structure but instead, some microfibrous accompanying with film-shape matrices were formed. When the chitosan was dipped into liquid nitrogen which made chitosan frozen in a second with a super fast cooling rate, the instant nucleation and superslow crystal growth rate resulted in the formation of nanofibrous structure with diameter from 50–500 nm. Therefore, it was hypothesized that surpercooling rate was required to induce phase separation of chitosan in acetic acid solvent.

Gelatin, as the basic or acidic hydrolysis product of collagen, has a very similar chemical composition to collagen, and therefore has been considered a good candidate for fabrication of 3D scaffolds. Phase separation is one of the popular methods to produce gelatin-based nanofibrous scaffolds. Liu and Ma developed a new processing technique to create 3D nanofibrous gelatin scaffolds, mimicking the natural collagen ECM in both physical architecture and chemcial composition, and its processing parameters were systematically investigated (Liu and Ma, 2009). This processing technique involved thermally induced phase separation combined with a porogen-leaching technique. In this method, the gelatin gel matrix phase separated at $-76°C$ was subjected to immersion in cold ethanol followed by the solvent exchange with 1,4-dioxane. The resulting 3D nanofibrous scaffold possessed high surface area, high porosities, and well-connected macropores. Its mechanical properties were controlled by polymer concentration and cross-linking density. The solvent composition played important roles in the phase separation process and gel porosity formation step. It was suggested that ethanol was required in the solvent composition to achieve the phase separation. If the aqueous solution of gelatin was freeze-dried directly, the gelatin foam with pores diameter of 100 μm without any nanofibrous architecture would be formed. However, when a certain concentration of ethanol or methanol was included in the aqueous solution as a nonsolvent, phase separation of gelatin can be induced thermally and nanofibrous structures were obtained easily. The study pointed out that the selection of proper solvent was critical to thermally induced phase separation, and that only ethanol- and methanol-aqueous solvent can be used to obtain nanostructured gelatin foam. Many other organic solvents, such as acetone, dioxane, tetrahydrofuran were unable to produce gelatin nanofibrous foam. The ethanol or methanol concentration was optimized to between 10–50%, otherwise gelatin beads would form during phase separation or gelatin cannot be dissolved completely. However, the specific mechanism underlying the interactions between gelatin molecules and the solvent to control the gelatin gel porosity is still unknown. Following the phase separation step, solvent exchange process was required in order to avoid shrinkage of the formed gelatin foam. The porogen-leaching technique was also introduced during the scaffolding fabrication process, using paraffin spheres with different pore sizes. The heat treatment to accelerate the dissolution of paraffin spheres were applied, showing that the smaller paraffin pore size and longer treatment time resulted in smaller dimension in gelatin foam porosity with higher interconnectivity. The prepared gelatin scaffold also showed much better dimensional stability in a tissue culture environment, compared with commercial gelatin foam (Gelfoam®).

4.3. *Applications of native polymer-based phase separation technique*

The biomimetic gelatin-based 3D nanofibrous scaffolds have been evaluated by osteoblasts cells for bone tissue engineering applications (Liu *et al.*, 2009). The 3D gelatin scaffolds were fabricated by thermally induced phase separation method with porogen leaching technique, using paraffin spheres as porogen. The scaffolds were further cross-linked chemically to obtain appropriate mechanical properties. The prepared nanofibrous gelatin scaffolds had similar porosity and pore size distribution of commercial product Gelfoam®, while the surface area and compressive modulus of the nanofibrous scaffolds were more than 700 and 10 times higher than that of Gelfoam®, respectively. The osteoblast cell adhesion profile was significantly improved on nanofibrous scaffolds, compared with Gelfoam.® The advantageous cell adhesion was attributed to both the higher surface area and higher porosity which promoted the cell-cell and cell-matrix interactions. Due to the excellent compressive modulus, the nanofibrous gelatin scaffolds were able to maintain its size after two weeks of pre-osteoblast cell culture, while the diameter of the Gelfoam® shrank to half the original size. To further explore its application in bone engineering, the mechanical strength was further enhanced by deposition of a bio-mimetic apatite layer throughout the porous structures of 3D scaffolds. After deposition, the compressive modulus was enhanced by 75% more than initial nanofibrous gelatin scaffolds, and also the cell differentiation was further enhanced as evidenced by two late osteogenic differentiation markers, i.e. bone sialoprotein and osteocalcin. Alternatively, instead of fabrication of nanofibrous scaffolds, the micro-sized scaffolds can be easily prepared from aqueous solution by freeze drying process, and then the hydroxyapatite nanoparticles coating can be further applied on the surface of the scaffolds to enhance functionality of the micro-sized scaffolds. Zandi and coworkers reported a fabrication of gelatin-based scaffolds with micro-sized pores by lyophilization of aqueous solution of gelatin premixed with hydroxyapatite (Zandi *et al.*, 2010). The obtained gelatin/hydroxyapatite scaffolds were subsequently coated by nano-size hydroxyapatite in nano-rod configuration, using chemical bath deposition method. The coated hydroxyapatite/gelatin scaffolds had significantly higher surface area and reactivity, resulting in not only improved mechanical strength that was comparable to human cancellous bone, but also more desirable biocompatibility for mesenchymal stem cells. Besides the pure gelatin scaffolds, hybrid scaffolds of gelatin with a secondary polymer has also received increasing attention. For instance, the fabrication of hydroxyapatite-chitosan-gelatin 3D scaffolds has been reported recently through phase-separation method. The pore size of the hybrid can be readily modulated by freezing conditions. The cellular evaluation using human mesenchymal stem cells was performed under two different conditions, i.e. static and perfusion conditions. It was pointed out that the perfusion preconditioning enhanced adsorption of ECM proteins and thus promoted the cell proliferation and osteogenic differentiation, compared with static preconditioning. The study highlighted the importance of convective flow in modulating the 3D scaffolds microenvironment, which provided new insight into the development of nanofibrous 3D substitutes.

In addition to biopolymer hybrid scaffolds, the incorporation of bioactive inorganic phase into polymeric nanofibers to more precisely mimic the natural bone ECM by phase separation technique has also emerged as a novel nanofabrication technique to fabricate 3D scaffolds. A novel way of producing nanofibrous gelatin-silica hybrid 3D scaffolds via thermally induced phase separation technique has been reported recently (Lei *et al.*, 2012). In this technique, the silica sol was prepared by conventional sol-gel process and then the various amount of silica sol was mixed with gelatin ethanol-aqueous solution using magnetic stirring at 40°C, followed by freezing at –70°C, solvent exchange and then freeze-drying. The gelatin and silica phases were found to be homogenously hybridized at molecular level due to their similar hydrophilicity. The mechanical properties of nanofibrous hybrid scaffolds were highly dependent on silica content, showing that higher silica content resulted in significant improvement of elastic response and compressive modulus. The silica gel also greatly delayed *in vitro* biodegradation rate, possibly as a result of the electrostatic and hydrogen interaction between silica hydroxyls and gelatin amino groups. More importantly, the incorporation of silica gel into gelatin scaffolds did not significantly affect the scaffold biocompatibility tested by cell viability experiments using pre-osteoblast cell line, as compared with pure gelatin scaffold. Therefore, it is feasible to fabricate native polymer-bioactive inorganic hybrid nanofibrous scaffolds by phase separation technique. However, in addition to evaluation of physical properties and biocompatibilities, further *in vitro* and *in vivo* evaluations are required to claim their potential applications in bone tissue engineering.

Nevertheless, the cellular evaluation of all-natural biopolymer-based 3D scaffolds fabricated by phase separation technique is quite limited. Most of the studies focus on the gelatin-based 3D scaffolds as discussed above. Instead of all-natural biopolymer components, many other studies have reported the phase separation technique on fabrication of synthetic/native polymer hybrid 3D scaffolds and their cellular evaluations, such as hyaluronic acid/collagen/poly-L-lactide hybrid polymer 3D scaffolds (Niu *et al.*, 2009).

4.4. *Challenges of phase separation technique*

Compared with electrospinning and self-assembly processes, this technique has been considered a simple and easy approach with consistent results. However several challenges still need to be overcome. Firstly, the variety of the polymers that can be used in this process is strictly limited, and only a handful of native polymers have been tested and reported. Secondly, some porogens are difficult to be completely extracted out. Thirdly, the industrialization of this technique on a large-scale is still difficult or even impractical to be realized by amenable process. Fourthly, the cellular evaluation of native polymer-based scaffolds prepared from phase separation technique is still lacking. So far, most of the literatures are focusing on the cellular evaluation of synthetic polymer based scaffolds that have similar dimensions to the natural collagen ECM, among which poly-L-lactide (PLLA) and poly(D,L-lactic-co-glycolic acid (PLGA) are the most investigated synthetic polymers evaluated at cellular level (Sun *et al.*, 2010) and *in vivo* animal model (Yu *et al.*, 2013).

5. Nano-Patterning Techniques using Native Polymers

In order to mimic the native ECM more accurately and improve the cell attachment and alignment, patterning techniques have been developed in recent years. Both micro-patterning and nano-patterning techniques are able to fabricate patterned topographies on 3D structured scaffolds. The conventional and standard micro-patterned topography technique is photolithography carried out on a silicon wafer with a photoresistant mask that selectively allows near-UV light source go through, and thereby recreates the patterns. However, micro-patterned topographies have been widely reported to produce unsatisfactory surfaces that may induce apoptosis after cells are seeded, mainly due to cell anisotropy and resultant inadequate cell attachment. Therefore, the development of new methods to fabricate nano-patterned topographies and 3D scaffolds has received increasing attention. The successful nano-patterning techniques include polymer demixing, biomolecules replication, colloidal lithography, and electron beam lithography, which are summarized in Table 1.

Table 1. Summary of nano-patterning techniques used for fabrication of tissue repair. Adapted and expanded from (Gadegard *et al.* 2006).

Nano-patterning method	Process illustration	Notes
Polymer demixing	Spin coating → Annealing	Various nanoscales can be created, e.g. pits, islands, ribbons; Quick and inexpensive; however, features are unordered ad hard to be controlled.
Biomolecule replication	Self-assembly → Replica	Native polymers can be used; dimensions with sub-10 nm scale can be achieved; quick process; however, accuracy is hard to control.
Colloidal lithography	Deposition → Dry etch	Large surface areas can be patterned; the heights and diameters of surfaces can be controlled; inexpensive; however, the features are unordered.
E-beam lithography	Spin coating → Exposure	Creation of ordered patterns of nanotopographical features, as small as 3–5 nm. However, it is time consuming and costly.
X-ray interference	Polymer Casting → Film peeled off	Most commonly studied method using native polymers, such as collagen. Ordered structures can be formed. However, it is costly by using X-ray.

Although many nano-patterning have been developed in recent decade, most of these techniques are studied with synthetic polymers due to their wide variety and modifiable physicochemical properties. Among these methods, X-ray interference technique is the only one that has been widely studied with native polymers, especially collagen. Therefore, only this technique will be discussed here. In this technique, X-ray is used to create a template with nanochannels using silicon wafer with a photoresistor. The template of pattern normally contains parallel channels of equal groove and ridge widths at nanoscale. Then, the biopolymer film is formed by casting method on the surface of the patterned template. After the film is dried, it is peeled off from the template and will have the patterned features with nano-groves. Films formed from collagen type I have been extensively investigated for its application in blood vessels repairing. The major limitation for its application is the mechanical properties. For the application in artificial blood vessels, the engineered construct needs to withstand certain load and pressure. However, the native proteins do not meet this requirement. A recent study showed that nano-patterning of collagen scaffolds improved the mechanical properties of tissue engineered vascular grafts (Zorlutuna et al., 2009). It was shown that vascular smooth muscle cells attached and aligned well on the nano-patterned collagen scaffolds in the same way as they are in the natural tissue. With cell attachment and alignment, the mechanical properties of the scaffolds were greatly improved, showing the similar ultimate tensile strength and Young's modulus to the natural vessels. With the aim to fabricate 3D substitutes for tissue engineering applications, the nano-patterned collagen films formed by X-ray interference can be shaped into different 3D shapes by bending and rolling. In this approach, it is best to minimize the volume of polymer to leave as much room for cells as possible (Gadegaard et al., 2006). The nano-grooved films can be rolled up into tubes with concentric layers of collagen to develop the vascular substitutes (Zorlutuna et al., 2009). The nano-patterned collagen tubular scaffold exhibited not only excellent cell retention, but also great molecular diffusivity of nutrients which is crucial to support cell growth, compared with unpatterned collagen films.

6. Concluding Remarks

The native polymer-based nanoscale 3D scaffolds can be prepared by various nanofabrication techniques, which may represent the future trend of tissue engineering. By adopting the appropriate native biopolymer and proper fabrication technique, the obtained scaffolds are able to precisely mimic the physical and chemical properties of natural ECM *in vivo*, including the topographical morphology, fiber diameter, mechanical properties, high surface area, cell-cell and cell-matrix interactions, etc. However, each nanofabrication technique discussed in this chapter has its own limitations, as well as the fact that only a selected variety of native polymers are currently used for these techniques. Because the complexities are often associated with natural polymers, such as complex structural composition, tedious purification process, possible immunogenicity and pathogen transmission problems, the development of nanofabrication techniques for 3D substitute using native polymers are still facing a lot of challenges.

Furthermore, although some of the nanofabrication techniques, such as self-assembly and phase separation, have been widely studied, the important parameters may vary with the variety of native polymers or their hybrid, and more cellular and *in vivo* evaluations of the prepared 3D scaffolds are required to better understand their potentials in tissue engineering applications. Even though a lot challenges remain on the road, there is no doubt that nanotechnology will play a critical role in the development of next generation scaffolds, especially in terms of fabrication component.

References

Ayres CE, Jha BS, Sell SA, Bowlin GL, Simpson DG. Nanotechnology in the design of soft tissue scaffolds: innovations in structure and function. *Wiley Interdiscip Rev Nanomed Nanobiotechnol* 2009;2:20–34.

Barnes CP, Sell SA, Boland ED, Simpson DG, Bowlin GL. Nanofiber technology: Designing the next generation of tissue engineering scaffolds. *Adv Drug Deliv Rev* 2007;59:1413–1433.

Beachley V, Wen X. Effect of electrospinning parameters on the nanofiber diameter and length. *Mater Sci Eng C* 2009;29:663–668.

Berndt P, Fields GB, Tirrell M. Synthetic lipidation of peptides and amino acids: Monolayer structure and properties. *J Am Chem Soc* 1995;117:9515–9522.

Boyer SA, Haudin J-M. Crystallization of polymers at constant and high cooling rates: A new hot-stage microscopy set-up. *Polym Test* 2010;29:445–452.

Chang J-J, Lee Y-H, Wu MH, Yang M-C, Chien CT. Preparation of electrospun alginate fibers with chitosan sheath. *Carbohyd Polym* 2012;87:2357–2361.

Chen Z, Wang P, Wei B, Mo X, Cui F. Electrospun collagen–chitosan nanofiber: A biomimetic extracellular matrix for endothelial cell and smooth muscle cell. *Acta Biomater* 2010;6:372–382.

Chew SY, Wen J, Yim EK, Leong KW. Sustained release of proteins from electrospun biodegradable fibers. *Biomacromolecules* 2005;6:2017–2024.

Coutinho D, Costa P, Neves N, Gomes ME, Reis RL. Micro- and nanotechnology in tissue engineering. *Tissue Eng* 2011;3–29.

Cui FZ, Li Y, Ge J. Self-assembly of mineralized collagen composites. *Mat Sci Eng R* 2007;57:1–27.

Davidenko N, Campbell J, Thian E, Watson C, Cameron R. Collagen-hyaluronic acid scaffolds for adipose tissue engineering. *Acta Biomater* 2010;6:3957–3968.

Ding B, Kimura E, Sato T, Fujita S, Shiratori S. Fabrication of blend biodegradable nanofibrous nonwoven mats via multi-jet electrospinning. *Polym* 2004;45:895–1902.

Douglas SM, Dietz H, Liedl T, Hogberg B, Graf F, Shih WM. Self-assembly of DNA into nanoscale three-dimensional shapes (vol 459, pg 414, 2009). *Nature* 2009;459:1154–1154.

Dvir T, Timko BP, Kohane DS, Langer R. Nanotechnological strategies for engineering complex tissues. *Nat Nanotechnol* 2011;6:13–22.

Ellis-Behnke RG, Liang Y-X, You S-W, Tay DK, Zhang S, So K-F, Schneider GE. Nano neuro knitting: Peptide nanofiber scaffold for brain repair and axon regeneration with functional return of vision. *Proc Natl Acad Sci USA* 2006;103:5054–5059.

Fang D, Chang C, Hsiao Benjamin S, Chu B. Development of Multiple-Jet Electrospinning Technology. In *Polymeric Nanofibers*, American Chemical Society: 2006;918:91–105.

Friess W. Collagen — biomaterial for drug delivery. *Eur J Pharm Biopharm* 1998;45:113–136.

Gadegaard N, Martines E, Riehle MO, Seunarine K, Wilkinson CDW. Applications of nano-patterning to tissue engineering. *Microelectron Eng* 2006;83:1577–1581.

Gelain F, Bottai D, Vescovi A, Zhang S. Designer self-assembling peptide nanofiber scaffolds for adult mouse neural stem cell 3-dimensional cultures. *PLoS One* 2006;1:e119.

Gelain F, Lomander A, Vescovi AL, Zhang SG. Systematic studies of a self-assembling peptide nanofiber scaffold with other scaffolds. *J Nanosci Nanotechno* 2007;7:424–434.

Hauser CA, Zhang S. Designer self-assembling peptide nanofiber biological materials. *Chem Soc Rev* 2010;39:2780–2790.

Homayoni H, Ravandi SAH, Valizadeh M. Electrospinning of chitosan nanofibers: Processing optimization. *Carbohyd Polym* 2009;77:656–661.

Hsiao H-Y, Huang C-M, Liu Y-Y, Kuo Y-C, Chen H. Effect of air blowing on the morphology and nanofiber properties of blowing-assisted electrospun polycarbonates. *J Appl Polym Sci* 2012;124:4904–4914.

Jayakumar R, Prabaharan M, Nair S, Tamura H. Novel chitin and chitosan nanofibers in biomedical applications. *Biotechnol Adv* 2010;28:142–150.

Jeong SI, Krebs MD, Bonino CA, Samorezov JE, Khan SA, Alsberg E. Electrospun Chitosan–Alginate Nanofibers with In Situ Polyelectrolyte Complexation for Use as Tissue Engineering Scaffolds. *Tissue Eng Part A* 2010; 17:59–70.

Kelleher CM, Vacanti JP. Engineering extracellular matrix through nanotechnology. *J R Soc Interface* 2010;7:S717–S729.

Kim GH, Yoon H. A direct-electrospinning process by combined electric field and air-blowing system for nanofibrous wound-dressings. *Appl Phys A* 2008; 90:389–394.

Kokini K, Sturgis JE, Robinson JP, Voytik-Harbin SL. Tensile mechanical properties of three-dimensional type I collagen extracellular matrices with varied microstructure. *J Biomech Eng* 2002;124:214–222.

Kopeček J, Yang J. Peptide-directed self-assembly of hydrogels. *Acta Biomater.* 2009;5:805–816.

Larsen G, Spretz R, Velarde-Ortiz R. Use of coaxial gas jackets to stabilize Taylor cones of volatile solutions and to induce particle-to-fiber transitions. *Adv Mater* 2004;16:166–169.

Lei B, Shin KH, Noh DY, Jo IH, Koh YH, Choi WY, Kim HE. *J Mater Chem* 2012;22:14133–14140.

Li D, Wang Y, Xia Y. Electrospinning nanofibers as uniaxially aligned arrays and layer_by_layer stacked films. *Adv Mater* 2004;16:361–366.

Li D, Xia YN. Electrospinning of nanofibers: Reinventing the wheel? *Adv Mater* 2004;16:1151–1170.

Liao S, Murugan R, Chan CK, Ramakrishna S. Processing nanoengineered scaffolds through electrospinning and mineralization suitable for biomimetic bone tissue engineering. *J Mech Behav Biomed Mater* 2008;1:252–260.

Liu X, Ma PX. Phase separation, pore structure, and properties of nanofibrous gelatin scaffolds. *Biomaterials* 2009;30:4094.

Liu X, Smith LA, Hu J, Ma PX. Biomimetic nanofibrous gelatin/apatite composite scaffolds for bone tissue engineering. *Biomaterials* 2009;30:2252–2258.

Lu QA, Wang XL, Lu SZ, Li MZ, Kaplan DL, Zhu, HS. Nanofibrous architecture of silk fibroin scaffolds prepared with a mild self-assembly process. *Biomaterials* 2011;32:1059–1067.

Lu T, Li Y, Chen T. Techniques for fabrication and construction of three-dimensional scaffolds for tissue engineering. *Int J Nanomedicine* 2013;8:337.

Madurantakam PA, Cost CP, Simpson DG, Bowlin GL. Science of nanofibrous scaffold fabrication: Strategies for next generation tissue-engineering scaffolds. *Nanomedicine* 2009;4:193–206.

McClure MJ, Sell SA, Simpson DG, Walpoth BH, Bowlin GL. A three-layered electrospun matrix to mimic native arterial architecture using polycaprolactone, elastin, and collagen: A preliminary study. *Acta Biomater* 2010;6:2422–2433.

McClure MJ, Simpson DG, Bowlin GL. Tri-layered vascular grafts composed of polycaprolactone, elastin, collagen, and silk: Optimization of graft properties. *J Mech Behav Biomed Mater* 2012;10:48–61.

McCullen SD, Autefage H, Callanan A, Gentleman E, Stevens MM. Anisotropic fibrous scaffolds for articular cartilage regeneration. *Tissue Eng — Part A* 2012;18:2073–2083.

Morton WJ. Method of dispersing fluids. 1902;No. 705691.

Ng R, Zhang R, Yang KK, Liu N, Yang S-T. Three-dimensional fibrous scaffolds with microstructures and nanotextures for tissue engineering. *RSC Advances* 2012.

Niu XF, Feng QL, Wang MB, Guo XD, Zheng QX. Porous nano-HA/collagen/PLLA scaffold containing chitosan microspheres for controlled delivery of synthetic peptide derived from BMP-2. *J Control Release* 2009;134:111–117.

Pan H, Li L, Hu L, Cui X. Continuous aligned polymer fibers produced by a modified electrospinning method. *Polym* 2006;47:4901–4904.

Panzavolta S, Gioffrè M, Focarete ML, Gualandi C, Foroni L, Bigi A. Electrospun gelatin nanofibers: Optimization of genipin cross-linking to preserve fiber morphology after exposure to water. *Acta Biomater* 2011;7:1702–1709.

Park CH, Kim CH, Pant HR, Tijing LD, Yu MH, Kim Y, Kim CS. An angled robotic dual-nozzle electrospinning set-up for preparing PU/PA6 composite fibers. *Text Res J* 2013;83:311–320.

Sell SA, McClure MJ, Garg K, Wolfe PS, Bowlin GL. Electrospinning of collagen/biopolymers for regenerative medicine and cardiovascular tissue engineering. *Adv Drug Deliv Rev* 2009;61:1007–1019.

Smith L, Ma P. Nano-fibrous scaffolds for tissue engineering. *Colloids Surf B Biointerfaces* 2004;39:125–131.

Stamov D, Grimmer M, Salchert K, Pompe T, Werner C. Heparin intercalation into reconstituted collagen I fibrils: Impact on growth kinetics and morphology. *Biomaterials* 2008;29:1–14.

Stamov D, Salchert K, Springer A, Werner C, Pompe T. Structural polymorphism of collagen type I-heparin cofibrils. *Soft Matter* 2009;5:3461–3468.

Stella JA, D'Amore A, Wagner WR, Sacks MS. On the biomechanical function of scaffolds for engineering load-bearing soft tissues. *Acta Biomater* 2010;6:2365–2381.

Stephanopoulos N, Ortony JH, Stupp SI. Self-assembly for the synthesis of functional biomaterials. *Acta Mater* 2013;61:912–930.

Sun H, Feng K, Hu J, Soker S, Atala A, Ma PX. Osteogenic differentiation of human amniotic fluid-derived stem cells induced by bone morphogenetic protein-7 and enhanced by nanofibrous scaffolds. *Biomaterials* 2010;31:1133–1139.

Telemeco T, Ayres C, Bowlin G, Wnek G, Boland E, Cohen N, Baumgarten C, Mathews J, Simpson, D. Regulation of cellular infiltration into tissue engineering scaffolds composed of submicron diameter fibrils produced by electrospinning. *Acta Biomater* 2005;1:377–385.

Tijing LD, Choi WL, Jiang Z, Amarjargal A, Park C-H, Pant HR, Im I-T, Kim CS. Two-nozzle electrospinning of (MWNT/PU)/PU nanofibrous composite mat with improved mechanical and thermal properties. *Curr App Phys* 2013.

Torres-Giner S, Gimenez E, Lagaron JM. Characterization of the morphology and thermal properties of Zein Prolamine nanostructures obtained by electrospinning. *Food Hydrocolloids* 2008;22:601–614.

Torres-Giner S, Gimeno-Alcañiz JV, Ocio MJ, Lagaron JM. Comparative performance of electrospun collagen nanofibers cross-linked by means of different methods. *ACS Appl Mater Interfaces* 2008;1:218–223.

Tsai SW, Liu RL, Hsu FY, Chen CC. A study of the influence of polysaccharides on collagen self-assembly: Nanostructure and kinetics. *Biopolymers* 2006;83:381–388.

Tu RS, Tirrell M. Bottom-up design of biomimetic assemblies. *Adv Drug Deliv Rev* 2004;56:1537–1563.

Um IC, Fang D, Hsiao BS, Okamoto A, Chu B. Electro-spinning and electro-blowing of hyaluronic acid. *Biomacromolecules* 2004;5:1428–1436.

Varesano A, Carletto RA, Mazzuchetti G. Experimental investigations on the multi-jet electrospinning process. *J Mater Process Technol* 2009;209:5178–5185.

Viry L, Moulton SE, Romeo T, Suhr C, Mawad D, Cook M, Wallace GG. Emulsion-coaxial electrospinning: Designing novel architectures for sustained release of highly soluble low molecular weight drugs. *J Mater Chem* 2012;22:11347–11353.

Wang Q, Zhang B. *Chapter 3: Self-assembled nanostructures. In: Nanotechnology research methods for foods and bioproducts. Padua GW Wang Q. (Edited).* Wiley-Blackwell: 2012.

Wang X, Sang L, Luo D, Li X. From collagen–chitosan blends to three-dimensional scaffolds: The influences of chitosan on collagen nanofibrillar structure and mechanical property. *Colloids Surf B Biointerfaces* 2011;82:233–240.

Wang X, Um IC, Fang D, Okamoto A, Hsiao BS, Chu B. Formation of water-resistant hyaluronic acid nanofibers by blowing-assisted electro-spinning and non-toxic post treatments. *Polym* 2005;46:4853–4867.

Wei M, Kang B, Sung C, Mead J. Core-sheath structure in electrospun nanofibers from polymer blends. *Macromol Mater Eng* 2006;291:1307–1314.

Yao C, Li X, Song T. Electrospinning and crosslinking of zein nanofiber mats. *J Appl Polym Sci* 2007;103:380–385.

Yarin AL. Coaxial electrospinning and emulsion electrospinning of core-shell fibers. *Polym Advan Technol* 2011;22:310–317.

Yu C-C, Chang J-J, Lee Y-H, Lin Y-C, Wu M-H, Yang M-C, Chien C-T. Electrospun scaffolds composing of alginate, chitosan, collagen and hydroxyapatite for applying in bone tissue engineering. *Materials Letters* 2013;93:133–136.

Yu NY, Schindeler A, Peacock L, Mikulec K, Fitzpatrick J, Ruys AJ, Cooper-White JJ, Little DG. Modulation of anabolic and catabolic responses via a porous polymer scaffold manufactured using thermally induced phase separation. *Eur Cells Mat* 2013;25.

Zandi M, Mirzadeh H, Mayer C, Urch H, Eslaminejad MB, Bagheri F, Mivehchi H. Biocompatibility evaluation of nano-rod hydroxyapatite/gelatin coated with nano-HAp as a novel scaffold using mesenchymal stem cells. *J Biomed Mater Res Part A* 2010;92:1244–1255.

Zhang S, Holmes T, Lockshin C, Rich A. Spontaneous assembly of a self-complementary oligopeptide to form a stable macroscopic membrane. *Proc Natl Acad Sci* 1993;90:3334–3338.

Zhang X, Reagan MR, Kaplan DL. Electrospun silk biomaterial scaffolds for regenerative medicine. *Adv Drug Deliv Rev* 2009;61:988–1006.

Zhao J, Han W, Chen H, Tu M, Zeng R, Shi Y, Cha Z, Zhou C. Preparation, structure and crystallinity of chitosan nano-fibers by a solid–liquid phase separation technique. *Carbohyd Polym* 2011;83:1541–1546.

Zhou M, Smith AM, Das AK, Hodson NW, Collins RF, Ulijn RV, Gough JE. Self-assembled peptide-based hydrogels as scaffolds for anchorage-dependent cells. *Biomaterials* 2009;30:2523–2530.

Zorlutuna P, Elsheikh A, Hasirci V. Nanopatterning of collagen scaffolds improve the mechanical properties of tissue engineered vascular grafts. *Biomacromolecules* 2009;10:814–821.

Zorlutuna P, Rong Z, Vadgama P, Hasirci V. Influence of nanopatterns on endothelial cell adhesion: Enhanced cell retention under shear stress. *Acta Biomater* 2009;5:2451–2459.

Chapter 7

Native Polymer-based 3D Substitutes as Alternatives with Slow-Release Functions

Dongwei Guo[†,2], *Benson J. Edagwa*[†,1], *Xin-Ming Liu*[*,1,2]

[1]*Department of Pharmacology and Experimental Neuroscience,*
[2]*Department of Pharmaceutical Sciences,*
University of Nebraska Medical Center,
Omaha, Nebraska, USA 68198.

1. Introduction

Tissue engineering has emerged as a promising biomedical technology and methodology to assist the regeneration and repair of defective tissues and organs. For this therapeutic strategy, scaffolds are needed to provide cells with physical support and a local environment that regulates cell proliferation and differentiation for tissue regeneration and repair. Many biodegradable and non-degradable polymeric biomaterials have been developed for tissue engineering, and only few of them have successfully been utilized in clinical application. Among them is a perfect example — synthetic poly-(lactic-co-glycolic acid) (PLGA). However, increased tissue inflammatory response caused by

*Corresponding author: Xinming Liu, Ph.D., Department of Pharmacology and Experimental Neuroscience, 985800 Nebraska Medical Center, Omaha, NE 68198-5800 USA; Email: xliu@unmc.edu; Phone: 01-402-559-4562
[†]Equal contributions

the byproducts of acid hydrolysis is still a major concern. Natural polymers have the potential to be an alternative and/or replacement biomaterial for tissue engineering due to their increased biocompatibility, biodegradability, and non-toxic degradation products. Many natural polymers, such as proteins and polysacchrides have been widely studied in tissue engineering. While great advances have been achieved, the use of natural polymers in tissue engineering still faces significant challenges. The extracellular matrix acts as a scaffold for cell attachment, proliferation and differentiation in tissue engineering. The limited ability to create the required microenvironment during engineered tissue development, as well as promoting accurate *in vivo* tissue repair and regeneration, presents a significant challenge. Biomaterials for tissue engineering are expected to produce scaffolds that have the ability to mimic *in vivo* tissue/organ developmental functions and provide the necessary signaling/therapeutic agents for cell attachment, proliferation and differentiation. Thus, encapsulation and efficient delivery of specific signaling/therapeutic molecules to their targets is essential to potentially achieve the required dynamic reciprocity in tissue engineering. Indeed, one goal of next generation tissue engineering is utilizing the controlled delivery and sustained release of signal/therapeutic agents in order to enhance tissue engineering efficacy. Increased effort has been placed into the development of drug delivery systems (DDS) for this purpose.

The delivery of signal/therapeutic agents and cells using natural polymers has the advantageous ability to function not only as drug delivery systems, but also structural scaffolds which may be applied for tissue engineering. The incorporation of signal/therapeutic agents into the natural polymer scaffolds has great potential in introducing signals and therapeutic agents to cells in a spatial and temporal manner, while enhancing the interplay between cells and the extracellular milieu for tissue growth and maintenance with neighbouring tissues. The combination of drug delivery and tissue engineering within a single system is recognized to be one future direction of tissue engineering for cell adhesion, proliferation, differentiation and synthesis of extracellular matrix. In addition, the combination of scaffold and DDS significantly contributes to the application in stem cell biology and medicine for obtaining a large number of high quality cells for cell transplantation therapy. Many natural polymers, such as proteins and polysaccharides, for tissue engineering have synergized the cell scaffold and DDS together to emphasize the significance of biomaterial technology in new therapeutic and research fields.

Natural polymers play important roles in their native forms. For example, proteins perform key roles as structural materials and enzymatic catalysts, while polysaccharides serve as key components of membranes and intracellular communication. From many points of view, the scaffolds for tissue engineering would ideally be made of biodegradable polymers with properties closely resembling the extracellular matrix (ECM). The current trend in tissue engineering is to mimic nature by using natural polymer scaffolds. Many natural polymers have the ability to form soft, tough and elastomeric networks which provide mechanical stability and structural integrity to tissues and organs. Many applications within the tissue engineering field can be extrapolated using natural polymers to repair and regenerate the tissue/organ. This chapter focuses mainly on the combination of drug delivery and tissue engineering within a single system using natural polymers and various well-studied natural protein and polysaccharide polymers in tissue engineering fields.

2. Proteins

Proteins are natural polymers composed of amino acid building blocks linked by amide bonds. The distribution of amino acids along the protein backbone renders proteins with distinct characteristics. Proteins are widely studied and used for drug delivery and tissue engineering because of their biodegradability, low toxicity, low antigenicity, high biocompatibility, high binding and encapsulation capacity of various therapeutic agents (Nitta and Numata, 2013; Uebersax *et al.*, 2009). Naturally derived proteins such as collagen, gelatin, elastin, albumin, fibronectin and zein have been widely used as biomaterials (Nitta and Numata, 2013; Uebersax *et al.*, 2009). For tissue engineering, hydrophilic or hydrophobic natural protein-based biomaterials with drug delivery capability serve as structural scaffolds and local bioreactors to stabilize encapsulated and transplanted cells, sustain release of encapsulated signal/therapeutic agents, and direct cell migration, growth, differentiation and organization by mimicking many features of the extracellular matrix.

2.1. *Collagen*

Collagen is a long and structural fibrous protein found in the extracellular matrices of connective tissues in mammals (Muyonga *et al.*, 2003). Type I collagen, the major component of tendons, bones and ligaments, is the most predominant in the collagen family. Collagen is also present in corneas, blood vessels and the gut (Brinckmann *et al.*, 2005). It plays a mechanical role in the body by supporting tissues and cell structures, while also protecting the body from external stimuli. The extracellular matrix provides space for cell replication and differentiation.

Collagen has garnered attention in tissue engineering and other biomedical applications due to its excellent biocompatibility, negligible immunogenicity and high bio-absorbability (Ramshaw *et al.*, 1996; Rao, 1995; Maeda *et al.*, 1999). Collagen-based biomaterials can be prepared through different methods. One of the techniques involves generation of a cellular collagen matrix by physical, chemical or enzymatic approaches (Gilbert *et al.*, 2006; Dong *et al.*, 2012). This method preserves the original tissue shape and extracellular matrix structure. Manufacturing collagen scaffolds is also accomplished by employing extraction, purification and polymerization. The scaffolds produced could be cross-linked to improve the mechanical strength and lower the biodegrading rate. Modification of collagen properties to generate various forms for different purposes have been achieved by incorporating other molecules such as glycosaminoglycans, elastin and chitosan (Parenteau-Bareil *et al.*, 2010).

Over the last decade, numerous reports from different research groups have explored the use of collagen-based biomaterials for cartilage regeneration, wound dressing, delivery of cells, drugs and nucleic acids, cardiovascular and plastic neurosurgery. Collagen-based implants have been used for skin replacement due to their mechanical strength.

Ma *et al.* (2003) reported the fabrication of porous collagen/chitosan scaffold cross-linked with glutaraldehyde for skin tissue engineering. In this study, Ma *et al.* (2003) investigated the morphology, swelling capacity and degradability of the cross-linked scaffold under different concentrations of glutaraldehyde. Both *in vivo* and *in vitro* studies found that the collagen/chitosan scaffolds were compatible with human dermal fibroblasts and could therefore sufficiently support their growth and infiltration from the surrounding tissue. The results from this study demonstrated the potential application of the collagen/chitosan scaffold with enhanced bio-stability and biocompatibility in skin tissue engineering. In a similar study, Tremblay *et al.* (2005) described human endothialized reconstructed skin that combined keratinocytes, fibroblasts and endothelial cells in a collagen sponge. *In vitro* data revealed the formation of capillary-like structures. This was further supported by studies in nude mice where the capillary like structures were formed underneath the epidermis in less than four days. This report demonstrates the potential application of collagen-based scaffolds in tissue engineered organs which require rapid vascularisation.

In 2009, Blais *et al.* sought to investigate the effect of Schwann cells on nerve regeneration and nerve function recovery using a chitosan-collagen sponge enriched with the cells. Both *in vivo* and *in vitro* studies demonstrated that the incorporation of Schwann cells into connective tissue enhanced nerve regeneration, thereby underscoring its important role in tissue engineered organs that require fast recovery after transplantation.

Collagen-based biomaterials have also been applied in bone and cartilage reconstruction. These scaffolds are mostly manufactured with either type I or II collagen, cross-linked with other materials such as hydroxyapatite. A report by Stone *et al.* (1997) described the design of a collagen scaffold that was later used as a template for the regeneration of meniscal cartilage in ten patients. This study was approved by the Food and Drug Administration (FDA) based on earlier *in vitro* and *in vivo* data conducted in dogs. The collagen-based scaffold was found to be safe and also supported tissue regeneration. In a different study, Tamimi *et al.* (2008) reported on the preparation of a brushite-collagen composite for bone regeneration. This biomaterial was assembled by combining citric acid and collagen type I solutions with a brushite cement powder. Collagen was found to enhance cell adhesion, thereby increasing the osteoconduction and osteoregenerative capacity of this biomaterial, properties which are important for materials designed for the treatment of bone defects. There have also been advances in the application of collagen-based biomaterials in cardiovascular regenerative medicine. Tissue engineered heart-like valves have been successively implanted in patients. In 2009, Tedder *et al.* evaluated the properties of moderately crosslinked collagen scaffolds prepared from porcine pericardium treated with penta-galloyl glucose. The aim of their study was to generate tissue engineered heart valves that would not only function immediately but also tolerate cell infiltration and gradual remodelling. *In vivo* studies indicated that the scaffold degraded slowly, supported infiltration by host fibroblasts and matrix remodelling, thereby demonstrating the importance of penta-galloyl glucose additive for heart valve tissue engineering biomaterials.

Recent studies have demonstrated the feasibility of collagen implants for drug delivery. Adhirajan *et al.* (2009) investigated the potential benefits of a previously reported collagen-based microsphere loaded with doxycycline in rats challenged with *Pseudomonas aeruginosa*. The healing process was evaluated based on wound reduction, matrix metalloprotease levels, the number of colony forming units and biochemical analysis. The authors noted that collagen dressing reduced both infection and metalloprotease levels, underscoring its potential application in wound healing. In a similar study, Adhirajan *et al.* (2009) reported the use of modified collagen microspheres in wound dressing to attenuate protease and bacterial growth. The collagen microspheres loaded with doxycycline inhibited matrix metalloproteinase production, enzymes were found to be persistently elevated, therefore retarding the healing process. A sustained drug release profile of the collagen microspheres was observed *in vitro*. In 2006, Shanmugasundaram *et al.* developed a collagen-based drug delivery system and loaded it with silver sulfadiazine. *In vitro* studies revealed a controlled release over 72 hours. The authors concluded that the microspheres could be used to control infection over an extended period of time with fewer dressing frequencies.

Collagen-based nanoparticles/nanospheres have been widely used in drug delivery. The biodegradable particles are thermally stable, therefore their sterilization is readily achieved. Berthold *et al.* (1998) evaluated collagen microparticles as a delivery system for glucocorticoids. Drug release from the particles was unaffected by the pH of the medium. Also, the encapsulated drugs remained stable, demonstrating the success of collagen-based particles as a carrier system for lipophilic steroids. In another study, Rossler *et al.* (1994) prepared collagen microparticles to investigate dermal delivery of all *trans* retinol. *In vitro* penetration of retinol into hairless mouse skin was also evaluated in comparison to freshly precipitated drug. The authors observed enhanced quantities of retinol in the skin when collagen-based microparticles were employed.

In summary, the use of collagen-based biomaterials in tissue engineering and drug delivery has been extensively studied both *in vivo* and *in vitro* by several research groups. Collagen has considerable applications because of its availability, biocompatibility and low immunogenicity. Current studies aim at enhancing its stability *in vivo*, mechanical strength and elasticity.

2.2. *Albumin*

Albumin is a class of hydrophilic globular protein. It is the major protein in serum, making up approximately 60% of the total protein. Albumin is also present in egg white, milk and plant tissues. The main functions of albumin include regulation of osmotic blood pressure and pH, maintenance of physiological functions by transporting hormones, drugs, fatty acids, inorganic ions, metabolites, vitamins and anticoagulants (Ellmerer *et al.*, 2000; Peters, 1995; Hostmark *et al.*, 2005).

There are only a few reports on the use of albumin in the manufacture of biomaterials for tissue engineering. A report by Lyu *et al.* (2012) investigated the application of albumin-grafted biomaterials as tissue engineering scaffolds to regenerate cartilaginous components. During their study, porcine knee chondrocytes were cultivated in a porous matrix consisting of polyethylene oxide, chitin and chitosan coated with albumin. The results demonstrated that the viability of the porcine knee chondrocytes in the graft was independent of the quantity of albumin. However, according to biochemical studies and staining images, increased concentration of albumin was found to favor adhesion of porcine knee chondrocytes, enhance glycosamino-glycan secretion, as well as collagen production. This report concluded that albumin scaffolds could promote the generation of tissue engineered cartilage for preclinical trials. In a similar study, Gallego *et al.* (2010) reported the ectopic bone formation from mandibular osteoblasts cultured in human serum derived albumin scaffold. *In vivo* studies with immunodeficient mice showed enhanced bone formation at ectopic sites of implantation.

In recent years, there has been increased interest in using albumin to prepare microspheres for targeted drug delivery with sustained and controlled release of the therapeutic agents. Albumin carriers are suitable for drug delivery since they are biodegradable, biocompatible, nontoxic, non-immunogenic and can be prepared efficiently over a wide range of particle sizes. Additionally, albumin is composed of charged amino acids that allow electrostatic adsorption of molecules carrying either positive or negative charge. Ghassabian *et al.* (1996) demonstrated that magnetically responsive albumin microspheres could be used to achieve drug targeting. During their study, magnetic albumin microspheres containing dexamethasone were prepared for *in vivo* and *in vitro* studies. The outcome of their study indicated that upon application of a magnetic field, magnetic albumin microspheres had good retention at their target site. The albumin microspheres were also found to release their drug over an extended period of time, underscoring their importance in delivery of chemotherapeutic agents.

Weil and co-workers (2012) reported the development of a nano-sized micellar drug delivery system derived from polycationic albumin protein precursor, cBSA-147. The micelles were subsequently loaded with an anticancer drug doxorubicin and subjected to *in vitro* studies. The results demonstrated that the micelles had high uptake into A549 cells after one hour of incubation and were also stable in various physiological buffers. According to the authors, cytotoxicity studies of the micelles revealed a fivefold increase in comparison with the free drug, indicating efficient intracellular drug release. This study opened new avenues for the application of a denatured albumin-based drug delivery system in cancer chemotherapy.

Chuo *et al.* (1996) reported the preparation of nifedipine loaded albumin microspheres cross-linked by varying concentrations of glutaraldehyde. *In vitro* studies revealed the release profile of nifedipine from the albumin microspheres, which was dependent upon the quantity of glutaraldehyde used for cross-linking. The authors of this report concluded that a variety of albumin-based microspheres with different release kinetics could be designed by controlling the extent of cross-linking.

Coombes *et al.* (2001) described a method for stabilizing albumin microspheres and nanospheres using lactic acid. The potential biomedical application was demonstrated by evaluating protein release from Dacron grafts coated with lactic acid stabilized albumin nanospheres. The interaction of albumin nanospheres with the Dacron graft was found to provide a surface that would promote endothelialisation.

In separate reports, bovine serum albumin has been used for drug delivery due to its abundance, low cost and ease of purification among other benefits. Gan *et al.* (2009) investigated the application and release profile of bovine serum beadlets generated by combined cross-linking of microbial transglutaminase and ribose. The beadlets were loaded with caffeine and subjected to *in vitro* release studies, in which the beadlets displayed sustained drug release profiles. In 2010, Yamazoe *et al.* reported the preparation of a water insoluble film, loaded with cilostazol by cross-linking recombinant human serum albumin. The generated film displayed drug binding ability and resistance to cell adhesion at levels consistent with that of native albumin. Furthermore, the release study of cilostazol albumin film in PBS containing Tween-80 was sustained over 144 hours.

Arnedo *et al.* (2002) evaluated albumin nanoparticles as drug delivery systems for antisense oligonucleatides. The nanoparticles, prepared by a coacervation process, were loaded with a phosphodiester oligonucleotide. The oligonucleotide incorporated into the nanoparticle matrix was protected against enzymatic degradation.

In 2002, Sahin *et al.* formulated terbutaline sulfate microspheres loaded with albumin aiming at passive lung targeting. Drug release from the microspheres was characterized by an initial fast release, followed by a subsequent slower release. The biodistribution data indicated a higher uptake by the lungs in comparison to the other organs, supporting the application of terbutaline sulfate loaded microspheres for drug delivery to the lungs. Lemma *et al.* (2006) reported the preparation of pH sensitive bovine serum albumin hydrophilic microspheres for oral drug delivery by free radical polymerization. According to the authors, drug release profiles in media were dependent on drug matrix interactions and the degree of cross-linking in the microspheres. The results demonstrated the potential application of the hydrogels in the delivery and release of acidic drugs to the intestines.

In summary, the application of albumin in drug delivery has been extensively investigated in literature. Even though there are only a few reports on the use of albumin-based biomaterials in the field of tissue engineering, this is a rapidly emerging area. Albumin's good biocompatibility, biodegradability and bioactivity make it a potential candidate for tissue engineering and drug delivery. Novel methods of stabilizing albumin for biomedical applications have been described by several research groups in literature.

2.3. *Gelatin*

Gelatin is a natural biopolymer formed by acid or base catalyzed thermal hydrolysis of insoluble collagen found in bones, connective tissues and skin. It contains oligopeptides with diverse physical and chemical properties depending on the parent collagen and the processing method used. Gelatin's physical properties depend on its amino acid composition and molecular weight distribution.

Gelatin has gained increased interest for its applications in tissue engineering and cell transplantation, not only due to its excellent biodegradability, but also its biocompatibility, mechanical strength and unique sequence of amino acids (Huang *et al.*, 2005). The repeating sequences of glycine-proline-hydroxyproline triplets and arginine-glycine-aspartic acid peptides are important for cell adhesion (Mao *et al.*, 2003). The drawback associated with gelatin dissolution at body temperature has been circumvented by establishing chemical cross-links through derivatization of functional group side chains, such as the amine of lysine and hydroxylysine. Several strategies for designing chemical cross-links between the gelatin chains reported in literature have utilized bi-functional reagents such as glutaraldehyde, diisocyanates, carbodiimides, polyepoxy compounds and acyl azides (Bigi *et al.*, 2002; Bigi *et al.*, 2001). To date, gelatin has been widely used for drug delivery applications, wound dressing and as a food additive. Several researchers have demonstrated that gelatin-based scaffolds are potentially useful for tissue engineering of damaged or lost structures such as peripheral nerves, muscle fibers, small blood vessels and kidney tubules.

In the field of cardiology, Langley *et al.* (1999) reported the replacement of the proximal aorta and aortic valve using a composite mechanical bileaflet valve incorporated into a gelatin polyester graft. Preliminary results indicated that this implantation technique can be undertaken with a relatively low early mortality and morbidity rate. According to this report, the intermediate-term benefits of this prosthesis include a low reoperation rate and high survival. The use of bioengineered cardiac grafts has been explored as an alternative to impaired myocardium. In 2005, Alperin *et al.* cultured embryonic stem cell-derived cardiomyocytes onto thin films coated with gelatin, laminin or collagen. The generated cells attached to the different types of substrates in a surface and protein dependent manner, highlighting their potential application for the repair of damaged heart tissue.

Due to their outstanding biodegradability, gelatin biomaterials have been utilized in wound dressing. Choi *et al.* (1999) reported the preparation of a cross-linked gelatin-alginate sponge and its application as a wound dressing material. In their study, sponges loaded with silver sulfadiazine or gentamicin sulphate were found to have sustained drug release profiles. Gelatin-based biomaterials have also been used as soft tissue adhesives. McDermott *et al.* (2004) demonstrated that adhesives based on gelatin and calcium independent microbial *trans* glutaminase would offer the benefits of fibrin sealants without the need for blood products. Cross-linked chitosan and gelatin blends have recently been developed for biomedical application. Chiono *et al.* (2008) showed that these cross-linked blends supported adhesion and proliferation of neuroblastoma cells, underscoring their potential use in nerve regeneration. Recent studies have also focused on the use of gelatin-based scaffolds for tissue engineering of cartilage. Huang and co-workers (2009) developed a cross-linking method for genipin and gelatin to improve the mechanical strength of gelatin. According to this work, this approach may potentially be applied to articular cartilage tissue engineering.

Guo *et al.* (2006) reported a technique that utilized combined tissue engineering and gene therapy to fabricate three-dimensional scaffolds. Chitosan-gelatin complex biomaterials were used to fabricate scaffolds incorporating plasmid DNA, which promoted cartilage regeneration. The plasmid DNA showed sustained release over a span of three weeks, demonstrating the potential application of a gene-activated chitosan-gelatin matrix in cartilage defect regeneration. Separately, Huang and co-workers (2009) reported the design of a ceramic-gelatin assembly for articular cartilage repair.

Gelatin has also been used in the delivery and controlled release of chemotherapeutic reagents. Park and Kim (1997) described the use of a fibrin-gelatin-antibiotic mixture to deliver antibiotics. The authors sought to compare the *in vitro* activity of the antibiotic released from either fibrin or the fibrin gelatin mixture. Three antibiotics, ampicillin, gentamicin and ofloxacin, were used in this study. Results indicated that the fibrin-gelatin-gentamicin mixture had a slower release profile while increasing the amount of gelatin in the mixture induced sustained delivery of the antibiotics. In a separate study, Bonfield and Silvio (1999) reported the use of a biodegradable gelatin drug delivery system for the treatment of bone infection and repair. The monolithic microspherical drug delivery system described by Bonfield and Silvio (1999) released both gentamicin and the growth hormone in a controlled manner.

Murata *et al.* (2012) developed oral gelatine beads, incorporating allopurinol, a xanthine oral inhibitor, for the prevention and treatment of mucositis. The authors evaluated rheological properties, as well as the beads release profiles. The beads were found to soften immediately upon contact with the dissolution media in a manner similar to allopurinol suspension. The gelatine beads are therefore a convenient dosage form for treating localized ailments in the oral cavity.

Jain *et al.* (2012) described the use of gelatin hybrid lipid nanoparticles for oral delivery of amphotericin B, a drug used against fungal infection and leishmaniasis. Amphotericin B is conventionally administered intravenously due to its poor solubility and low intestinal permeability. Intravenous administration causes severe hemolytic toxicity and nephrotoxicity prompting exploration of other methods of delivery. The gelatin nano-formulation prepared by Jain *et al.* (2012) exhibited a sustained drug release profile, increased oral bioavailability and intestinal permeability.

In summary, the application of gelatin-based scaffolds for drug delivery has been explored for many years. However, its use in tissue engineering, albeit relatively new, has witnessed significant advances. Several research groups have evaluated gelatin-based materials because of their unique properties which include biocompatibility, non-immunogenicity, and its versatile functionality based upon differential cross-linking. It is important to note that the structural integrity of gelatin-based biomaterials is enhanced by cross-linking with other molecules. Also, several researchers have demonstrated that a combination of gelatin with other biopolymers generates scaffolds may be applied broadly to fields of hard and soft tissue engineering.

2.4. *Zein*

Zein is a natural protein that is mainly stored in corn kernels, accounting for 40–50% of the total corn proteins. The molecular weight of zein is about 44 kDa. Zein is water insoluble, yet soluble in an alcohol solution. The high nonpolar amino acid composition of zein is responsible for its hydrophobicity. The molecular structure of zein is a helical wheel conformation that contains nine homologous repeating units arranged in an anti-parallel form stabilized by hydrogen bonds (Argos *et al.*, 1982). Zein can form tough, glossy, hydrophobic coatings and has anti-photo-degradation, as well as anti-bacterial activity. Due to its good biocompatibility, low water uptake value, high thermal resistance, and good mechanical property, zein has been used as edible coating for food, FDA approved pharmaceutical excipients and promising biomaterials for tissue engineering (Gong *et al.*, 2006; Gong *et al.*, 2011; Lai and Guo, 2011; Liu *et al.*, 2005; Qu *et al.*, 2008; Wang *et al.*, 2009; Wang *et al.*, 2007; Dong *et al.*, 2004; Tu *et al.*, 2009; Wang *et al.*, 2005).

Large bone defect regeneration is a substantial medical challenge. Various three-dimensional (3D) scaffolds of metals, ceramics, synthetic and natural polymers have been applied for bone tissue engineering in an attempt to resolve this unmet medical need. However, challenges remain, and each individual material has its own inherent limitations: the non-biodegradability of metals, the brittle and unpredictable degradation of ceramics, low mechanical properties and poor biocompatibility of synthetic polymers, and the vascularization failure of most materials. As a natural polymer, zein and its degraded products show good biocompatibility (Dong *et al.*, 2004; Wang *et al.*, 2005).

Dr. Wang's group developed 3D porous zein scaffolds for bone tissue engineering. The results indicated that the controllable highly porous structures of zein scaffolds provided appropriate space for cell migration and tissue ingrowth. Additionally, the mechanical properties and positive *in vivo* tissue compatibility of zein scaffolds imply that the use of zein in bone-grafting substitution has its potential advantages. It was also proven that zein scaffolds could allow rat and human mesenchymal stem cells to adhere, proliferate and undergo osteogenic differentiation in the presence of dexamethasone (Gong *et al.*, 2006; Qu *et al.*, 2008), which clearly demonstrates the potential of using zein for bone tissue engineering (Tu *et al.*, 2009).

Dr. Wang's *in vivo* studies showed that implantation of a 3D zein scaffold itself only resulted in small amount of bone formation in the thigh muscle pouches of nude mice, without any evidence of blood vessel formation (Qu *et al.*, 2008; Tu *et al.*, 2009). To improve the bone formation efficacy, rat mesenchymal stem cells (MSCs) were seeded into zein scaffords to form the complexes between zein scaffolds and rat MSCs. These complexes were then implanted in the thigh muscle of mice. Significantly high amounts of ectopic bone formation and new blood vessels were found at the implantation sites. Most importantly, the zein scaffolds showed obvious degradation, and the pores were filled with mature bone tissue and blood vessels at week 12. The zein and MSC complexes were then evaluated in a critical-sized bone defect rabbit model. It was found that the bone defects were almost fully repaired with high quality new bone formation at week 12, and the zein scaffolds had completely disappeared and were replaced with new bones. On the contrary, only sporadic neo-bone formed and the bone defects remained unrepaired when the bone defects were implanted with zein scaffolds alone (Tu *et al.*, 2009). Clearly, zein scaffolds and MSCs worked synergistically to effectively repair the critical-sized radius defects of rabbits. MSCs not only provided an osteogenic cell source for new bone formation, but also secreted and sustained growth factors in the zein scaffolds necessary to recruit native cells to migrate to the defect site (Frank *et al.*, 2002). Zein scaffolds complexed with MSCs serve as localized delivery systems of growth factors.

Zein has been well-studied as drug delivery vehicles for encapsulation and controlled release of various hydrophilic and hydrophobic compounds such as insulin, heparin, and ivermectin. Zein microspheres and nanoparticles have been produced by many techniques including liquid-liquid dispersion, phase separation and nanoprecipitation (Nitta and Numata, 2013). Chemical cross-linking of zein solution containing the drug has also been applied for the preparation of zein microspheres (Suzuki *et al.*, 1989). To enhance drug encapsulation efficiency, hollow zein nanoparticles were developed by introducing sodium carbonate as a template (Reddy *et al.*, 2011). The encapsulated agents could slowly be released into phosphate buffer (pH 7.4) from zein particles for up to one–two weeks (Nitta and Numata, 2013; Liu *et al.*, 2005; Wang *et al.*, 2005). Hollow zein nanoparticles showed a more sustained and controlled release profile than that of solid zein nanoparticles. To combine drug delivery and

tissue engineering into one system, 2D films composed of zein particles suitable for *in vitro* cell cuture were prepared. Further studies also
showed that 3D zein scaffolds could be successfully developed using
zein particles. The encapsulated agents were slowly released from 3D
zein scaffolds, which had a strong correlation with the degradation
of zein scaffolds (Gong *et al.*, 2006; Liu *et al.*, 2005).

In summary, there is great potential that 3D zein scaffolds in
desired shapes can be developed using zein particles preloaded with
various bioactive agents. As the next generation of 3D scaffolds for
tissue engineering, the integrity of zein particles in the scaffolds will
not be destroyed in the preparation process, and the scaffold can sustain release of bioactive components for stimulating cell proliferation/
differentiation and tissue repairing/regeneration.

2.5. *Recombinant proteins and peptides*

One of the drawbacks of natural proteins is their batch-to-batch variation. With advances in recombinant technology, protein-based polymers that allow fine control over molecular structure have been
sought to resolve this limitation. These recombinant proteins and
peptides have an amino acid sequence controlled at the DNA level.
Proteins and peptides prepared by this promising technology have the
advantages of monodispersed molecular distribution, precisely defined
physicochemical properties, programmable distribution of amino
acids from the first to the last along the polypeptide chain, predictable
degradation rates and better biocompatibility. These properties make
them very attractive and useful for drug delivery and tissue engineering. Elastin-like proteins and polypeptides are the most commonly
studied genetically engineered proteins (Werkmeister and Ramshaw,
2012; Kim and Evans, 2012).

Elastin-like polymers (ELPs) have very good biocompatibility, are
easily biodegradable and do not elicit any immunogenic response.
Synthesis of ELPs is genetically encoded and precise control over
amino acid sequence and molecular weight can be easily achieved.
Also, these polymers can be purified by taking advantage of their
inverse transition temperature. The ELPs are soluble in the presence

of water below the transition temperature and conversely segregate because of hydrophobic polymer chains above the transition temperature. Elastin is a 68 KDa protein with several repeating amino acid sequences. Depending on the amino acid sequences, properties like temperature sensitivity are sought. ELPs are a class of recombinant polymers with a pentapeptide motif VPGVG that can be modified to generate functionalized scaffolds, which are temperature sensitive and allow regeneration of natural tissue. A simple change in the fourth amino acid "V" can fine-tune the characteristic transition temperature of the polymer. Such modifications can generate a variety of gels, fibers, foams etc. Recent advances also generated block copolymers of ELPs with variations in physicochemical properties like swelling, degradation etc. Alternatively, insertion of RGD or REDV sequences for cell recognition alters cell adhesive properties of ELPs. A variety of applications that include cartilage, vascular grafting, liver tissue and ocular tissue engineering have been achieved (Nettles *et al.*, 2010).

ELP polymer has amino acid sequence VPAVG that showed formation of spherical nanoparticles above 330C. These particles are stable, non-toxic, biocompatible and does not elicit inflammatory response. The inverse phase transition property of ELPs allows for controlled drug delivery. Also, because of the thermo sensitivity of ELPs, it is easy to purify the polymer. This polymer has been used to encapsulate bone morphogenetic proteins (BMPs). BMPs are essential for the formation of cartilage and bone. Therefore, these nanoparticles could clinically demonstrate bone regeneration and healing. The ELPs are thus explored for the encapsulation of proteins. Alternatively, ELPS are also used with radioligands for intratumoral injections in brachytherapy (Liu *et al.*, 2012).

Growth factor-based recombinant fusion protein is the bioactive material that contains growth factors and could both have cell-binding and recognition properties. Fibroblast growth factor (FGF) based recombinant bioactive material is currently in use. Fibroblast, being a vascular endothelial cell modulator, has shown potential in tissue repair. These recombinant proteins are usually multifunctional and are called fusion proteins. Fusion proteins with constructs of human fibronectin and human basic FGF have shown cell proliferation and adhesion. A multifunctional recombinant protein ERE-EGF also showed cell adhesion and tissue growth properties. Additionally, fusion protein of collagen binding domain and bFGF showed wound healing capacity. Alternately, GST fusion with bFGF enhanced growth of human umbilical vein endothelial cells. Epidermal growth factor fusion proteins are also used because the peptide chain shows a variety of properties like cell replication, DNA synthesis activation, wound healing etc. The recombinant fusion proteins with collagen binding domain, fibronectin and immunoglobulin G Fc region etc. have been made for a variety of functions. However, their potential in tissue engineering is still under research. Platelet derived growth factor (PDGF) is another growth factor that has been researched for generation of recombinant fusion proteins (Takashima and Tanaka, 2010). The use of fnCBD-ERE-EGF is researched for both collagen binding and cell growth-promoting activity (Nakamura *et al.*, 2009). On the other hand, fibrin hydrogels are under research to study their ability to encapsulate bone-marrow-derived mesenchymal stem cells. Their applicability is tested in facial reconstruction. Also, Tβ4 formulation was developed using a chitosan-collagen hydrogel to increase cellular growth, tube formation and angiogenesis with a subcutaneous injection (Chiu and Radisic, 2011). The ratio between chitosan and collagen was 2:1. Because of gelation, Tβ4 was encapsulated in this hydrogel. There was an enhanced vascularization at seven days in rats and the effectiveness of the hydrogel was assessed by vessel density.

2.6. *Silk fibroin*

Silk fibroin is obtained from silkworms and is biocompatible, has optimum mechanical strength and is biodegradable. This allows for implantable systems of silk fibroin. The advantage of silk is that it is easily purified using degumming process. The purification step is not expensive compared to other synthetic polymers. Silk fibroin has large hydrophobic domains intertwined with simple hydrophilic moieties. The hydrophobic domains in H-chain form β sheets while L-chain is flexible owing to the hydrophilic moieties. Additionally, the polymer complex integrity of silk fibroin is because of the P25 protein. Silk fibroin can be easily functionalized because the chemical moieties are readily accessible. The mechanical strength, elongation and ductility balance made silk more attractive as a biomaterial (Kundu *et al.*, 2013). However, to address the brittleness of silk materials, structural manipulations are researched. Silk scaffolds gave mild immune reactions with a subcutaneous injection implantation but no significant toxicities were found. The immune reactions may be due to degraded products from silk biomaterials that were found after implantation in rats. The mechanism for mild toxicities as found in *in vitro* studies is because of the presence of either metalloproteinases or integrins (MMPs). Owing to water solubility of silk fibroin, it has been manufactured under mild conditions that allow it to be used for a variety of applications.

Silk has been used as a vascular stent device with promising results in patients. It has been in use for the abdominal aortic blood vessels in rats, as flow devices in aneurysms. Blood pressure and compatibility are the two main challenges for these kinds of vascular applications. In bone regeneration, the challenges are the toughness and formation of matrix deposition. Silk fibroin is employed for tendon regeneration, as found by Achilles tendon regeneration. The applicability in ocular, tracheal, eardrum and spinal cord regeneration makes it popular. The biodegradability, very low immune reactions and compatibility with biological systems complement the silk-based systems.

As a drug delivery vehicle, silk fibroin has two essential properties, minimal toxic byproducts unlike synthetic polymers and the ability of silk fibroin to enhance uptake into the biological tissues/cells. Silk fibroin is formulated as a microsphere in controlled drug delivery. Also, drug loaded liposomes of silk fibroin have been evaluated for targeted-drug delivery. PLGA and alginate were formulated as microspheres and coated with silk to enhance the protein delivery. Compounds like Rhodamine B, azoalbumin were encapsulated with silk fibroin coatings. In insulin delivery, crystalline silk nanoparticles were prepared. In this preparation, silk fibroin was solubilized in a calcium chloride solution and then desalted to generate polypeptides. The half-life of silk fibroin nanoparticles increased about 2.5 times over native insulin. Silk fibroin has been used as a nanosphere for drug delivery because of its excellent biocompatibility. Silk spheres were prepared with polyvinyl alcohol (PVA). The microsphere size was consistent and controllable and was insoluble in water. These nanospheres can easily cross physiological barriers or pass through capillaries and therefore can be used for targeted delivery in tissues. Functionalization of these nanospheres allows them to target for tissue engineering. Other applications include modifying silk fibroin for ligament and bone deformations. Low crystallinity hydroxyapatite (LHA) is used here to modify silk fibroin for cell proliferation and differentiation functions. Here, the knitted silk was degummed in $NaHCO_3$ solution at high temperatures to prepare sericin-free silk. The central part was immersed in silk solution and ended in LHA/silk and lyophilized to produce silk sponges. The β sheets were formed after methanol-water immersion and then air-dried and rolled. The formulation was tested in rabbits in which enhanced expression of RUNX2 and osteocalcin were found, signifying compatibility. The bone volume and strength were regained after LHA/silk fibroin administration.

2.7. *Fibrin*

Fibrin is a non-globular clotting protein that is important for repair and healing wounds and homeostasis. It is a biocompatible polymer that even supports growth of new blood vessels. The properties like biodegradability, ease of modifying polymerization, compatibility made it attractive for tissue engineering. The polymeric network has nano-scale structures and is closely related to ECM. In biological systems, the fibrin is formed from fibrinogen enzymatically in the presence of thrombin. The thickness, fiber length and pore, branching can be easily altered during polymerization. The fibrin network can be modified to incorporate growth factors and heparin, because of the presence of RGD motifs, to closely resemble ECM (Spicer and Mikos, 2010). One drawback is degradation of fibrin that could be avoided either with stabilisers, such as apoprotinin and factor XIII, or the presence of thrombin and calcium. Fibrin hydrogels are mainly used in cartilage (Chrondrocytes), ligament and tendon tissue engineering. An interesting character of fibrin is that its degradation products like E fragment, peptide from N-terminus show healing and myocardial reperfusion properties. These properties could be harnessed for tissue regeneration. Fibrin scaffolds can be used as vehicles for injection in fibrin grafts, and can be used for preparation of *in-vitro* drug and disease testing models. The fibrin products can be obtained from the patients to manufacture personalized or autologous drug delivery systems. Also, the choice of clinical application can determine the functionalization of fibrin scaffold (incorporation of growth factors, strength and cost) (Hunt and Grover, 2010).

As an injectable vehicle for cardiac bioengineering, different biomaterials like collagen-based scaffolds were used. However, these scaffolds were found to contract during encapsulation stage. Fibrin was used in these cardiac applications (Barsotti *et al.*, 2011). The property of fibrin to allow modifications for targeted delivery, using growth factors, is harnessed for this application. The resemblance of fibrin to ECM, high elasticity, expedite healing are other essential features of fibrin gels. One study used fibrin sealant for dermal augmentation. In this study, hyaluronic acid gel solution (20%) was mixed with fibrinogen (from fibrin). The mixture was added to a 0.5 ml thrombin. Fibroblasts were suspended in the mixture to produce implants. These implants were then tested for subdermal tissue augmentation. Another study evaluated the use of fibrin glue in cartilage regeneration in goats. Here, the chondrocytes from a rabbit were suspended in fibrin glue and xenografted (van Susante *et al.*, 1999). The fibrin gel was extensively researched for its applicability in cardiac tissue engineering. Endothelial cells were suspended in fibrin matrix and were tested in sheep ischemic myocardia. There was a sign of generation of new blood vessels and improved blood flow in myocardia. Another study injected skeletal myoblasts in fibrin glue to rats that had left coronary occlusion. The study showed increased arteriole density and decreased infarct size. Other studies include the use of cardiac monocytes, stem cells from heart and bone marrow, cardiac fibroblasts injections/implantations with fibrin gel in various myocardial models. Studies also involved the use of bioactive factors like bFGF and VEGF along with fibrin glue in tissue engineering (Spicer and Mikos, 2010; Wong *et al.*, 2003). It was also studied that interaction between fibrin and bioactive factors like TGF-β1 caused slow release. This signified that alteration in initial fibrinogen component could alter the release parameters of encapsulated factors. Also, a study showed that controlled release of dual factors TGF-β1 and PDGF was possible using poly(ethylene glycol) with fibrin glue. This study showed that by varying PEG, release of factors from the scaffold also varied. The fibrin allowed one factor to be covalently linked whereas the other to be entrapped. Such scaffolds could be used in vasculogenisis and angiogenesis (Spicer and Mikos, 2010; Drinnan et al., 2010; Huang et al., 2002). A group of studies used fibrin glue with several biological active factors that act as signaling molecules. One such signaling molecule is heparin sulfate which showed increased bone regeneration in rats (Wong et al., 2003; Arkudas et al., 2009; Arkudas et al., 2009). Some studies modified affinity of factors to fibrin gel, like bone morphogenetic protein-2 which increased retention in the gel. These researchers found elevated bone formation with the BMP-2 that was modified (Schmoekel et al., 2004).

3. Polysaccharides

Polysaccharides are natural polymers of monosaccharides that are chemically linked by O-glycosidic bonds between the hydroxyl groups of the monosaccharide. Polysaccharides are abundantly found in nature (microbial, animal and vegetal) with minimal cost. Polysaccharides have a large number of reactive groups and varying chemical compositions on molecular chains. These can be easily modified chemically and biochemically to form several kinds of linear and branched polysaccharide derivatives with unique structures and properties. These include positively charged polyelectrolytes, negatively charged polyelectrolytes and non-polyelectrolytes. As natural polymers, polysaccharides are hydrophilic, highly stable, safe, biocompatible and biodegradable, which endow polysaccharides promising biomaterials. A large number of polysaccharides and their derivatives have been widely studied as scaffolds for tissue engineering as well as drug delivery systems which are described in the following sections.

3.1. *Chitosan*

Chitosan (CHT) is a polysaccharide consisting of varying fractions of randomly distributed β-(1-4)-linked D-glucosamine (deacetylated unit) and N-acetyl-D-glucosamine (acetylated unit). It is obtained by deacetylation of chitin, a natural polysaccharide found in the exoskeletons in crustaceans. The molecular weight of commercially available CHT is usually between 10 and 1000 kDa. The properties such as biodegradability and biological ability of chitosan depend on the degree of deacetylation. Chitosan molecule has amino and hydroxyl groups that can be chemically modified to provide various versatile applications. CHT has good solubility in water at pH less than ~6.5, while it is insoluble at higher pH ranges (Jain *et al.*, 2013). CHT is biodegradable and bioadhesive, and it is ideal for tissue engineering. This pH-controlled swelling property of CHT has been widely studied as a delivery matrix in tissue engineering. For tissue engineering applications, researchers have either chemically modified CHT, or combined CHT with a variety of biomaterials, such as collagen, poly(lactide-co-glycolide) (PLGA), HA, calcium phosphate, poly(methyl methacrylate), PLLA. These CHT-based matrixes have been demonstrated as promising scaffolds for skin, bone, and cartilage tissue engineering.

Bilayer CHT was used as skin substitute with great biocompatibility and stability (Ma *et al.*, 2001). It significantly accelerates skin tissue regeneration, promotes collagen synthesis, and stimulates various cytokine productions. For bone tissue engineering, CHT was proven to enhance bone formation both *in vitro* and *in vivo* (Di Martino *et al.*, 2005; Muzzarelli *et al.*, 1994). However, due to mechanical weakness and instability, chitosan is usually combined with other materials for better mechanical properties. CHT is also an attractive natural-origin polymer for cartilage engineering because its structure is similar to that of glycosaminoglycans (GAGs) — a native articular cartilage that is found in extracellular matrices and is vital for chondrocytes regulation. Although CHT is an ideal material for tissue engineering, CHT-based biomaterials alone cannot meet the needs for rapid wound healing and tissue regeneration. For bone tissue engineering, bone formation required a long period of time (several months or years) after implantation of CHT-based matrices. For skin tissue engineering, scarring is an inevitable problem. In order to solve these problems and maximize the application of CHT, various drugs, cytokines, and recently exogenous small interfering RNA (siRNA) have been encapsulated into CHT-based scaffolds to achieve quicker and better outcomes of tissue regeneration.

Epidermal growth factor (EGF) plays important roles in the stimulation of cell growth, proliferation and differentiation. EGF has been loaded into CHT-based materials for liver, nerve, and skin tissue regeneration for fast wound healing and tissue regeneration. Park et al. (2003) fabricated galactosylated chitosan extracellular matrix (ECM) as a replacement of natural ECM to keep cells into a three-dimensional architecture, provide mechanical integrity to the new tissue, and provide a 3D space for the nutrients and metabolites exchange. EGF was also successfully encapsulated into galactosylated chitosan ECM. The ECM and EGF combination system demonstrated enhanced hepatocyte adhesion and DNA synthesis, indicating its increased potency in liver tissue engineering (Park et al., 2003). For nerve regeneration, Goraltchouk et al. (2006) used CHT-based tubes loaded with poly(lactide-co-glycolide) (PLGA) microspheres to locally deliver EGF or other therapeutic molecules to the injured peripheral nerve or spinal cord. EGF was slowly released from this combined system for more than one month with good bioactivity (Goraltchouk et al., 2006). Alemdaroğlu et al. (2006) developed CHT gel formulation containing EGF, which significantly accelerated burn wound healing. Also, EGF-immobilized CHT scaffolds promote chondrogenesis and have potential for reticular cartilage regeneration (Tigli and Gumusderelioglu, 2008).

Transforming growth factor-β1 (TGF-β1) has the functions of regulating cell proliferation, differentiation and apoptosis. It was also loaded into CHT-based materials, and found its application in tissue engineering. Kim's group (2003) developed scaffolds using chitosan microsphere loaded with TGF-β1 to enhance chondrogenesis for cartilage tissue engineering. Clearly, chitosan microspheres loaded with TGF-β1 significantly augmented the cell proliferation and production of ECM (Kim *et al.*, 2003; Lee *et al.*, 2004). Gene delivery is recognized as a very promising strategy for various incurable human diseases, and TGF-β1 DNA loaded CHT-based scaffolds have also been developed for tissue engineering. Zhang *et al.* (2006) fabricated porous chitosan/collagen scaffolds via freeze-drying process, and plasmid and adenoviral vector encoding TGF-β1 was then loaded into the scaffolds. These CHT scaffold/DNA complexes resulted in periodontal tissue regeneration as promising substrate candidates (Zhang *et al.*, 2006). The similar TGF-β1 plasmid DNA loaded chitosan-gelatin scaffold was also applied for chondrocytes regeneration (Guo *et al.*, 2006). Furthermore, recent studies showed that TGF-β1 played a vital role in the formation of scars in skin tissue regeneration. Liu *et al.* (2013) incorporated trimetylchitosan (TMC)/siRNA complexes targeting TGF-β1 into a collagen-chitosan/silicone membrane bilayer dermal equivalent (BDE). The complex could sustain and interfere TGF-β1 signal pathway for the inhibition of scar formation (Liu *et al.*, 2013).

Platelet derived growth factor (PDGF) is a key regulator for angiogenesis. Liao *et al.* (2005) encapsulated chitosan–alginate fibers with different components, including dexamethasone, BSA, PDGF. Depending on the component properties, the release time of encapsulated components ranged from hours to weeks. Generally, charged components such as BSA, PDGF, and avidin exhibited sustained release for three weeks. It's noteworthy that the bioactivity of PDGF was retained over the release period (Liao *et al.*, 2005). To achieve better outcomes, PDGF DNA plasmid was loaded into various CHT-based materials for bone and periodontal bone engineering (Lee *et al.*, 2002; Zhang *et al.*, 2007; Jeong *et al.*, 2000; Peng *et al.*, 2009).

Fibroblast growth factors (FGFs) are involved in angiogenesis, wound healing, and embryonic development. Mizuno *et al.* (2003) encapsulated basic FGF (bFGF) into chitosan film. bFGF showed sustained release from the film and accelerated the rate of skin wound healing. Fujita *et al.* (2004) developed an injectable FGF-2 immobilized CHT-based hydrogel for soft tissue engineering. The controlled release of biologically active FGF-2 molecules was regulated by the biodegradation of hydrogel. *In vivo* studies demonstrated that this hydrogel/FGF complex induced vascularization formation (Fujita *et al.,* 2004). Further studies from the same group showed that the hydrogel/FGF complex significantly induced angiogenesis and enhanced collateral circulation in ischemic myocardium, thereby protecting the myocardium (Fujita *et al.,* 2005).

Recently, with the maturation of encapsulation techniques, dual/ multiple growth factors have been encapsulated into one CHT-based scaffold system to mimic *in vivo* conditions and maximize synergistic effects. Dyondi *et al.* (2013) successfully encapsulated bFGF and bone morphogenetic protein 7 (BMP7) into gellan xanthan gel along with chitosan nanoparticles to develop the complex of scaffolds and DDS with growth factor combination for the accelerated differentiation of human fetal osteoblasts.

In summary, the complexes of chitosan scaffolds and DDS significantly improved the efficacy of tissue engineering through the sustained release of growth factors and other key agents for cell proliferation, differentiation and apoptosis. Currently, some chitosan-based systems are advanced from benchside to bedside, and they are commercially available as novel pharmaceutical products/devices with enhanced efficacy.

3.2. *Starch*

Starch is a carbohydrate consisting of a large number of glucose units joined by glycosidic bonds. It is a white, tasteless and odorless powder that is insoluble in cold water or alcohol. Starch is synthesized by all green plants and deposited in the chloroplasts as their energy store. It contains two types of molecules: the linear α-amylose composed of thousands of glucose residues linked by α $(1{\to}4)$ bonds, and the branched amylopectin, which mainly consists of α $(1{\to}4)$-linked glucose residues but has α $(1{\to}6)$ branch points every 24 to 30 glucose residues. Generally, starch contains 20–30% α-amylose and 70–80% amylopectin.

Because of its inherent biodegradability, overwhelming abundance and renewability, starch is considered as one of the most promising natural materials for tissue engineering. However, the intrinsic properties such as thermal, mechanical, and biological properties of starch limit its direct applications. Typically, starch is modified or blended with other materials to achieve better chemical and physical properties. Early in 1990s, Reis's group proposed the potential of starch-based materials for biomedical applications (Reis *et al.*, 1995; Gomes *et al.*, 2001). Since then, many studies have demonstrated the capability of starch-based materials for tissue engineering applications. Starch-based porous scaffolds have shown great potential for tissue engineering (Lam *et al.*, 2002; Torres *et al.*, 2007), especially for bone tissue engineering (Salgado *et al.*, 2007; Gomes *et al.*, 2008; Martins *et al.*, 2009). These materials are biocompatible *in vivo* (Marques *et al.*, 2002), and have been shown to facilitate vascularization (Santos *et al.*, 2007). To achieve better tissue repair, surface modifications have been applied. These modifications either increase the adhesion to the tissue, or increase the targeting ability to specific tissues (Pashkuleva *et al.*, 2005; Oliveira and Reis, 2004). To control the degradation speed, alpha-amylase was encapsulated (Azevedo and Reis, 2009; Martins *et al.*, 2009). In order to meet the need of fast tissue regeneration, drugs and growth factors were encapsulated in these starch-based materials.

Reis's group has been working on the development and characterization of starch-based materials to be used as drug delivery carriers in tissue engineering applications. They have developed starch-based microparticles, and loaded the microparticles with different corticosteroids. It has been shown that the starch-based microparticles were able to sustain a controlled release of the entrapped corticosteroids over 30 days, which confirmed the potential of these systems to be used as carriers for the delivery of bioactive agents (Silva *et al.*, 2005). They also fabricated starch microspheres, and encapsulated the microspheres with nonsteroidal anti-inflammatory drug-NSAID. Because of the mild processing conditions, these starch microspheres were expected to encapsulate bioactive growth factors and even live cells (Malafaya *et al.*, 2006). To control the release speed of the drugs, an enzymatic-mediated degradation carrier system has been developed for bone tissue engineering application. In this study, a microparticle carrier which consists of starch and polycaprolactone (SPCL) was encapsulated with dexamethasone (DEX). The DEX was initially released by diffusion until a gradual degradation of the carrier took place, which allowed the remaining drug in the polymeric matrix to release. The enzymes, including α-amylase and lipase, regulated the degradation rate of the matrix (Balmayor *et al.*, 2008).

To obtain faster tissue repair, different growth factors have been encapsulated with starch-based materials. Based on the study mentioned above, the starch-based microparticles have been further encapsulated with platelet-derived growth factor (PDGF). The release profile has indicated that the microparticles can constantly release PDGF for eight weeks. Moreover, the incorporation and release did not affect the biological activity of PDGF. The result has demonstrated that the starch-based microparticles are suitable vehicles for the incorporation and release of bioactive growth factors, and can enhance tissue regeneration (Silva *et al.*, 2007). Bone morphogenetic protein-2 (BMP2) is a protein that belongs to a superfamily called transforming growth factor beta (TGF-beta). It induces the formation of bone and cartilage, and thus plays an important role in bone tissue regeneration. A novel injectable drug carrier consisting of starch-poly-epsilon-caprolactone microparticles have been used to encapsulate BMP2 to induce osteogenesis. BMP2 was rapidly released during the first 12 hours, followed by sustained release for up to 10 days.

Recently, a blend of corn starch with polycaprolactone (30/70 wt.%, SPCL) has been combined with human adipose stem cells (hASC) for bone tissue engineering application. SPCL was functionalized by introduction of silanol (Si–OH) groups that promote bone-related gene expression, and stimulate osteoblast proliferation and differentiation. The result demonstrated that under the presence of Si–OH groups, the osteogenic development was enhanced. The scaffolds can sustain hASC proliferation and differentiation into the osteogenic lineage (Rodrigues *et al.*, 2012). These bioactive materials combined with differentiable hASC have great potential for bone tissue engineering application, and were considered a promising biomaterial for bone tissue replacement. Based on the above result, a combination of SPCL scaffolds, hASC, and growth factors has been developed for bone tissue engineering. It is known that vascularization is vital for bone growth, healing, and consequent bone tissue regeneration. Therefore, SPCL scaffolds have been encapsulated with vascular endothelial growth factor (VEGF) and fibroblast growth factor-2 (FGF-2) to stimulate the angiogenesis of injured bone tissue. The animal study on nude mice has demonstrated that the assembled tissue engineering constructs composed of fibrin sealant, hASCs, VEGF, and FGF-2 have great biocompatibility. After implantation, the release of VEGF and FGF-2 from the constructs has upregulated the expression of vascular endothelial growth factor receptor 2 (VEGFR2) and other important mediators in neovascularization (Santos *et al.*, 2013).

3.3. *Alginate*

Alginates are naturally occurring polysaccharides derived from brown seaweeds. These unbranched polysaccharides are comprised of $1\rightarrow4$ linked β-D-mannuronic acid (M) and α-L-guluronic acid epimer (G) monomers that vary along the polymer chain (Pawar and Edgar, 2012). Alginate is also classified as a block copolymer comprised of sequences of M block and G block units interconnected with M-G blocks (Haug *et al.*, 1966; Haug *et al.*, 1967). The copolymer composition, sequence and molecular weights vary with the algal or bacterial source. Alginates have the ability to bind divalent cations to form soft-stable hydrogels that have found wide applications. Industrial production of alginate involves extraction from brown algae by treatment with aqueous alkali solution (Smidsroed and Skjaak-Braek, 1999). The molecular weight of commercially available alginate is in the range of 32,000 and 40,000 g/mol (Rinaudo, 1992). It is also important to note that the viscosity of alginate solutions increases with a decrease in the pH. Maximum viscosity is attained in the pH range of 3 and 3.5 at which point protonation of the carboxylate groups initiate hydrogen bonding.

Alginate has been used for biomedical application due to its biocompatibility, relatively low cost, low toxicity and ability to form mild gels by addition of divalent cations (Lee and Mooney, 2012). The scaffolds are currently used as dressing materials in wound healing, delivery of bioactive agents and cell transplantation in tissue engineering. Alginate gels have also been used to encapsulate islets of Langerhans for diabetes treatment. The gels also play a significant role in pharmaceutical industry since they can be orally administered or injected into the body in a minimally invasive manner.

Treatment of bone injuries presents a major challenge to orthopaedic surgeons due to the poor healing process. Alginate gels have been used in bone regeneration since they can be introduced into the body in a minimally invasive manner. Kolambkar *et al.* (2011) reported a hybrid growth factor delivery system combined with a peptide modified alginate gel for sustained growth factor release. In this study, the researchers investigated the ability of the system to deliver recombinant bone morphogenetic protein-2(rhBMP-2) for the repair of bone defects in a rat model. The hybrid growth factor delivery system was found to generate a significant repair response. In a similar study elsewhere, Lópiz-Morales *et al.* (2010) compared the treatment of osteochondral defects by employing either bone morphogenetic proteins (rhBMP-2, rhBMP-4) alone or in combination with cells delivered in a calcium alginate gel. Morphological assessment demonstrated that the alginate gel carrier was more effective in the restoration of the subchondral bone. Basmanav *et al.* (2008) designed an alginate scaffold capable of sequentially delivering two bone morphogenetic proteins (BMP) for bone tissue engineering. The delivery system consisted of microspheres of polyelectrolyte complexes of poly(4-vinyl pyridine) and alginic acid loaded with growth factors BMP-2 and BMP-7. Co-administration of BMP using the alginate scaffold was found to enhance osteogenic differentiation in stem cells derived from the bone marrow of rats. Similarly, Kanczler *et al.* (2010) utlilized alginate-VEGF165/BM-2 scaffolds to deliver human bone marrow stromal cells. The scaffolds were implanted into a mouse femur for 28 days and found to increase the level of bone repair.

Alginate scaffolds have also shown potential application in the repair of damaged cartilage. The gels have been used in the transplantation of chondrogenic cells to restore damaged cartilage. Chang *et al.* (2001) prepared molds of facial implants incorporating chondrocytes gelled in 2% alginate. The constructs of different cell concentrations were then implanted in the dorsal aspect of nude mice and harvested at various time points. The results indicated progressive cartilage formation with time. In a different study, Thornton *et al.* (2004) prepared macroporous alginate hydrogel scaffolds and implanted them into immunocompromised mice by means of minimally invasive surgical delivery through a small catheter. The scaffolds were rehydrated with primary bovine articular chondrocytes. Specimens were then harvested to evaluate the efficacy of cell delivery, recovery of scaffold geometry and neotissue formation. The study demonstrated high recovery of the scaffolds into their original shape and size within a short time. The delivery procedure reported by Thornton *et al.* (2004) generated cartilage structures with similar geometry to the original scaffold.

Alginate scaffolds have also been exploited in the formation of blood vessels. Neovascularization is a promising alternative to patients suffering from coronary and peripheral arterial diseases. Alginate gels have been used as a delivery vehicle of angiogenic molecules necessary for neovascularization. Gu *et al.* (2004) designed a vascular endothelial growth factor (VEGF) delivery system using calcium alginate matrix with the aim of optimizing VEGF encapsulation and good release profile from the matrix *in vitro*. Vascular endothelial growth factor is a signal transduction molecule that acts on vascular endothelial cells. The rate of VEGF release from alginate beads into fetal calf serum was found to be higher than that in PBS. The authors noted that the presence of $CaCl_2$ in the released supernatant not only preserved the alginate integrity but also ensured sustained release of VEGF. Gu *et al.* (2004) concluded that the alginate-VEGF delivery system could be utilized in vascular tissue engineering and wound healing. Similarly, Jay *et al.* (2009) developed methods for sustained release of VEGF from small alginate microparticles cross-linked with divalent cations. This strategy generated non-toxic microparticles and the bioactivity of VEGF was unaffected after release. In another study, Lee *et al.* (2010) prepared PLGA microspheres incorporating alginate hydrogels loaded with rhVEGF and evaluated the formation of new blood vessels after injection into the ischemic hindlimb site of a mouse model. The microsphere/hydrogel combination system released rhVEGF at a controlled rate, significantly increasing the number of newly formed blood vessels.

Alginate scaffolds are also under investigation for the repair of the nervous system. A report by Prang *et al.* (2006) described the promotion of oriented axonal regrowth in the injured spinal cord using alginate-based anisotropic capillary hydrogels. In this work, alginate-based hydrogels were introduced into an axon outgrowth assay *in vitro* and adult rat spinal cord lesions *in vivo*. The alginate scaffolds were found to promote linear axon regrowth and targeted neuron reinnervation. The alginate-based anisotropic capillary hydrogels described by Prang *et al.* (2006) therefore have the potential to induce directed nerve regrowth following spinal cord injury.

In summary, alginate scaffolds have been widely explored as biomaterials for many biomedical applications. Specifically, alginates biomaterials have been utilized in wound healing, drug delivery and tissue engineering. Alginates have often been used since they are biocompatible and can easily be formulated into a variety of fibers, foams and elastic gels under mild gelation conditions. Chemical derivatization generates scaffolds with new properties. Alginate scaffolds also retain the structural similarity to the extracellular matrices in tissues.

3.4. *Hyaluronan*

Hyaluronan (HA) is a high-molar-mass linear glycosaminoglycan (GAG) found in the extracellular matrix (ECM). It is a linear polysaccharide that can reach a size of 6 to 8 MDa (Stem *et al.*, 2007). The polymer has the structure of poly [(1~3)-2-acetamido-2-deoxy-β-D-glucose-(1~4)-β-D-glucopyranosyluronic acid]. It possesses one carboxyl group per disaccharide repeating unit, and is therefore a polyelectrolyte with negative charge at neutral pH. HA is produced by a membrane-bound group of enzymes named hyaluronan synthases. This organizes D-glucuronic acid and N-acetyl-D-glucosamine disaccharides into a non-branched linear polymer that can range between 1 MDa to 4 MDa in size, which is generally known as high molecular weight (HMW) HA. HMW HA is very flexible and can readily form into coils, nets or fibers. HMW HA is strongly negatively charged and therefore absorbs up to 10–10,000 times its weight water. It is generally a bio-inert molecule that acts to maintain a hydrated and porous environment, which absorbs mechanical shock and regulate osmotic balance. It can also sequester and gradually release growth factors and other bioactive molecules to communicate a local biological influence over cells. From a tissue-engineering standpoint, HMW HA is specifically interesting because it can diminish the interaction of encapsulated cells with other cells and growth factors. Therefore, HMW HA conceals them from local harmful signals that is realized by modulating inflammatory response of macrophages. HMW HA can degrade into highly bioactive oligomers by enzymes (hyaluronidases). HA degradation is facilitated in inflammation and injury by the production of reactive oxygen and nitrogen species (Stem *et al.,* 2006; Moshayedi *et al.,* 2013).

The size of HA polymers used in bioengineering and biotechnology is obviously of critical importance. High-molar-mass forms of HA are reflections of intact normal tissue while fragmented forms of HA are usually indications of stress, for various sizes. HA fractions have an enormous repertoire of functions and constitute an information-rich system. Large matrix polymers of HA are space-filling, anti-angiogenic, and immunosuppressive, whereas the intermediate-sized polymers comprising 25 to 50 disaccharides are inflammatory, immunostimulatory, and highly angiogenic. These low-molar-mass oligosaccharides (Oss) appear to function as endogenous danger signals.

Due to its novel characteristics of biocompatibility and rheological properties, HA is being increasingly used as a component of artificial matrices and in bioengineering for tissue scaffolding. HA finds a wide range of biomedical applications, the overwhelming majority of which is because of its viscoelastic properties. HA is present in all soft tissues of higher organisms and extracellular matrix of articular cartilage and in the mesenchyme of the developing embryo. It plays different biological roles depending on its physical and chemical properties. It interacts with binding proteins, proteoglycans and active molecules, such as growth factors, which influence matrix structure, water balance lubrication and cell interaction (Tognana *et al.,* 2007). Due to its large number of hyaluronan-binding proteins that exhibit significant differences in tissue expression, cellular localization, specificity, affinity and regulation, HA can influence cell movement through its ability to alter the osmotic pressure which would lead to the formation of hydrated pathways.

Chemical modification of HA is widely used to produce engineered tissues, which are controllably degradable, by introducing new properties while maintaining its unique properties. These novel biomaterials can facilitate angiogenesis, osteointegration and cell phenotype preservation. Such derivatives can be the basis of novel biochemical markers or drug delivery methods. The intrinsic biocompatibility of HA provides great potential for the development of engineered tissues and biomaterials on various purposes, such as orthopedic, cardiovascular, pharmacologic and oncologic applications (Almond, 2007).

HA can elicit a wide variety of responses from the cellular microenvironment, and HA can be introduced into a tissue engineering construct through overexpression of endogenous cellular HA synthases or through exogenous addition. Cross-linking methods have been used to stabilize the HA polymers including water-soluble carbodiimide, polyvalent hydrazide, divinyl sulfone, disulfide, and photo cross-linking hydrogels through glycidyl methacrylate-HA conjugates. Covalent cross-linking of HA provides the ability to combine the desirable biological properties of HA with the properties of a mechanically stronger molecule (Volpi *et al.*, 2009). Also, cross-linking HA results in a range of pore sizes as well as prolonged degradation profiles and long-term stability. A novel hydrogel was prepared by cross-linking hyaluronan with glycol chitosan in aqueous solution using water-soluble carbodiimide at nearly neutral pH and room temperature. The products can be easily formulated into injectable gels, various films, membranes and sponges for soft tissue augmentation, viscosupplementation, drug delivery, preventing adhesion of post operation, wound dressing and tissue engineering scaffolds (Wang, 2006). Other studies have shown that HA was immobilized onto the surface of macroporous biodegradable poly (D,L-lactic acid-coglycolic acid) scaffolds to enhance the attachment, proliferation, and differentiation of chondrocytes for cartilage tissue engineering. It was also observed that collagen II expression, a marker of healthy cartilage phenotype, and DNA synthesis were significantly increased in polymers that incorporated HA (Yoo *et al.*, 2005).

In addition, HA-polymer scaffolds can be engineered for the controlled release of drugs, DNA and other molecular agents (Allison and Grande-Allen, 2006). Hyaluronan nanocapsules were reported as new drug nanocarriers for the intracellular delivery of hydrophobic drugs. Nanocapsules were produced by modified solvent displacement technique, which allows the formation of the polymer shell around the oily core using a cationic surfactant as an interphase bridge (Oyarzun-Ampuero *et al.*, 2013). HA hydrogel has been used for delivery of DNA into the cellular microenvironment. Hyaluronidase degradable HA can facilitate the DNA release, endosome escape and transfection efficacy. Segura *et al.* (2005) developed procedures to immobilize non-viral DNA complexes to hyaluronic acid-collagen hydrogel. They found that DNA was released slowly from the cross-linked HA with poly (ethylene glycol) diglycidyl ether hydrogel substrate during 48 hours incubation in PBS or hyaluronidase and taken up by adherent cells (Segura *et al.*, 2005). Liu's work (2005) described nanoparticles made of oleoyl-carboxymethy-chitosan/HA using coacervation process as novel potential carriers for gene delivery. Zhang *et al.* (2013) found that PEGylated HA would be a good candidate to increase the uptake of non-viral vectors and the low transfection efficacy. The strong electrostatic interaction between the polyelectrolytes inhibited the release of inner layer of nucleic acid (Zhang *et al.*, 2013).

In summary, HA has had a profound impact on the field of tissue engineering. The biomaterials incorporating HA have shown biocompatible, controllable and biodegradable properties, and great potentials on remodeling of engineered tissues as well as preservation of cell phenotypes. HA, as a powerful molecular delivery vehicle for pharmacological and oncological applications, has the potential to transform the clinical future of tissue engineering. However, more applications like harnessing the endogenous activity of the hyaluronan synthases can be achieved with this novel macromolecule.

3.5. *Chondroitin sulphate*

Chondroitin sulfate (CS) is a glycosaminoglycan (GAG), usually linked to proteins to form proteoglycans. CS is a linear polysaccharide composed of sulfated N-acetyl-galactosamine and glucuronic acid units. CS is composed of repeating disaccharide units of glucuronic acid N-acetylgalactosamine. The hydroxyl groups on these disaccharide units can be differentially sulphated, yielding considerable structural heterogeneity within the CS chains.

The different isoforms of CS are defined by the position of the sulfated groups. The two most common are chondroitin-4-sulfate and chondroitin-6-sulfate depending on the position of the sulfate group in the pyranose ring. *In vivo*, sulfated GAGs are structural components abundant in the extracellular matrix (ECM), which have been proposed to play a role in wound repair, morphogenesis, and growth factor signaling (Hansson *et al.*, 2012). Due to its polyanionic nature and high charge density, CS is able to form stable polyelectrolyte complexes (PEC) with polycationic polymers either as a major component or as a cross-linker connecting the chains of the cationic polymer.

Because of their high content of sulphate and carboxyl groups, CS chains have a high negative charge density, thus an inherent ability to attract positively charged matrix molecules through electrostatic interaction. As a result, they contribute to the water-binding capacity of proteoglycans (PGs) via their ionic, carboxyl and sulphate radicals and, therefore, are important to the hydrodynamic properties of the disc in resisting biomechanical load (Hayes *et al.*, 2011). Also, CS is the major glycosaminoglycan component of native cartilage tissue. Other reported benefits include applications for osteoarthritis, ability to increase the production of ECM matrix, and capacity to induce the differentiation of multipotent stromal cells (Silva *et al.*, 2013). In pharmaceutical and biomedical applications, CS has been investigated in nanoparticles intended for drug delivery, wound dressings, and scaffolds in tissue engineering.

Nowadays, there is a significantly growing need for tissue integration and cartilage repair is a uniquely challenging application as well. Cartilage is made of thick matrix of proteoglycans and collagen, which makes it challenging for usage of adhesive and bonding strategies. Clinically, a decrease in the rate of GAG synthesis is indicated in the later stages of osteoarthritis. In such cases, supplementing these joints with CS, the most abundant GAG in articular cartilage, may prove beneficial. CS has been proven to reduce the friction coefficient of native articular cartilage tissue while articulating against polished glass (Katta *et al.*, 2009; Basalo *et al.*, 2007).

Cytokine and growth factors (GFs) play an indispensable role in regenerative medicine and tissue engineering. Tissue healing cascades involve the combination of several GFs and cytokines, expressed and secreted in a spatiotemporal manner. Platelet lysates (PLs) has been studied as a promising alternative source of bioactive agents that can be released at physiological concentrations. PLs has been loaded into CS-based biomaterials for drug delivery (Ehrlich *et al.*, 1998). A study shows that a new formulation of nanoparticles (NPs) based on the electrostatic interaction between chitosan and chondroitin sulphate (CH–CS NPs) is proposed for the controlled release of proteins and growth factors (GFs), specifically platelet lysates (PLs). Spherical non-cytotoxic NPs were successfully produced, exhibiting high encapsulation efficiency for physiological levels of GFs with a controlled protein release profile for longer than a month. Moreover, it was also observed that these NPs could be taken up by human adipose-derived stem cells (hASCs), depending on the concentration of NPs in the culture medium and incubation time. This shows the versatility of the developed NPs, which, besides acting as a protein delivery system, can also be used in the future as intracellular carriers for bioactive agents, such as nucleotides (Santo *et al.*, 2012; Venkatesan *et al.*, 2012).

Researchers also showed that CS could be chemically functionalized with methacrylate and aldehyde groups on the polysaccharide backbone to chemically bridge biomaterials and tissue proteins via a twofold covalent link (Wang *et al.*, 2007). The multifunctional CS served as a 'tissue primer' that formed a covalent bridge between an acrylate-based polymer and a tissue surface. The application of the CS adhesive can be compared to a paint primer that integrates the polymeric grafts to a host tissue. Wiltsey (2013) developed an injectable nucleus pulposus tissue engineering scaffold with the ability to form an adhesive interface with surrounding disc tissue. A family of in situ forming hydrogels based on poly(N-isopropylacrylamide)-graft-chondroitin sulfate (PNIPAAm-g-CS) were evaluated for their mechanical properties, bioadhesive strength, and cytocompatibility. It was shown experimentally and computationally with the Neo-hookean hyperelastic model that increasing the cross-link density and decreasing the CS concentration increased mechanical properties at 37°C. This generated several hydrogel formulations with unconfined compressive modulus values similar to the values reported for the native nucleus pulposus (Wiltsey *et al.*, 2013). Another nanoparticulate system was developed by Hansson (2012) from an RGD-functionalized chitosan derivative by complexation with chondroitin sulfate. These bioactive complexes were developed to promote wound healing by inducing adhesion and subsequently migration of skin cells.

Hydrogels have been widely studied as vehicles for delivery and maintenance of cells or bioactive factors for cartilage repair. The injectable and hydrophilic nature of hydrogels makes them suitable candidates for cartilage tissue engineering. The application of CS adhesive neither penetrates nor damages the cartilage tissue and is easy to administer. The CS adhesive is applied to the surface of cartilage by a cotton swab or brush. A pre gel solution containing a biocompatible initiator can then be applied to the tissue. The hydrogels can be formed with or without suspended cells to generate cellular or acellular implants. The resulting hydrogel biomaterial is covalently bound to the cartilage surface via the CS-adhesive bridge. Cells surgically stimulated from the marrow or exogenously added to the liquid pregel can be incorporated into the hydrogel layer. The CS hydrogels for tissue repair are designed to support cell viability and phenotype, and nutrient-waste flow. Furthermore, they have the potential to support neocartilage formation. Hydrogels can be formed from or blended with cartilage extracellular matrix (ECM) components by methods such as temperature change, chemical cross-linking and self-assembly (Wang *et al.*, 2010). To realize these characteristics, the CS hydrogels have a high porosity or lower cross-linking density and limited physical strength.

4. Conclusion

Natural polymers play an indispensable role in drug delivery and tissue engineering application. However, it is in some way limited because of the availability of natural products with clinical relevance.

Drug delivery applications in the field of tissue engineering are still in its infancy. Although the natural polymers present some drawbacks like difficulties in controlling the variability from batch to batch, mechanical properties that lead to limited processability, their advantages are self-evident. The degradability, biocompatibility, availability, low cost, similarity with the extracellular matrix and intrinsic cellular interaction are essential properties for potential biomedical applications, especially as drug delivery systems for tissue engineering. It is expected that more research will be conducted to obtain commercially available and clinically successful natural polymer-based scaffolds with slow release properties for tissue engineering.

References

Adhirajan N, Shanmugasundaram N, Shanmuganathan S, Babu M. Functionally modified gelatin microspheres impregnated collagen scaffold as novel wound dressing to attenuate the proteases and bacterial growth. *Eur J Pharm Sci* 2009;36:235–245.

Alemdaroglu C, Degim Z, Celebi N, Zor F, Ozturk S, Erdogan D. An investigation on burn wound healing in rats with chitosan gel formulation containing epidermal growth factor. Burns: *Burns* 2006;32:319–327.

Allison DD, Grande-Allen KJ. Review. Hyaluronan: A powerful tissue engineering tool. *Tissue Eng* 2006;12:2131–2140.

Almond A. Hyaluronan. *Cell Mol Life Sci* 2007;64:1591–6.

Alperin C, Zandstra PW, Woodhouse KA. Polyurethane films seeded with embryonic stem cell-derived cardiomyocytes for use in cardiac tissue engineering applications. *Biomaterials* 2005;26:7377–7386.

Argos P, Pedersen K, Marks MD, Larkins BA. A structural model for maize zein proteins. *J Bio chem* 1982;257:9984–9990.

Arkudas A, Pryymachuk G, Hoereth T, Beier JP, Polykandriotis E, Bleiziffer O *et al.* Dose-finding study of fibrin gel-immobilized vascular endothelial growth factor 165 and basic fibroblast growth factor in the arteriovenous loop rat model. *Tissue Eng Part A* 2009;15:2501–2511.

Arkudas A, Tjiawi J, Saumweber A, Beier JP, Polykandriotis E, Bleiziffer O *et al.* Evaluation of blood vessel ingrowth in fibrin gel subject to type and concentration of growth factors. *J Cell Mol Med* 2009;13:2864–2874.

Arnedo A, Espuelas S, Irache JM. Albumin nanoparticles as carriers for a phosphodiester oligonucleotide. *Int J Pharm* 2002;244:59–72.

Azevedo HS, Reis RL. Encapsulation of alpha-amylase into starch-based biomaterials: an enzymatic approach to tailor their degradation rate. *Acta Biomateri* 2009;5:3021–3030.

Balmayor ER, Tuzlakoglu K, Marques AP, Azevedo HS, Reis RL. A novel enzymatically-mediated drug delivery carrier for bone tissue engineering applications: Combining biodegradable starch-based microparticles and differentiation agents. *J Mater Sci Mater in Medi* 2008;19:1617–1623.

Barsotti MC, Felice F, Balbarini A, Di Stefano R. Fibrin as a scaffold for cardiac tissue engineering. *Biotechnol Appl Biochem* 2011;58:301–310.

Basalo IM, Chahine NO, Kaplun M, Chen FH, Hung CT, Ateshian GA. Chondroitin sulfate reduces the friction coefficient of articular cartilage. *J Biomech* 2007;40:1847–1854.

Berthold A, Cremer K, Kreuter J. Collagen microparticles: Carriers for glucocorticosteroids. *Eur J Pharm Biopharm* 1998;45:23–29.

Bigi A, Cojazzi G, Panzavolta S, Roveri N, Rubini K. Stabilization of gelatin films by crosslinking with genipin. *Biomaterials* 2002;23:4827–4832.

Bigi A, Cojazzi G, Panzavolta S, Rubini K, Roveri N. Mechanical and thermal properties of gelatin films at different degrees of glutaraldehyde crosslinking. *Biomaterials* 2001;22:763–768.

Blais M, Grenier M, Berthod F. Improvement of nerve regeneration in tissue-engineered skin enriched with schwann cells. *J Invest Dermatol* 2009;129: 2895–2900.

Brinckmann J, Notbohm H, Mueller PK, Editors. Collagen: Primer in Structure, Processing and AssemblyIn: Top. Curr. Chem.; 2005, 247: *Springer GmbH*; 2005.

Buket BF, Kose GT, Hasirci V. Sequential growth factor delivery from complexed microspheres for bone tissue engineering. *Biomaterials* 2008;29:4195–4204.

Chang SCN, Rowley JA, Tobias G, Genes NG, Roy AK, Mooney DJ, *et al.* Injection molding of chondrocyte/alginate constructs in the shape of facial implants. *J Biomed Mater Res* 2001;55:503–511.

Chiono V, Pulieri E, Vozzi G, Ciardelli G, Ahluwalia A, Giusti P. Genipin-crosslinked chitosan/gelatin blends for biomedical applications. *J Mater Sci Mater Med* 2008;19:889–898.

Chiu LL, Radisic M. Controlled release of thymosin beta4 using collagen-chitosan composite hydrogels promotes epicardial cell migration and angiogenesis. *J Control Release* 2011;155:376–385.

Choi YS, Hong SR, Lee YM, Song KW, Park MH, Nam YS. Study on gelatin-containing artificial skin: I. Preparation and characteristics of novel gelatin-alginate sponge. *Biomaterials* 1999;20:409–417.

Chuo W-H, Tsai T-R, Hsu S-H, Cham T-M. Preparation and *in-vitro* evaluation of nifedipine loaded albumin microspheres cross-linked by different glutaraldehyde concentrations. *Int J Pharm* 1996;144:241–245.

Coombes AGA, Breeze V, Lin W, Gray T, Parker KG, Parker T. Lactic acid-stabilised albumin for microsphere formulation and biomedical coatings. *Biomaterials* 2001;22:1–8.

Di Martino A, Sittinger M, Risbud MV. Chitosan: A versatile biopolymer for orthopaedic tissue-engineering. *Biomaterials* 2005;26:5983–5990.

Di Silvio L, Bonfield W. Biodegradable drug delivery system for the treatment of bone infection and repair. *J Mater Sci Mater Med* 1999;10:653–658.

Dong J, Mo X, Li Y, Chen D. Recent research progress of decellularization of native tissues. *Shengwu Yixue Gongchengxue Zazhi* 2012;29:1007–1013.

Dong J, Sun Q, Wang JY. Basic study of corn protein, zein, as a biomaterial in tissue engineering, surface morphology and biocompatibility. *Biomaterials* 2004;25:4691–4697.

Drinnan CT, Zhang G, Alexander MA, Pulido AS, Suggs LJ. Multimodal release of transforming growth factor-beta1 and the BB isoform of platelet derived growth factor from PEGylated fibrin gels. *J Control Release* 2010;147:180–186.

Dyondi D, Webster TJ, Banerjee R. A nanoparticulate injectable hydrogel as a tissue engineering scaffold for multiple growth factor delivery for bone regeneration. *Int J Nanomed* 2013;8:47–59.

Ehrlich S, Wolff N, Schneiderman R, Maroudas A, Parker KH, Winlove CP. The osmotic pressure of chondroitin sulphate solutions: experimental measurements and theoretical analysis. *Biorheology* 1998;35:383–397.

Ellmerer M, Schaupp L, Brunner GA, Sendlhofer G, Wutte A, Wach P *et al.* Measurement of interstitial albumin in human skeletal muscle and adipose tissue by open-flow microperfusion. *Am J Physiol* 2000;278:E352–E6.

Frank O, Heim M, Jakob M, Barbero A, Schafer D, Bendik I *et al.* Real-time quantitative RT-PCR analysis of human bone marrow stromal cells during osteogenic differentiation *in vitro*. *J Cell Biochem* 2002;85:737–746.

Fujita M, Ishihara M, Morimoto Y, Simizu M, Saito Y, Yura H *et al.* Efficacy of photocrosslinkable chitosan hydrogel containing fibroblast growth factor-2 in a rabbit model of chronic myocardial infarction. *J Surg Research* 2005;126:27–33.

Fujita M, Ishihara M, Simizu M, Obara K, Ishizuka T, Saito Y *et al.* Vascularization *in vivo* caused by the controlled release of fibroblast growth factor-2 from an injectable chitosan/non-anticoagulant heparin hydrogel. *Biomaterials* 2004; 25:699–706.

Gallego L, Junquera L, Meana A, Alvarez-Viejo M, Fresno M. Ectopic bone formation from mandibular osteoblasts cultured in a novel human serum-derived albumin scaffold. *J Biomater Appl* 2010;25:367–381.

Gan C-Y, Cheng L-H, Phuah E-T, Chin P-N, AlKarkhi AFM, Easa AM. Combined cross-linking treatments of bovine serum albumin gel beadlets for controlled-delivery of caffeine. *Food Hydrocolloids* 2009;23:1398–1405.

Ghassabian S, Ehtezazi T, Forutan SM, Mortazavi SA. Dexamethasone-loaded magnetic albumin microspheres: preparation and *in vitro* release. *Int J Pharm* 1996;130:49–55.

Gilbert TW, Sellaro TL, Badylak SF. Decellularization of tissues and organs. *Biomaterials* 2006;27:3675–3683.

Gomes M, Ribeiro A, Malafaya P, Reis R, Cunha A. A new approach based on injection moulding to produce biodegradable starch-based polymeric scaffolds: morphology, mechanical and degradation behaviour. *Biomaterials* 2001;22:883–889.

Gomes ME, Azevedo HS, Moreira AR, Ella V, Kellomaki M, Reis RL. Starch-poly(epsilon-caprolactone) and starch-poly(lactic acid) fibre-mesh scaffolds for bone tissue engineering applications: structure, mechanical properties and degradation behaviour. *J Tissue Eng Regenerative med* 2008;2:243–252.

Gong S, Wang H, Sun Q, Xue ST, Wang JY. Mechanical properties and *in vitro* biocompatibility of porous zein scaffolds. *Biomaterials* 2006;27:3793–3799.

Gong SJ, Sun SX, Sun QS, Wang JY, Liu XM, Liu GY. Tablets based on compressed zein microspheres for sustained oral administration: design, pharmacokinetics, and clinical study. *J Biomater appl* 2011;26:195–208.

Goraltchouk A, Scanga V, Morshead CM, Shoichet MS. Incorporation of protein-eluting microspheres into biodegradable nerve guidance channels for controlled release. *J Control Release*: official journal of the Controlled Release Society. 2006;110:400–407.

Gu F, Amsden B, Neufeld R. Sustained delivery of vascular endothelial growth factor with alginate beads. *J Control Release* 2004;96:463–472.

Guo T, Zhao J, Chang J, Ding Z, Hong H, Chen J, *et al.* Porous chitosan-gelatin scaffold containing plasmid DNA encoding transforming growth factor-beta1 for chondrocytes proliferation. *Biomaterials* 2006;27:1095–1103.

Hansson A, Di Francesco T, Falson F, Rousselle P, Jordan O, Borchard G. Preparation and evaluation of nanoparticles for directed tissue engineering. *Int J Pharm* 2012;439:73–80.

Haug A, Larsen B, Smidsroed O. A study of the constitution of alginic acid by partial acid hydrolysis. *Acta Chem Scand* 1966;20:183–190.

Haug A, Larsen B, Smidsroed O. Sequence of uronic acid residues in alginic acid. *Acta Chem Scand* 1967;21:691–704.

Hayes AJ, Hughes CE, Ralphs JR, Caterson B. Chondroitin sulphate sulphation motif expression in the ontogeny of the intervertebral disc. *Eur Cells Mater* 2011;21:1–14.

Hostmark AT, Tomten SE, Berg JE. Serum albumin and blood pressure: a population-based, cross-sectional study. *J Hypertens* 2005;23:725–730.

Huang Q, Goh JC, Hutmacher DW, Lee EH. *In vivo* mesenchymal cell recruitment by a scaffold loaded with transforming growth factor beta1 and the potential for in situ chondrogenesis. *Tissue Eng* 2002;8:469–482.

Huang Y, Onyeri S, Siewe M, Moshfeghian A, Madihally SV. *In vitro* characterization of chitosan-gelatin scaffolds for tissue engineering. *Biomaterials* 2005;26:7616–7627.

Hunt NC, Grover LM. Cell encapsulation using biopolymer gels for regenerative medicine. *Biotechnol Lett* 2010;32:733–742.

Iemma F, Spizzirri UG, Puoci F, Muzzalupo R, Trombino S, Cassano R, *et al.* pH-sensitive hydrogels based on bovine serum albumin for oral drug delivery *Int J Pharm* 2006;312:151–157.

Jain A, Gulbake A, Shilpi S, Jain A, Hurkat P, Jain SK. A new horizon in modifications of chitosan: syntheses and applications. *Crit Rev Ther Drug Carrier Syst* 2013;30:91–181.

Jain S, Valvi PU, Swarnakar NK, Thanki K. Gelatin coated hybrid lipid nanoparticles for oral delivery of amphotericin *B Mol Pharm* 2012;9:2542–2553.

Jay SM, Saltzman WM. Controlled delivery of VEGF via modulation of alginate microparticle ionic crosslinking. *J Control Release* 2009;134:26–34.

Jeong Park Y, Moo Lee Y, Nae Park S, Yoon Sheen S, Pyoung Chung C, Lee SJ. Platelet derived growth factor releasing chitosan sponge for periodontal bone regeneration. *Biomaterials* 2000;21:153–159.

Kanczler JM, Ginty PJ, White L, Clarke NMP, Howdle SM, Shakesheff KM, *et al.* The effect of the delivery of vascular endothelial growth factor and bone morphogenic protein-2 to osteoprogenitor cell populations on bone formation. *Biomaterials* 2010;31:1242–1250.

Katta J, Jin Z, Ingham E, Fisher J. Chondroitin sulphate: an effective joint lubricant? Osteoarthritis and cartilage / OARS, *Osteoarthritis Research Society* 2009;17: 1001–1008.

Kim JJ, Evans GR. Applications of biomaterials in plastic surgery. *Clin Plast Surg* 2012;39:359–376.

Kim SE, Park JH, Cho YW, Chung H, Jeong SY, Lee EB *et al.* Porous chitosan scaffold containing microspheres loaded with transforming growth factor-β1: Implications for cartilage tissue engineering. *J Control Release* 2003;91:365–374.

Kolambkar YM, Dupont KM, Boerckel JD, Huebsch N, Mooney DJ, Hutmacher DW, *et al.* An alginate-based hybrid system for growth factor delivery in the functional repair of large bone defects. *Biomaterials* 2011;32:65–74.

Kundu B, Rajkhowa R, Kundu SC, Wang X. Silk fibroin biomaterials for tissue regenerations. *Adv Drug Deliv Rev* 2013;65:457–470.

Lai LF, Guo HX. Preparation of new 5-fluorouracil-loaded zein nanoparticles for liver targeting. *Int J Pharm* 2011;404:317–323.

Lam CXF, Mo XM, Teoh SH, Hutmacher DW. Scaffold development using 3D printing with a starch-based polymer. *Mater Sci Eng: C* 2002;20:49–56.

Langley SM, Rooney SJ, Dalrymple-Hay MJ, Spencer JM, Lewis ME, Pagano D *et al.* Replacement of the proximal aorta and aortic valve using a composite bileaflet prosthesis and gelatin-impregnated polyester graft (Carbo-Seal): early results in 143 patients. *J Thorac Cardiovasc Surg* 1999;118:1014–1020.

Lee J, Bhang SH, Park H, Kim B-S, Lee KY. Active Blood Vessel Formation in the Ischemic Hindlimb Mouse Model Using a Microsphere/Hydrogel Combination System. *Pharm Res* 2010;27:767–774.

Lee JE, Kim KE, Kwon IC, Ahn HJ, Lee SH, Cho H, *et al.* Effects of the controlled-released TGF-beta 1 from chitosan microspheres on chondrocytes cultured in a collagen/chitosan/glycosaminoglycan scaffold. *Biomaterials* 2004;25:4163–4173.

Lee J-Y, Nam S-H, Im S-Y, Park Y-J, Lee Y-M, Seol Y-J *et al.* Enhanced bone formation by controlled growth factor delivery from chitosan-based biomaterials. *J Control Release* 2002;78:187–197.

Lee KY, Mooney DJ. Alginate: Properties and biomedical applications. *Prog Polym Sci* 2012;37:106–126.

Liao IC, Wan AC, Yim EK, Leong KW. Controlled release from fibers of polyelectrolyte complexes. *J Control Release*: official journal of the Controlled Release Society. 2005;104:347–358.

Lien SM, Ko LY, Huang TJ. Effect of pore size on ECM secretion and cell growth in gelatin scaffold for articular cartilage tissue engineering. *Acta Biomater* 2009;5:670–679.

Liu W, McDaniel J, Li X, Asai D, Quiroz FG, Schaal J, *et al.* Brachytherapy using injectable seeds that are self-assembled from genetically encoded polypeptides in situ. *Cancer Res* 2012;72:5956–5965.

Liu X, Ma L, Liang J, Zhang B, Teng J, Gao C. RNAi functionalized collagen-chitosan/silicone membrane bilayer dermal equivalent for full-thickness skin regeneration with inhibited scarring. *Biomaterials* 2013;34:2038–2048.

Liu X, Sun Q, Wang H, Zhang L, Wang JY. Microspheres of corn protein, zein, for an ivermectin drug delivery system. *Biomaterials* 2005;26:109–115.

Liu Y, Li H, Shu XZ, Gray SD, Prestwich GD. Crosslinked hyaluronan hydrogels containing mitomycin C reduce postoperative abdominal adhesions. *Fertility and sterility* 2005;83 Suppl 1:1275–1283.

Lopiz-Morales Y, Abarrategi A, Ramos V, Moreno-Vicente C, Lopez-Duran L, Lopez-Lacomba JL, *et al.* *In vivo* comparison of the effects of RHBMP-2 and RHBMP-4 in osteochondral tissue regeneration. *Eur Cells Mater* 2010; 20:367–378.

Lyu SR, Kuo YC, Lin MH, Hsieh WH, Chuang CW. Application of albumin-grafted scaffolds to promote neocartilage formation. *Colloids Surf B Biointerfaces* 2012;91:296–301.

Ma J, Wang H, He B, Chen J. A preliminary *in vitro* study on the fabrication and tissue engineering applications of a novel chitosan bilayer material as a scaffold of human neofetal dermal fibroblasts. *Biomaterials* 2001;22:331–336.

Ma L, Gao C, Mao Z, Zhou J, Shen J, Hu X, *et al.* Collagen/chitosan porous scaffolds with improved biostability for skin tissue engineering. *Biomaterials* 2003;24:4833–4841.

Maeda M, Tani S, Sano A, Fujioka K. Microstructure and release characteristics of the minipellet, a collagen-based drug delivery system for controlled release of protein drugs. *J Controll Release* 1999;62:313–324.

Malafaya PB, Stappers F, Reis RL. Starch-based microspheres produced by emulsion crosslinking with a potential media dependent responsive behavior to be used as drug delivery carriers. *J Materials Sci Mat in Med* 2006;17:371–377.

Mao J, Zhao L, Kang DY, Shang Q, Yang G, Cao Y. Study of novel chitosan-gelatin artificial skin *in vitro J Biomed Mater Res, Part A* 2003;64A:301–308.

Marques AP, Reis RL, Hunt JA. The biocompatibility of novel starch-based polymers and composites: *in vitro* studies. *Biomaterials* 2002;23:1471–1478.

Martins A, Chung S, Pedro AJ, Sousa RA, Marques AP, Reis RL *et al*. Hierarchical starch-based fibrous scaffold for bone tissue engineering applications. *J Tissue Eng Regenerative Med* 2009;3:37–42.

Martins AM, Pham QP, Malafaya PB, Sousa RA, Gomes ME, Raphael RM, *et al*. The role of lipase and alpha-amylase in the degradation of starch/poly(epsilon-caprolactone) fiber meshes and the osteogenic differentiation of cultured marrow stromal cells. *Tissue Engin Part A* 2009;15:295–305.

McDermott MK, Chen T, Williams CM, Markley KM, Payne GF. Mechanical properties of biomimetic tissue adhesive based on the microbial transglutaminase-catalyzed crosslinking of gelatin. *Biomacromolecules* 2004;5:1270–1279.

Mizuno K, Yamamura K, Yano K, Osada T, Saeki S, Takimoto N, *et al*. Effect of chitosan film containing basic fibroblast growth factor on wound healing in genetically diabetic mice. *J Biomed Mater Research Part A* 2003;64A:177–181.

Moshayedi P, Carmichael ST. Hyaluronan, neural stem cells and tissue reconstruction after acute ischemic stroke. *Biomatter* 2013;3.

Murata Y, Kofuji K, Nishida N, Kamaguchi R. Development of film dosage form containing allopurinol for prevention and treatment of oral mucositis. *ISRN Pharm* 2012;764510:15.

Muyonga JH, Cole CGB, Duodu KG. Characterisation of acid soluble collagen from skins of young and adult Nile perch (Lates niloticus). *Food Chem* 2003;85:81–9.

Muzzarelli RAA, Mattioli-Belmonte M, Tietz C, Biagini R, Ferioli G, Brunelli MA, *et al*. Stimulatory effect on bone formation exerted by a modified chitosan. *Biomaterials* 1994;15:1075–1081.

Nakamura M, Mie M, Mihara H, Nakamura M, Kobatake E. Construction of a multi-functional extracellular matrix protein that increases number of N1E-115 neuroblast cells having neurites. *J Biomed Mater Res B Appl Biomater* 2009;91:425–432.

Nettles DL, Chilkoti A, Setton LA. Applications of elastin-like polypeptides in tissue engineering. *Adv Drug Deliv Rev* 2010;62:1479–1485.

Nitta SK, Numata K. Biopolymer-based nanoparticles for drug/gene delivery and tissue engineering. *Int J Mol Sci* 2013;14:1629–1654.

Oliveira AL, Reis RL. Pre-mineralisation of starch/polycrapolactone bone tissue engineering scaffolds by a calcium-silicate-based process. *J Mater Sci Mater in Med* 2004;15:533–540.

Oyarzun-Ampuero FA, Rivera-Rodriguez GR, Alonso MJ, Torres D. Hyaluronan nanocapsules as a new vehicle for intracellular drug delivery. *Eur J Pharm Sci*: official journal of the European Federation for Pharmaceutical Sciences. 2013.

Parenteau-Bareil R, Gauvin R, Berthod F. Collagen-based biomaterials for tissue engineering applications. *Materials* 2010;3:1863–1887.

Park I-K, Yang J, Jeong H-J, Bom H-S, Harada I, Akaike T *et al*. Galactosylated chitosan as a synthetic extracellular matrix for hepatocytes attachment. *Biomaterials* 2003;24:2331–2337.

Park MS, Kim YB. Sustained release of antibiotic from a fibrin-gelatin-antibiotic mixture. *Laryngoscope* 1997;107:1378–1381.

Pashkuleva I, Marques AP, Vaz F, Reis RL. Surface modification of starch based blends using potassium permanganate-nitric acid system and its effect on the adhesion and proliferation of osteoblast-like cells. *J Mater Sci Mater in Med* 2005;16:81–92.

Pawar SN, Edgar KJ. Alginate derivatization: A review of chemistry, properties and applications. *Biomaterials* 2012;33:3279–3305.

Peng L, Cheng X, Zhuo R, Lan J, Wang Y, Shi B, *et al*. Novel gene-activated matrix with embedded chitosan/plasmid DNA nanoparticles encoding PDGF for periodontal tissue engineering. *J Biomed Mater Research Part A* 2009;90:564–576.

Peters Jt. All About Albumin: Biochemistry, Genetics, and Medical Applications. *Academic* 1995.

Prang P, Mueller R, Eljaouhari A, Heckmann K, Kunz W, Weber T, *et al*. The promotion of oriented axonal regrowth in the injured spinal cord by alginate-based anisotropic capillary hydrogels. *Biomaterials* 2006;27:3560–3569.

Qu ZH, Wang HJ, Tang TT, Zhang XL, Wang JY, Dai KR. Evaluation of the zein/ inorganics composite on biocompatibility and osteoblastic differentiation. *Acta Biomater* 2008;4:1360–1368.

Ramshaw JAM, Werkmeister JA, Glattauer V. Collagen-based biomaterials. *Biotechnol Genet Eng Rev* 1996;13:335–382.

Rao KP. Recent developments of collagen-based materials for medical applications and drug delivery systems. *J Biomater Sci Polym Ed* 1995;7:623–645.

Reddy N, Xu H, Yang Y. Unique natural-protein hollow-nanofiber membranes produced by weaver ants for medical applications. *Biotech bioeng* 2011;108:1726–1733.

Reis R, Cunha A. Characterization of two biodegradable polymers of potential application within the biomaterials field. *J Mater Sci: Mater Med* 1995;6:786–792.

Rinaudo M. On the abnormal exponents aη and aD in Mark Houwink type equations for worm like chain polysaccharides. *Polym Bull (Berlin)* 1992;27:585–589.

Rodrigues AI, Gomes ME, Leonor IB, Reis RL. Bioactive starch-based scaffolds and human adipose stem cells are a good combination for bone tissue engineering. *Acta Biomater* 2012;8:3765–3776.

Rossler B, Kreuter J, Ross G. Effect of collagen microparticles on the stability of retinol and its absorption into hairless mouse skin *in vitro*. *Pharmazie* 1994;49:175–179.

Sahin S, Selek H, Ponchel G, Ercan MT, Sargon M, Hincal AA *et al.* Preparation, characterization and *in vivo* distribution of terbutaline sulfate loaded albumin microspheres. *J Control Release* 2002;82:345–358.

Salgado AJ, Coutinho OP, Reis RL, Davies JE. *In vivo* response to starch-based scaffolds designed for bone tissue engineering applications. *J Biomed Mater Res Part A* 2007;80A:983–989.

Santo VE, Gomes ME, Mano JF, Reis RL. Chitosan-chondroitin sulphate nanoparticles for controlled delivery of platelet lysates in bone regenerative medicine. *J Tissue Eng Regenerative Med* 2012;6 Suppl 3:S47–S59.

Santos MI, Fuchs S, Gomes ME, Unger RE, Reis RL, Kirkpatrick CJ. Response of micro- and macrovascular endothelial cells to starch-based fiber meshes for bone tissue engineering. *Biomaterials* 2007;28:240–248.

Santos TC, Morton TJ, Moritz M, Pfeifer S, Reise K, Marques AP, *et al.* Vascular endothelial growth factor and fibroblast growth factor-2 incorporation in starch-based bone tissue-engineered constructs promote the *in vivo* expression of neovascularization mediators. *Tissue Eng Part A* 2013;19:834–848.

Schmoekel H, Schense JC, Weber FE, Gratz KW, Gnagi D, Muller R *et al.* Bone healing in the rat and dog with nonglycosylated BMP-2 demonstrating low solubility in fibrin matrices. *J Orthop Res* 2004;22:376–381.

Segura T, Chung PH, Shea LD. DNA delivery from hyaluronic acid-collagen hydrogels via a substrate-mediated approach. *Biomaterials* 2005;26:1575–1584.

Shanmugasundaram N, Sundaraseelan J, Uma S, Selvaraj D, Babu M. Design and delivery of silver sulfadiazine from alginate microspheres-impregnated collagen scaffold. *J Biomed Mater Res B Appl Biomater* 2006;77:378–388.

Silva GA, Costa FJ, Neves NM, Coutinho OP, Dias AC, Reis RL. Entrapment ability and release profile of corticosteroids from starch-based microparticles. *J Biomed Mater Research Part A* 2005;73:234–243.

Silva GA, Coutinho OP, Ducheyne P, Shapiro IM, Reis RL. Starch-based microparticles as vehicles for the delivery of active platelet-derived growth factor. *Tissue Eng* 2007;13:1259–1268.

Silva JM, Georgi N, Costa R, Sher P, Reis RL, Van Blitterswijk CA *et al.* Nanostructured 3D constructs based on chitosan and chondroitin sulphate multilayers for cartilage tissue engineering. *PloS one* 2013;8:e55451.

Smidsroed O, Skjaak-Braek G. Alginate as immobilization matrix for cells. *Trends Biotechnol* 1990;8:71–78.

Spicer PP, Mikos AG. Fibrin glue as a drug delivery system. *J Control Release* 2010;148:49–55.

Stern R, Jedrzejas MJ. Hyaluronidases: their genomics, structures, and mechanisms of action. *Chem Rev* 2006;106:818–839.

Stern R, Kogan G, Jedrzejas MJ, Soltes L. The many ways to cleave hyaluronan. *Biotechnol Adv* 2007;25:537–557.

Stone KR, Steadman JR, Rodkey WG, Li ST. Regeneration of meniscal cartilage with use of a collagen scaffold. Analysis of preliminary data. *J Bone Joint Surg Am* 1997;79:1770–1777.

Suzuki T, Sato E, Matsuda Y, Tada H, Unno K, Kato T. Preparation of zein microspheres conjugated with antitumor drugs available for selective cancer chemotherapy and development of a simple colorimetric determination of drugs in microspheres. *Chem Pharm Bull* 1989;37:1051–1054.

Takashima M, Tanaka Y. Application of acidic thermal treatment for one- and two-stage anaerobic digestion of sewage sludge. *Water Sci Technol* 2010;62:2647–2654.

Tamimi F, Kumarasami B, Doillon C, Gbureck U, Le Nihouannen D, Cabarcos EL, *et al.* Brushite-collagen composites for bone regeneration. *Acta Biomater* 2008;4:1315–1321.

Tedder ME, Liao J, Weed B, Stabler C, Zhang H, Simionescu A, *et al.* Stabilized collagen scaffolds for heart valve tissue engineering. *Tissue Eng Part A* 2009;15:1257–1268.

Thornton AJ, Alsberg E, Albertelli M, Mooney DJ. Shape-defining scaffolds for minimally invasive tissue engineering. *Transplantation* 2004;77:1798–1803.

Tigli RS, Gumusderelioglu M. Evaluation of RGD- or EGF-immobilized chitosan scaffolds for chondrogenic activity. *Int J Bio Macromolecules* 2008;43:121–128.

Tognana E, Borrione A, De Luca C, Pavesio A. Hyalograft C: Hyaluronan-based scaffolds in tissue-engineered cartilage. *Cells Tissues Organs* 2007;186:97–103.

Torres FG, Boccaccini AR, Troncoso OP. Microwave processing of starch-based porous structures for tissue engineering scaffolds. *J Appl Polym Sci* 2007;103:1332–1339.

Tremblay PL, Hudon V, Berthod F, Germain L, Auger FA. Inosculation of tissue-engineered capillaries with the host's vasculature in a reconstructed skin transplanted on mice. *Am J Transplant* 2005;5:1002–1010.

Tu J, Wang H, Li H, Dai K, Wang J, Zhang X. The *in vivo* bone formation by mesenchymal stem cells in zein scaffolds. *Biomaterials* 2009;30:4369–4376.

Uebersax L, Merkle HP, Meinel L. Biopolymer-based growth factor delivery for tissue repair: from natural concepts to engineered systems. *Tissue Eng Part B Rev* 2009;15:263–289.

van Susante JL, Buma P, Schuman L, Homminga GN, van den Berg WB, Veth RP. Resurfacing potential of heterologous chondrocytes suspended in fibrin glue in large full-thickness defects of femoral articular cartilage: an experimental study in the goat. *Biomaterials* 1999;20:1167–1175.

Venkatesan J, Pallela R, Bhatnagar I, Kim SK. Chitosan-amylopectin/hydroxyapatite and chitosan-chondroitin sulphate/hydroxyapatite composite scaffolds for bone tissue engineering. *Int J Bio Macromolecules* 2012;51:1033–1042.

Volpi N, Schiller J, Stern R, Soltes L. Role, metabolism, chemical modifications and applications of hyaluronan. *Curr Med Chem* 2009;16:1718–1745.

Wang DA, Varghese S, Sharma B, Strehin I, Fermanian S, Gorham J, *et al*. Multifunctional chondroitin sulphate for cartilage tissue-biomaterial integration. *Nature Mater* 2007;6:385–392.

Wang HJ, Fu JX, Wang JY. Effect of water vapor on the surface characteristics and cell compatibility of zein films. Colloids and surfaces B, *Biointerfaces* 2009;69:109–115.

Wang HJ, Gong SJ, Lin ZX, Fu JX, Xue ST, Huang JC, *et al. In vivo* biocompatibility and mechanical properties of porous zein scaffolds. *Biomaterials* 2007;28:3952–3964.

Wang HJ, Lin ZX, Liu XM, Sheng SY, Wang JY. Heparin-loaded zein microsphere film and hemocompatibility. *J Control Release*: Official journal of the Controlled Release Society. 2005;105:120–131.

Wang QG, Hughes N, Cartmell SH, Kuiper NJ. The composition of hydrogels for cartilage tissue engineering can influence glycosaminoglycan profile. *Eur Cells Mater* 2010;19:86–95.

Wang W. A novel hydrogel crosslinked hyaluronan with glycol chitosan. *J Mater Sci Mater in Med* 2006;17:1259–1265.

Werkmeister JA, Ramshaw JA. Recombinant protein scaffolds for tissue engineering. *Biomed Mater* 2012;7:012002.

Wiltsey C, Kubinski P, Christiani T, Toomer K, Sheehan J, Branda A *et al*. Characterization of injectable hydrogels based on poly(N-isopropylacrylamide)-g-chondroitin sulfate with adhesive properties for nucleus pulposus tissue engineering. *J Mater Sci Mater in Med* 2013;24:837–847.

Wong C, Inman E, Spaethe R, Helgerson S. Fibrin-based biomaterials to deliver human growth factors. *Thromb Haemost* 2003;89:573–582.

Wu Y, Shih EK, Ramanathan A, Vasudevan S, Weil T. Nano-sized albumin-copolymer micelles for efficient doxorubicin delivery. *Biointerphases* 2012;7:011–0005.

Yamazoe H, Tanabe T. Drug-carrying albumin film for blood-contacting biomaterials. *J Biomater Sci Polym Ed* 2010;21:647–657.

Yoo HS, Lee EA, Yoon JJ, Park TG. Hyaluronic acid modified biodegradable scaffolds for cartilage tissue engineering. *Biomaterials* 2005;26:1925–1933.

Zhang W, Cheng Q, Guo S, Lin D, Huang P, Liu J *et al*. Gene transfection efficacy and biocompatibility of polycation/DNA complexes coated with enzyme degradable PEGylated hyaluronic acid. *Biomaterials* 2013;34:6495–6503.

Zhang Y, Cheng X, Wang J, Wang Y, Shi B, Huang C, *et al*. Novel chitosan/collagen scaffold containing transforming growth factor-beta1 DNA for periodontal tissue engineering. *Biochem Biophys Res Commun* 2006;344:362–369.

Zhang Y, Wang Y, Shi B, Cheng X. A platelet-derived growth factor releasing chitosan/coral composite scaffold for periodontal tissue engineering. *Biomaterials* 2007;28:1515–1522.

Conclusions and Future Outlook

Generally speaking, the prospect of using native polymer-based 3D tissue engineered substitutes to generate living tissues is logical and exciting. The field of native polymer-based 3D tissue engineering has been evolving rapidly with significant advances in scaffold fabrication and design techniques. However, efforts to create native polymer-based 3D tissue engineered substitutes remain in an early developmental stage and native polymers are not without their limitations as 3D tissue engineering scaffolds. Disease transmission, immunogenic responses *in vivo* studies, as well as batch-to-batch variability, and poor mechanical properties are being recognized and becoming the major blocks for the applications of native polymer-based 3D tissue engineering substitutes in clinic (Malafaya *et al.*, 2007). In the future, further combinations of advanced techniques may be considered in the production and design of native polymer-based tissue engineered substitutes. Undoubtedly, much progress are likely to be achieved with more in–depth research and continuous optimization of native polymer-based tissue engineered scaffolds.

The first is the combination of tissue engineering and slow/controllable release technique. Of particular interest in the complications arising from currently available native polymer-based tissue engineered substitutes is the lack of growth and remodeling of replacement tissues after implantation. It is well known that cellular attachment, growth, proliferation, differentiation and cell-cell communication are always stimulated by different growth factors *in vivo*. Moreover, different growth factors will function at different stages of tissue regeneration. The combination with the slow/controllable release technique is a promising route to endow the bio-functions to native polymers (Liang and Andreadis, 2011; Pang *et al.*, 2010).

The second is the combination of tissue engineering with protein or genetic engineering. Protein or genetic engineering allows us to design and synthesize new protein sequences, new self-assembling peptides or fusions of different bioactive domains or protein motifs (Romano *et al.*, 2010). That is to say, there will be a chance to generate a multifunctional biomaterial with better biological activity, as well as mechanical properties that are absent in native polymers (Gomes *et al.*, 2012). Consequently, protein or genetic engineering provides a promising alternative for new biomaterials with ideal properties. How to succeed in fulfilling the potential of protein or genetic engineering in the field of tissue engineering is becoming another future direction.

The third is the combination of tissue engineering and nano-techniques. It is widely accepted that the organization of extracellular matrix is frequently hierarchical from nano to macro (Kim *et al.*, 2012). Moreover, a handful of studies have demonstrated that a microenvironment similar to native tissue is capable to guide cell growth and promote tissue regeneration (Holzapfel *et al.*, 2012). Thus, nano-techniques have lately been applied to produce biomimetic scaffolds, which will possess one or multiple characteristics of the natural extracellular matrix, such as biomimetic mechanical properties, biological activities, topographical features, etc. (Arafat *et al.*, 2011; Fleischer and Dvir, 2012; Tan *et al.*, 2012). What is more important, these nano-scaffolds displayed superior properties to bulk scaffolds at improving cell functions and inducing tissue regeneration (Barnes *et al.*, 2007; Cao *et al.*, 2009; Holzwarth and Ma, 2011).

References

Arafat MT, Lam CXF, Ekaputra AK, Wong SY, Li X, Gibson I. Biomimetic composite coating on rapid prototyped scaffolds for bone tissue engineering. *Acta Biomater* 2011;7:809–820.

Barnes C, Sell SA, Boland ED, Simpson DG, Bowlin GL. Nanofiber technology: Designing the next generation of tissue engineering scaffolds. *Adv Drug Deliv Rev* 2007;59(14):1413–1433.

Cao H, Liu T, Chew SY. The application of nanofibrous scaffolds in neural tissue engineering. *Adv Drug Deliv Rev* 2009;61(12):1055–1064.

Fleischer S, Dvir T. Tissue engineering on the nanoscale: lessons from the heart. *Curr Opinion Biotech* 2012;24:1–8.

Gomes S, Leonor IB, Mano JF, Reis RL, Kaplan DL. Natural and genetically engineered proteins for tissue engineering. *Prog Polym Sci* 2012;37:1–17.

Holzapfel BM, Reichert JC, Schantz JT, Gbureck U, Rackwitz L, Noth U, Jakob F, Rudert M, Groll J, Hutmacher DW. How smart do biomaterials need to be? A translational science and clinical point of view. *Adv Drug Deliv Rev* 2013;65:581–603.

Holzwarth JM, Ma PX. Biomimetic nanofibrous scaffolds for bone tissue engineering. *Biomaterials* 2011;32:9622–9629.

Kim HN, Jiao A, Hwang NS, Kim MS, Kang DH, Kim DH, Suh KY. Nanotopography-guided tissue engineering and regenerative medicine. *Adv Deliv Rev* 2013;65:536–558.

Liang MS, Andreadis ST. Engineering fibrin-binding TGF-β1 for sustained signaling and contractile function of MSC based vascular constructs. *Biomaterials* 2011;32:8684–8693.

Malafaya PB, Silva GA, Reis RL. Natural-origin polymers as carriers and scaffolds for biomolecules and cell delivery in tissue engineering applications. *Adv Drug Deliv Rev* 2007;59:207–233.

Pang Y, Wang X, Ucuzian AA, Brey EM, Burgess WH, Jones KJ, Alexander TD, Greisler HP. Local delivery of a collagen-binding FGF-1 chimera to smooth muscle cells in collagen scaffolds for vascular tissue engineering. *Biomaterials* 2010;31:878–885.

Romano NH, Sengupta D, Chung C, Heilshorn SC. Protein-engineered biomaterials: Nanoscale mimics of the extracellular matrix. *Biochim Biophys Acta* 2010;1810:339–349.

Tan A, Alavijeh MS, Seifalian AM. Next generation stent coatings: Convergence of biotechnology and nanotechnology. *Trends in Biotechnology* 2012;30(8):406–409.

Index

3D scaffolds   107, 130, 147, 222, 228–230, 233, 234, 236, 237, 239–241, 244, 246–248, 251

acellular matrix   91–93, 95
agarose   83, 85, 118
air-blowing electrospinning   226, 227
albumin   259, 263–266
alginate   4, 8–10, 17–20, 84, 107, 116–119, 121, 122, 124, 147, 160, 161, 163 164, 231–233, 235, 267, 275, 281,285–288
Artefill   194–196, 209, 210
articular cartilage   76, 78, 79, 81, 82, 87, 88, 90, 91, 100, 101, 111, 118, 119, 123–125, 127, 128

beads   116, 124
bioceramics   150, 172
biocompatibility   8, 13–15, 37–39, 44, 49, 52, 56, 60
biodegradability   5, 8, 13, 14, 39, 49, 52
bone defect   145, 146, 150, 156, 157, 160, 161, 164, 165, 169, 172, 173
bone graft   146, 162, 172

cardiovascular   1, 2, 20, 22
cartilage regeneration   76, 77, 80, 83, 85, 87, 89, 91–93, 101, 105, 110, 112, 114, 119, 120, 123–125, 129, 130
cartilage repair   75, 77, 79, 80, 85, 95, 99, 101, 102, 113, 114, 123, 125, 127, 128
cell-cell interaction   221
cellulose   147, 166–169
chitosan   4, 8–10, 15–18, 22, 49, 51–55, 86, 88, 102–109, 114, 117, 124, 222, 229–233, 235–238, 243, 244, 246, 260, 261, 264, 268, 274, 279–282, 290, 291, 293, 294
chondrocytes   76, 79, 82, 83, 85–87, 89, 90, 101, 111, 115, 117–119, 123–125, 130
chondroitin sulfate   291, 293, 294
coaxial electrospinning   225, 226
cofibrillogenesis   237–239
collagen   4–6, 10, 11, 13–15, 17, 19–21, 78, 79, 81, 83, 84, 86, 88, 95–97, 99–102, 112, 117–120, 124, 127, 147, 148–153, 155, 160, 161, 169, 186–191, 193–196, 202–205,

222, 230–239, 244, 247, 248, 250, 259–264, 266, 267, 274, 277, 279, 281, 290–292
controlled delivery 258
cross-linking 148, 150, 162, 163, 170

dextran 170
DNA 268, 272, 274, 280, 281, 290, 291

elastin 11–13, 15, 21
electrospinning 222–234, 236, 240, 242, 248
encapsulation 258, 259, 271, 273, 277, 282, 287, 293
Extracellular matrix (ECM) 37–39, 41, 43, 48, 112, 119, 123, 221, 234

facial rejuvenation 187, 213
fibrin 4–6, 10–12, 17, 43–48, 115, 116, 119, 120, 124
fibronectin 6, 15, 38, 43, 44, 46
freeze-dried alginate sponge 61
functional recovery 37, 40–42, 49, 53, 57, 58, 61, 62

gelatin 80, 83, 107, 119, 120, 127, 222, 230, 244–247
graft 264, 265, 267, 277, 293
growth factor 271, 273, 274, 276, 277, 280–287, 289, 292, 293

heart valve graft 3
hierarchical structures 241
hyaluronan 288–291

hyaluronic acid 4, 6, 8, 10, 79, 85, 109, 110, 112, 161, 162, 186–189, 197–199, 202, 204, 205, 207, 208, 212
hydrogel 83–87, 90, 101, 102, 106, 113, 114, 116, 118, 120, 124

implant 2, 19
injectable filler 186, 187, 189

laminin 38, 41–43, 46
Lip enhancement 204

mechanical property 41, 49, 53, 60
Mesenchymal stem cell (MSC) 154, 156, 160, 161, 164, 173
micelles 264, 265
microspheres 262–266, 271, 275, 280, 283, 286, 287
multi-channel conduits 40, 42, 53, 55

nanofabrication 221, 222, 233, 241, 247, 251
nanofibrous scaffolds 226, 229, 230, 233, 237, 242–247
nanoparticles 263, 265, 269, 271, 273, 275, 282, 291–293
nano-patterning 248, 250
nasojugal groove 198, 206
nasolabial fold 189, 197, 200–202, 208–210
native polymer 1, 2, 4, 5, 8–10, 15–17, 19, 22, 75, 77–82, 84–86, 88, 91, 114, 119, 121–127, 129, 222, 228–230,

232, 233, 237, 239–242, 248, 250, 251
natural materials 38, 61
nerve conduits 37–42, 43, 49, 53, 61, 62
nerve regeneration 35, 37, 43, 46–49, 52, 56, 57, 60–62

osteoconductivity 146, 151, 155
osteogenicity 146
osteoinductivity 146, 155

peripheral nerve injury 35, 36, 44, 46
phase separation 230, 242–245, 247, 248, 251
polyester 147, 171
polylactic acid 80, 196
polypeptides 272, 275
polysaccharides 147, 157, 159, 166, 259, 278, 285
porogen leaching 245
proteins 258, 259, 269, 272–274, 286, 289, 291, 293

Restylane 188, 189, 191, 205, 212
rhinoplasty 198, 199
rhytids 203

scaffold 2–16, 19–22, 257–264, 267–273, 275, 277–288, 290–293, 295
sciatic nerve 40, 52, 53, 55, 57
self-assembly 230, 233–237, 239–241, 248, 251

serial puncture 202, 204, 208, 210
signal/therapeutic agents 258, 260
silk 114, 147, 152–155
silk fibroin 13, 14, 19, 20, 22, 55–57, 274–276
silk protein 233, 236
soft augmentation 213
sponge 99–101, 115, 120, 125
starch 147, 164–166, 282–284
sustained release 258, 268, 281, 282, 284, 287
synthetic polymers 150

targeted 264, 275–277, 288
tear trough 197, 198, 206–208
tissue-engineered blood vessel 10, 11, 13
Tissue-engineered myocardium 16, 17, 19, 20, 22
tissue engineering 221, 222, 226, 228–231, 233, 234, 237, 239–241, 245, 247, 250, 251, 257–263, 266–270, 272–274, 276–288, 290–295
Tracheal cartilage 125
Tunneling technique 209

X-ray interference 249, 250

zein 156, 259, 269–272
Zyplast 189, 191, 193